Genomics of Bacterial Metal Resistance

Genomics of Bacterial Metal Resistance

Editors

Alessio Mengoni
Carlo Viti
Raymond J. Turner
Li-Nan Huang

MDPI • Basel • Beijing • Wuhan • Barcelona • Belgrade • Manchester • Tokyo • Cluj • Tianjin

Editors

Alessio Mengoni
Università degli Studi di Firenze
Italy

Carlo Viti
University of Florence
Italy

Raymond J. Turner
University of Calgary
Canada

Li-Nan Huang
Sun Yat-Sen University
China

Editorial Office
MDPI
St. Alban-Anlage 66
4052 Basel, Switzerland

This is a reprint of articles from the Special Issue published online in the open access journal *Genes* (ISSN 2073-4425) (available at: https://www.mdpi.com/journal/genes/special_issues/genomics_bacterial_metal_resistance).

For citation purposes, cite each article independently as indicated on the article page online and as indicated below:

LastName, A.A.; LastName, B.B.; LastName, C.C. Article Title. *Journal Name* **Year**, *Volume Number*, Page Range.

ISBN 978-3-0365-0390-5 (Hbk)
ISBN 978-3-0365-0391-2 (PDF)

© 2021 by the authors. Articles in this book are Open Access and distributed under the Creative Commons Attribution (CC BY) license, which allows users to download, copy and build upon published articles, as long as the author and publisher are properly credited, which ensures maximum dissemination and a wider impact of our publications.

The book as a whole is distributed by MDPI under the terms and conditions of the Creative Commons license CC BY-NC-ND.

Contents

About the Editors vii

Raymond J. Turner, Li-Nan Huang, Carlo Viti and Alessio Mengoni
Metal-Resistance in Bacteria: Why Care?
Reprinted from: *Genes* 2020, *11*, 1470, doi:10.3390/genes11121470 1

Rob Van Houdt, Ann Provoost, Ado Van Assche, Natalie Leys, Bart Lievens, Kristel Mijnendonckx and Pieter Monsieurs
Cupriavidus metallidurans Strains with Different Mobilomes and from Distinct Environments Have Comparable Phenomes
Reprinted from: *Genes* 2018, *9*, 507, doi:10.3390/genes9100507 5

Felipe A. Millacura, Paul J. Janssen, Pieter Monsieurs, Ann Janssen, Ann Provoost, Rob Van Houdt and Luis A. Rojas
Unintentional Genomic Changes Endow *Cupriavidus metallidurans* with an Augmented Heavy-Metal Resistance
Reprinted from: *Genes* 2018, *9*, 551, doi:10.3390/genes9110551 29

Md Muntasir Ali, Ann Provoost, Laurens Maertens, Natalie Leys, Pieter Monsieurs, Daniel Charlier and Rob Van Houdt
Genomic and Transcriptomic Changes That Mediate Increased Platinum Resistance in *Cupriavidus metallidurans*
Reprinted from: *Genes* 2019, *10*, 63, doi:10.3390/genes10010063 47

Natalie Gugala, Joe Lemire, Kate Chatfield-Reed, Ying Yan, Gordon Chua and Raymond J. Turner
Using a Chemical Genetic Screen to Enhance Our Understanding of the Antibacterial Properties of Silver
Reprinted from: *Genes* 2018, *9*, 344, doi:10.3390/genes9070344 69

Natalie Gugala, Kate Chatfield-Reed, Raymond J. Turner and Gordon Chua
Using a Chemical Genetic Screen to Enhance Our Understanding of the Antimicrobial Properties of Gallium against *Escherichia coli*
Reprinted from: *Genes* 2019, *10*, 34, doi:10.3390/genes10010034 91

Zaaima AL-Jabri, Roxana Zamudio, Eva Horvath-Papp, Joseph D. Ralph, Zakariya AL-Muharrami, Kumar Rajakumar and Marco R. Oggioni
Integrase-Controlled Excision of Metal-Resistance Genomic Islands in *Acinetobacter baumannii*
Reprinted from: *Genes* 2018, *9*, 366, doi:10.3390/genes9070366 115

Yuan Ping Li, Nicolas Carraro, Nan Yang, Bixiu Liu, Xian Xia, Renwei Feng, Quaiser Saquib, Hend A Al-Wathnani, Jan Roelof van der Meer and Christopher Rensing
Genomic Islands Confer Heavy Metal Resistance in *Mucilaginibacter kameinonensis* and *Mucilaginibacter rubeus* Isolated from a Gold/Copper Mine
Reprinted from: *Genes* 2018, *9*, 573, doi:10.3390/genes9120573 129

Cameron Parsons, Sangmi Lee and Sophia Kathariou
Heavy Metal Resistance Determinants of the Foodborne Pathogen *Listeria monocytogenes*
Reprinted from: *Genes* 2019, *10*, 11, doi:10.3390/genes10010011 143

Gabhan Chalmers, Kelly M. Rozas, Raghavendra G. Amachawadi, Harvey Morgan Scott, Keri N. Norman, Tiruvoor G. Nagaraja, Mike D. Tokach and Patrick Boerlin
Distribution of the *pco* Gene Cluster and Associated Genetic Determinants among Swine *Escherichia coli* from a Controlled Feeding Trial
Reprinted from: *Genes* **2018**, *9*, 504, doi:10.3390/genes9100504 . . . . **157**

Camilla Fagorzi, Alice Checcucci, George C. diCenzo, Klaudia Debiec-Andrzejewska, Lukasz Dziewit, Francesco Pini and Alessio Mengoni
Harnessing Rhizobia to Improve Heavy-Metal Phytoremediation by Legumes
Reprinted from: *Genes* **2018**, *9*, 542, doi:10.3390/genes9110542 . . . . **173**

George C diCenzo, Klaudia Debiec, Jan Krzysztoforski, Witold Uhrynowski, Alessio Mengoni, Camilla Fagorzi, Adrian Gorecki, Lukasz Dziewit, Tomasz Bajda, Grzegorz Rzepa and Lukasz Drewniak
Genomic and Biotechnological Characterization of the Heavy-Metal Resistant, Arsenic-Oxidizing Bacterium *Ensifer* sp. M14
Reprinted from: *Genes* **2018**, *9*, 379, doi:10.3390/genes9080379 . . . . **189**

Nia Oetiker, Rodrigo Norambuena, Cristóbal Martínez-Bussenius, Claudio A. Navarro, Fernando Amaya, Sergio A. Álvarez, Alberto Paradela and Carlos A. Jerez
Possible Role of Envelope Components in the Extreme Copper Resistance of the Biomining *Acidithiobacillus ferrooxidans*
Reprinted from: *Genes* **2018**, *9*, 347, doi:10.3390/genes9070347 . . . . **213**

About the Editors

Alessio Mengoni is a Professor of genetics at the University of Florence, Italy, and visiting professor at the Intercollegiate Faculty of Biotechnology, University of Gdansk, Poland. He is interested in understanding the evolution of bacterial genomes and the dynamics of microbiota in relation to symbiotic interactions and adaptations to heavy-metal contaminated sites.

Carlo Viti is a Full Professor of Microbiology at the University of Florence with a PhD in Soil Science. His main research topics include the ecology and taxonomy of bacteria, dynamics of microbiota in soils, rumen and the phycosphere. Carlo Viti pioneered the use of electrical signalling as a possible driver in bacterial biofilm sociomicrobiology.

Raymond J. Turner Ph.D. (Professor of Microbiology and Biochemistry), received a Ph.D. in Physical Biochemistry and a PDF in Microbial Biochemistry. He has been based at the Department of Biological Sciences, University of Calgary, since 1998, where he provides lectures in courses including introductory biology and biochemistry, biomembranes, molecular and biochemical advanced techniques, environmental chemistry, and biochemical toxicology. He has held visiting professorships at the University of Bologna and the University of Verona, Italy. His research interests include: bacterial resistance mechanisms, including metal and metalloid toxicity and tolerance as well as multidrug resistance efflux pumps; bioremediation of metals and organic pollutants; protein translocation by the Tat system and the accessory proteins involved.

Li-Nan Huang, of the School of Life Sciences at San Yat-sen University, has made significant contributions to research focussed on the microbiology of mine tailings and acid mine drainage. He is at the forefront of understanding how stressors affect microbial community composition and their variations. He also manages a productive group funded by the National Natural Science Foundation of China.

Editorial

Metal-Resistance in Bacteria: Why Care?

Raymond J. Turner [1,*], Li-Nan Huang [2], Carlo Viti [3] and Alessio Mengoni [4,*]

1. Department of Biological Sciences, Faculty of Science, University of Calgary, Calgary, AB T2N 1N4, Canada
2. School of Life Sciences, Sun Yat-Sen University, Guangzhou 510275, China; eseshln@mail.sysu.edu.cn
3. Laboratorio Genexpress, Dipartimento di Scienze e Tecnologie Agrarie, Alimentari, Ambientali e Forestali, Università di Firenze, 50144 Florence, Italy; carlo.viti@unifi.it
4. Laboratorio di Genetica Microbica, Dipartimento di Biologia, Università di Firenze, 50019 Florence, Italy
* Correspondence: turnerr@ucalgary.ca (R.J.T.); alessio.mengoni@unifi.it (A.M.)

Received: 26 November 2020; Accepted: 3 December 2020; Published: 8 December 2020

Heavy metal resistance is more than the tolerance one has towards a particular music genera. The study of metal resistance mechanisms in bacteria traces back to the 1970s and through to the mid 1990s. During these early days, specific metal or metalloid ion resistance determinants, consisting of single metal(loid) resistance genes (MRGs) to large complex operons, were being identified on large conjugative plasmids and other mobile genetic elements. These determinants were often used to classify these accessory plasmid components of genomes. Thinking back to a conference on this topic with speakers from our ancestors of this field, such as Simon Silver, Ann Summers, Barry Rosen, Diane Taylor, Geoff Gadd, Dietrich Nies, and Max Mergeay, who were presenting their work of cloning, sequencing, and characterizing metal resistance in microbes at this time. This early work performed in the pre-omics' era made great strides in exploring bacteria response to silver, nickel, cadmium, mercury, copper, arsenite/arsenate, and tellurite. It was not that long before it was realized that metal resistance in bacteria essentially follows a limited number of biochemical processes [1–3], e.g., prevention of metals' uptake; if it gets in; efflux it back out again; sequestration through metal binding proteins or chelating metabolites; oxidation-reduction to change redox state or other chemical modification (either removal or addition of organic constituents) to change the metal's speciation; sequestration through precipitation to metal crystal form or the production of metal binding proteins or chelating metabolites. These seem like trivial statements to say today, but a remarkable amount of work has been put forward to understand such processes at the genetic, biochemical, and structural biology levels. Yet, even with the power of omics approaches, there are still many metal-microbe interaction puzzles left to solve.

The work exploring specific metal-resistance determinants has been complemented by those researchers exploring the ability of various bacterial species to respire using different metal(loids) as electron donors or acceptors [4,5]. Additionally, the studies on metal resistance over the past 50 years, derived primarily from the clinical environment, are complemented by the work evaluating the microbiology of extreme environments, from deep sea vents to mine drainage/tailings and industrial sites, evaluating bacteria's role in geochemistry. This work gave us the multi-metal resistant *Cupriavidus metallidurans*, which has become an important model organism in this regard [6]. The genomics of this species strains have provided us amazing insight into how bacteria can survive high metal loads and how bacteria can survive anthropogenically abused environments. It would be impossible to deny the advances in knowledge that the genomic revolution has given our field. Even as early as the late 2000s, it took 3 years of work to sequence a strain to obtain a rough draft of an aluminum resistant polychlorinated biphenyl (PCBs) degrading strain [7]. At the time, getting this information was remarkable, and this sequenced genome sparked new hypothesis and important findings. On reflection, it could take 6 months to sequence an operon in 1990. Now, of course, sequencing and assembling a genome can be done in a week. Multiple strains can be sequenced and their genomes compared for unique single nucleotide polymorphisms (SNPs) and gene operon changes. Bioinformatic mining of genomes allows for an understanding of specific genetic traits related to metals [8]. Beyond sequencing, other omic approaches have evolved and been applied to the field of metal resistance in bacteria, including proteomic (example [9]), metabolomic [10,11], and comparative genomics approaches [12], methods of chemical genomics [13], or the comprehensive approach of resistance metalloproteomics [14]. Combining various omics together to look at the response of the transcriptome, proteome, and metabolome by metals is referred to as metallomics [15].

So why do we care, or why should we care, about metal-resistance in bacteria? The research directions described above still continue in labs around the world, but now more often focus on the advent of biotechnological or bioremediation advances. Over the past decade, we have seen the knowledge of metal resistance in bacteria be used in the eco-friendly production of a wide variety of metal nanomaterials [16]. The appreciation of the normal sensitivity of most bacteria to several metals has led to a resurgence of their use as metal(loid)-based antimicrobials [17,18] as a result of moving into the antimicrobial resistance era and the need for new and novel antimicrobials. As such, we have also seen an exponential use of metal(loid)-based nanoparticles used as antimicrobial agents [19]. Of course, resistance has already started to develop against different metal nanomaterial formulations [20].

Through the journey from the 1970s, we have obtained a good view of the acquired MRGs. It is now reasonably well established that many are found on mobile genetic elements and genomic islands similar to antibiotic resistant genes (ARGs). Using modern day genomics, we can see beyond the specific gene determinants and toward the full system responses of metal challenges to bacteria. We have begun to see various global regulator systems, such as MarR [21], providing regulated tolerance to both antibiotics and metals. Similarly, we see multidrug resistance efflux pumps providing co-resistance to metals, antiseptics, and antibiotics [22]. This also helps us to understand the link between the use of metal ions in agriculture practices and its influence on the world's antimicrobial resistance challenges [23].

It was through the variety of genomic approaches that we found the genes, metabolic pathways, and key enzymes involved in resistance and tolerance mechanisms in bacteria. Yet knowledge gaps exist in our understanding of bacterial sensitivity to metal challenges. How do naïve bacterial species respond to metal stress? Can we see metal resistance develop in real-time? Our various anthropogenic activities have led to metal resistance bacteria in aquatic and marine environments [24]. This is beginning to allow us to understand how bacteria survive acute metal ion challenges as well as chronically living under constant metal exposed aggression.

We have learned a lot about metal-resistance to date. What does the future hold in this field that genomics tools will feed? As metabolic modeling of microbes improves [25], how will our view and use of metal-resistance in bacteria change? Pontification here gives possibilities of novel metal(loid) respiring species, bioremediation strategies for the many metal polluted sites world-wide, novel metal-based antimicrobial treatments, biocatalysts in green chemistry, understanding of bacterial evolution in relationship to the Earth's geological history, and modelling natural selection of microbial communities and microbial strains.

The present Special Issue, which includes two reviews [26,27], two featured papers [28,29], and eight original manuscripts [30–37] covers many of the above-mentioned aspects of genomics in bacteria resistance. The review papers discuss the knowledge and perspective of heavy-metal resistance in human pathogens and in the challenge of plant symbiotic microbiome exploitation in phytoremediation of heavy-metal polluted soils. The research papers present novel data on the genetics of resistance in model and pathogenic species and in biotechnologically relevant strains. Witnessing the need to still fully understand the genetics and evolution of heavy-metal resistance novel work on the previously mentioned model bacterium *C. metallidurans* is also presented.

Conflicts of Interest: The authors declare no conflict of interest.

References

1. Summers, A.O.; Silver, S. Microbial Transformations of Metals. *Annu. Rev. Microbiol.* **1978**, *32*, 637–672. [CrossRef] [PubMed]
2. Silver, S.; Phung, L.T. A bacterial view of the periodic table: Genes and proteins for toxic inorganic ions. *J. Ind. Microbiol. Biotechnol.* **2005**, *32*, 587–605. [CrossRef] [PubMed]
3. Hobman, J.L.; Crossman, L.C. Bacterial antimicrobial metal ion resistance. *J. Med. Microbiol.* **2015**, *64*, 471–497. [CrossRef] [PubMed]
4. Fredrickson, J.K.; Romine, M.F. Genome-assisted analysis of dissimilatory metal-reducing bacteria. *Curr. Opin. Biotechnol.* **2005**, *16*, 269–274. [CrossRef]
5. Csotonyi, J.T.; Stackebrandt, E.; Yurkov, V.V. Anaerobic Respiration on Tellurate and Other Metalloids in Bacteria from Hydrothermal Vent Fields in the Eastern Pacific Ocean. *Appl. Environ. Microbiol.* **2006**, *72*, 4950–4956. [CrossRef]
6. Mergeay, M.; Van Houdt, R. (Eds.) *Metal Response in Cupriavidus metallidurans: Volume I: From Habitats to Genes and Proteins*; Springer Science and Business Media LLC.: Cham, Switzerland, 2015. [CrossRef]
7. Triscari-Barberi, T.; Simone, D.; Calabrese, F.M.; Attimonelli, M.; Hahn, K.R.; Amoako, K.K.; Turner, R.J.; Fedi, S.; Zannoni, D. Genome Sequence of the Polychlorinated-Biphenyl Degrader *Pseudomonas pseudoalcaligenes* KF707. *J. Bacteriol.* **2012**, *194*, 4426–4427. [CrossRef]
8. Bini, E. Archaeal transformation of metals in the environment. *FEMS Microbiol. Ecol.* **2010**, *73*, 1–16. [CrossRef]
9. Zammit, C.M.; Weiland, F.; Brugger, J.; Wade, B.; Winderbaum, L.J.; Nies, D.H.; Southam, G.; Hoffmann, P.; Reith, F. Proteomic responses to gold(iii)-toxicity in the bacterium *Cupriavidus metallidurans* CH34. *Metallomics* **2016**, *8*, 1204–1216. [CrossRef]
10. Tremaroli, V.; Workentine, M.L.; Weljie, A.M.; Vogel, H.J.; Ceri, H.; Viti, C.; Tatti, E.; Zhang, P.; Hynes, A.P.; Turner, R.J.; et al. Metabolomic Investigation of the Bacterial Response to a Metal Challenge. *Appl. Environ. Microbiol.* **2009**, *75*, 719–728. [CrossRef]
11. Booth, S.C.; Workentine, M.L.; Weljie, A.M.; Turner, R.J. Metabolomics and its application to studying metal toxicity. *Metals* **2011**, *3*, 1142–1152. [CrossRef]
12. Permina, E.A.; Kazakov, A.E.; Kalinina, O.V.; Gelfand, M.S. Comparative genomics of regulation of heavy metal resistance in Eubacteria. *BMC Microbiol.* **2006**, *6*, 49. [CrossRef] [PubMed]
13. Zheng, X.F.; Chan, T.F. Chemical Genomics: A Systematic Approach in Biological Research and Drug Discovery. *Curr. Issues Mol. Biol.* **2002**, *4*, 33–43. [CrossRef] [PubMed]
14. Wang, H.; Yan, A.; Liu, Z.; Yang, X.; Xu, Z.; Wang, Y.; Wang, R.; Koohi-Moghadam, M.; Hu, L.; Xia, W.; et al. Deciphering molecular mechanism of silver by integrated omic approaches enables enhancing its antimicrobial efficacy in *E. coli*. *PLoS Biol.* **2019**, *17*, e3000292. [CrossRef] [PubMed]
15. Haferburg, G.; Kothe, E. Metallomics: Lessons for metalliferous soil remediation. *Appl. Microbiol. Biotechnol.* **2010**, *87*, 1271–1280. [CrossRef] [PubMed]
16. Choi, Y.; Lee, S.Y. Biosynthesis of inorganic nanomaterials using microbial cells and bacteriophages. *Nat. Rev. Chem.* **2020**, 1–19. [CrossRef]
17. Lemire, J.A.; Harrison, J.J.; Turner, R.J. Antimicrobial activity of metals: Mechanisms, molecular targets and applications. *Nat. Rev. Genet.* **2013**, *11*, 371–384. [CrossRef] [PubMed]
18. Turner, R.J. Metal-based antimicrobial strategies. *Microb. Biotechnol.* **2017**, *10*, 1062–1065. [CrossRef]
19. Sánchez-López, E.; Gomes, D.; Esteruelas, G.; Bonilla, L.; Machado, A.L.; Galindo, R.; Cano, A.; Espina, M.; Ettcheto, M.; Camins, A.; et al. Metal-Based Nanoparticles as Antimicrobial Agents: An Overview. *Nanomaterials* **2020**, *10*, 292. [CrossRef]
20. Niño-Martínez, N.; Salas-Orozco, M.; Martinez-Castañon, G.A.; Méndez, F.T.; Ruiz, F. Molecular Mechanisms of Bacterial Resistance to Metal and Metal Oxide Nanoparticles. *Int. J. Mol. Sci.* **2019**, *20*, 2808. [CrossRef]
21. Chen, S.; Li, X.; Sun, G.-X.; Zhang, Y.; Su, J.-Q.; Ye, J. Heavy Metal Induced Antibiotic Resistance in Bacterium LSJC7. *Int. J. Mol. Sci.* **2015**, *16*, 23390–23404. [CrossRef]
22. Yu, Z.; Gunn, L.; Wall, P.; Fanning, S. Antimicrobial resistance and its association with tolerance to heavy metals in agriculture production. *Food Microbiol.* **2017**, *64*, 23–32. [CrossRef] [PubMed]
23. Baker-Austin, C.; Wright, M.S.; Stepanauskas, R.; McArthur, J. Co-selection of antibiotic and metal resistance. *Trends Microbiol.* **2006**, *14*, 176–182. [CrossRef] [PubMed]

24. Squadrone, S. Water environments: Metal-tolerant and antibiotic-resistant bacteria. *Environ. Monit. Assess.* **2020**, *192*, 238. [CrossRef] [PubMed]
25. Kim, W.J.; Kim, H.U.; Lee, S.Y. Current state and applications of microbial genome-scale metabolic models. *Curr. Opin. Syst. Biol.* **2017**, *2*, 10–18. [CrossRef]
26. Fagorzi, C.; Checcucci, A.; DiCenzo, G.C.; Debiec-Andrzejewska, K.; Dziewit, L.; Pini, F.; Mengoni, A. Harnessing Rhizobia to Improve Heavy-Metal Phytoremediation by Legumes. *Genes* **2018**, *9*, 542. [CrossRef]
27. Parsons, C.; Lee, S.; Kathariou, S. Heavy Metal Resistance Determinants of the Foodborne Pathogen *Listeria monocytogenes*. *Genes* **2018**, *10*, 11. [CrossRef]
28. Gugala, N.; Lemire, J.; Chatfield-Reed, K.; Yan, Y.; Chua, G.; Turner, R.J. Using a Chemical Genetic Screen to Enhance Our Understanding of the Antibacterial Properties of Silver. *Genes* **2018**, *9*, 344. [CrossRef]
29. Oetiker, N.; Norambuena, R.; Martínez-Bussenius, C.; Navarro, C.A.; Amaya, F.; Álvarez, S.A.; Paradela, A.; Jerez, C.A. Possible Role of Envelope Components in the Extreme Copper Resistance of the Biomining *Acidithiobacillus ferrooxidans*. *Genes* **2018**, *9*, 347. [CrossRef]
30. Ali, M.; Provoost, A.; Maertens, L.; Leys, N.; Monsieurs, P.; Charlier, D.; Van Houdt, R. Genomic and Transcriptomic Changes That Mediate Increased Platinum Resistance in *Cupriavidus metallidurans*. *Genes* **2019**, *10*, 63. [CrossRef]
31. Al-Jabri, Z.; Zamudio, R.; Horvath-Papp, E.; Ralph, J.D.; Al-Muharrami, Z.; Rajakumar, K.; Oggioni, M.R. Integrase-Controlled Excision of Metal-Resistance Genomic Islands in *Acinetobacter baumannii*. *Genes* **2018**, *9*, 366. [CrossRef]
32. Chalmers, G.; Rozas, K.M.; Amachawadi, R.G.; Scott, H.M.; Norman, K.N.; Nagaraja, T.G.; Tokach, M.D.; Boerlin, P. Distribution of the *pco* Gene Cluster and Associated Genetic Determinants among Swine *Escherichia coli* from a Controlled Feeding Trial. *Genes* **2018**, *9*, 504. [CrossRef] [PubMed]
33. Van Houdt, R.; Provoost, A.; Van Assche, A.; Leys, N.; Lievens, B.; Mijnendonckx, K.; Monsieurs, P. *Cupriavidus metallidurans* Strains with Different Mobilomes and from Distinct Environments Have Comparable Phenomes. *Genes* **2018**, *9*, 507. [CrossRef] [PubMed]
34. DiCenzo, G.C.; Debiec-Andrzejewska, K.; Krzysztoforski, J.; Uhrynowski, W.; Mengoni, A.; Fagorzi, C.; Gorecki, A.; Dziewit, L.; Bajda, T.; Rzepa, G.; et al. Genomic and Biotechnological Characterization of the Heavy-Metal Resistant, Arsenic-Oxidizing Bacterium *Ensifer* sp. M14. *Genes* **2018**, *9*, 379. [CrossRef] [PubMed]
35. Gugala, N.; Chatfield-Reed, K.; Turner, R.J.; Chua, G. Using a Chemical Genetic Screen to Enhance Our Understanding of the Antimicrobial Properties of Gallium against *Escherichia coli*. *Genes* **2019**, *10*, 34. [CrossRef] [PubMed]
36. Li, Y.P.; Carraro, N.; Yang, N.; Liu, B.; Xia, X.; Feng, R.; Saquib, Q.; Al-Wathnani, H.A.; Van Der Meer, J.R.; Rensing, C. Genomic Islands Confer Heavy Metal Resistance in *Mucilaginibacter kameinonensis* and *Mucilaginibacter rubeus* Isolated from a Gold/Copper Mine. *Genes* **2018**, *9*, 573. [CrossRef]
37. Millacura, F.A.; Janssen, P.J.; Monsieurs, P.; Janssen, A.; Provoost, A.; Van Houdt, R.; Rojas, L.A. Unintentional Genomic Changes Endow *Cupriavidus metallidurans* with an Augmented Heavy-Metal Resistance. *Genes* **2018**, *9*, 551. [CrossRef]

Publisher's Note: MDPI stays neutral with regard to jurisdictional claims in published maps and institutional affiliations.

© 2020 by the authors. Licensee MDPI, Basel, Switzerland. This article is an open access article distributed under the terms and conditions of the Creative Commons Attribution (CC BY) license (http://creativecommons.org/licenses/by/4.0/).

Article

Cupriavidus metallidurans Strains with Different Mobilomes and from Distinct Environments Have Comparable Phenomes

Rob Van Houdt [1,*], Ann Provoost [1], Ado Van Assche [2], Natalie Leys [1], Bart Lievens [2], Kristel Mijnendonckx [1] and Pieter Monsieurs [1]

1   Microbiology Unit, Belgian Nuclear Research Centre (SCK•CEN), B-2400 Mol, Belgium; aprovoos@sckcen.be (A.P.); nleys@sckcen.be (N.L.); kmijnend@sckcen.be (K.M.); pmonsieu@sckcen.be (P.M.)
2   Laboratory for Process Microbial Ecology and Bioinspirational Management, KU Leuven, B-2860 Sint-Katelijne-Waver, Belgium; ado.vanassche@kuleuven.be (A.V.A.); bart.lievens@kuleuven.be (B.L.)
*   Correspondence: rvhoudto@sckcen.be

Received: 21 September 2018; Accepted: 15 October 2018; Published: 18 October 2018

Abstract: *Cupriavidus metallidurans* has been mostly studied because of its resistance to numerous heavy metals and is increasingly being recovered from other environments not typified by metal contamination. They host a large and diverse mobile gene pool, next to their native megaplasmids. Here, we used comparative genomics and global metabolic comparison to assess the impact of the mobilome on growth capabilities, nutrient utilization, and sensitivity to chemicals of type strain CH34 and three isolates (NA1, NA4 and H1130). The latter were isolated from water sources aboard the International Space Station (NA1 and NA4) and from an invasive human infection (H1130). The mobilome was expanded as prophages were predicted in NA4 and H1130, and a genomic island putatively involved in abietane diterpenoids metabolism was identified in H1130. An active CRISPR-Cas system was identified in strain NA4, providing immunity to a plasmid that integrated in CH34 and NA1. No correlation between the mobilome and isolation environment was found. In addition, our comparison indicated that the metal resistance determinants and properties are conserved among these strains and thus maintained in these environments. Furthermore, all strains were highly resistant to a wide variety of chemicals, much broader than metals. Only minor differences were observed in the phenomes (measured by phenotype microarrays), despite the large difference in mobilomes and the variable (shared by two or three strains) and strain-specific genomes.

Keywords: phenotype microarray; mobile genetic elements; *Cupriavidus*; metal; resistance

1. Introduction

Cupriavidus metallidurans type strain CH34, which was isolated from a decantation basin in the non-ferrous metallurgical factory at Engis, Belgium [1], has been mostly studied because of its resistance to numerous heavy metals [2]. It tolerates high concentrations of metal (oxyan)ions, including Cu^+, Cu^{2+}, Ni^{2+}, Zn^{2+}, Co^{2+}, Cd^{2+}, CrO_4^{2-}, Pb^{2+}, Ag^+, Au^+, Au^{3+}, $HAsO_4^{2-}$, AsO^{2-}, Hg^{2+}, Cs^+, Bi^{3+}, Tl^+, SeO_3^{2-}, SeO_4^{2-} and Sr^{2+} [2,3]. Metal detoxification is encoded by at least 24 gene clusters and many of them are carried by its two megaplasmids pMOL28 and pMOL30 [4]. Resistance to metal ions is mediated by multiple systems, including transporters belonging to the resistance nodulation cell division (RND), the cation diffusion facilitator (CDF) and the P-type ATPase families [2,5].

Cupriavidus metallidurans strains have characteristically been isolated from metal-contaminated industrial environments such as soils around metallurgical factories in the Congo (Katanga) and North-Eastern Belgium [6,7], as well as from contaminated soils in Japan [8] and gold mining sites in

Queensland (Australia) [9]. Other environments include sewage plants [10], laboratory wastewater (Okayama University, Okayama, Japan) [11] and spacecraft assembly cleanrooms [12]. In addition, *C. metallidurans* strains were also found in the drinking water and dust collected from the International Space Station (ISS) [12,13].

Remarkably, more and more reports describe the isolation of *C. metallidurans* strains from medically-relevant settings and sources such as the pharmaceutical industry, human cerebrospinal fluid and cystic fibrosis patients [14]. It remains to be elucidated if the isolates caused the active infection or only intruded as secondary opportunistic pathogens [14]. Nevertheless, an invasive human infection and four cases of catheter-related infections caused by *C. metallidurans* were recently reported [15,16].

All *Cupriavidus* genomes characteristically carry, next to their chromosome, a second large replicon. This 2 to 3 Mb-sized replicon has recently been coined chromid as it neither fully fits the term chromosome nor plasmid [17,18]. In addition to the chromid, most *Cupriavidus* strains harbor one or more megaplasmids (100 kb or larger in size), which probably mediate the adaptation to certain ecological niches by the particular functions they encode (see [19] for detailed review). For instance, pMOL28 and pMOL30 from *C. metallidurans* CH34 are pivotal in metal ion resistance [4]; hydrogenotrophic and chemolithotrophic metabolism are encoded by pHG1 from *Cupriavidus necator* H16 [20], and pRALTA from *Cupriavidus taiwanensis* LMG19424 codes for nitrogen fixation and legume symbiosis functions [21]. Next to these megaplasmids, other plasmids (mostly broad host range) can be present. One example is pJP4 from *Cupriavidus pinatubonensis* JMP134, which is a broad host range IncP-1β plasmid involved in the degradation of substituted aromatic pollutants [22].

The *C. metallidurans* mobilome is completed with a large diversity of genomic islands (GIs), integrative and conjugative elements, transposons and insertion sequence (IS) elements [7,23–25]. Many mobile genetic elements (MGEs) carry accessory genes beneficial for adaptation to particular niches (resistance, virulence, catabolic genes), but acquired genes may also impact the host by cross-talk to host global regulatory networks [26]. In addition, without accessory genes, MGEs such as IS elements can have an impact on genome plasticity and concomitant adaptability of phenotypic traits, including resistance to antibacterial agents, virulence, pathogenicity and catabolism [27]. Finally, the presence of prophages, until now not identified in *C. metallidurans*, may also affect many different traits and lead to phenotypic changes in the host [28,29].

Recently, we showed that *C. metallidurans* strains share most metal resistance determinants irrespective of their isolation type and place [7]. In contrast, significant differences in the size and diversity of their mobilome was observed. However, our comparison was based on whole-genome hybridization to microarrays containing oligonucleotide probes present on the CH34 microarray. These observations triggered us to further study the diversity of the mobilome, its relation to the environment and impact on the host's global phenome. Therefore, we inventoried the mobilomes and compared the global metabolic capabilities of type strain CH34, strains NA1 and NA4 isolated from water sources aboard ISS [12], and H1130 isolated from an invasive human infection [15]. The global metabolic activities were assessed by employing phenotype microarrays (PMs), which highlight differences in growth requirements, nutrient utilization and sensitivity to chemicals [30].

2. Materials and Methods

2.1. Strains, Media and Culture Conditions

Bacterial strains and plasmids used in this study are summarized in Table 1. *Cupriavidus metallidurans* strains were routinely cultured at 30 °C in lysogeny broth (LB) or tris-buffered mineral medium (MM284) supplemented with 0.2% (*w/v*) gluconate [1]. *Escherichia coli* strains were routinely cultured at 37 °C in LB. Liquid cultures were grown in the dark on a rotary shaker at 150 rpm. For culturing on agar plates, 1.5% agar (Thermo Scientific, Oxoid, Hampshire, UK) was added. When appropriate, the following chemicals (Sigma-Aldrich (Overijse, Belgium) or Fisher Scientific (Merelbeke, Belgium)) were added to the growth medium at the indicated final concentrations: kanamycin (50 µg/mL for *E. coli* or 1500 µg/mL for *C. metallidurans*), tetracycline (20 µg/mL), 5-bromo-4-chloro-3-indolyl-β-D-galactopyranoside (X-Gal; 40 µg/mL), isopropyl-β-D-thiogalactopyranoside (IPTG; 0.1 mM) and diaminopimelic acid (DAP; 1 mM).

Table 1. Strains and plasmids used in this study.

Strain or Plasmid	Genotype/Relevant Characteristics	Reference
STRAIN		
Cupriavidus metallidurans		
CH34^T	Type strain	[31]
NA1	Isolated from a water sample, ISS	[12]
NA4	Isolated from a water sample, ISS	[12]
NA4 ΔCRISPR	ΔCRISPR::*tet*, Tc^R	This study
H1130	Isolated from invasive human infection	[15]
Escherichia coli		
DG1	*mcrA* Δ*mrr-hsdRMS-mcrBC* (r$_B^-$ m$_B^-$) Φ80*lacZ*Δ*M15* Δ*lacX74 recA1 araD139* Δ*(ara-leu)7697 galU galK rpsL endA1 nupG*	Eurogentec
MFDpir	MG1655 RP4-2-Tc::[Δ*Mu1*::*aac(3)IV*-Δ*aphA*-Δ*nic35*-Δ*Mu2*::*zeo*] Δ*dapA*::(*erm-pir*) Δ*recA*	[32]
PLASMID		
pK18mob	pMB1 ori, *mob+, lacZ*, Km^R	[33]
pK18mob-CRISPR	CRISPR region of NA4 in pK18mob, Km^R	This study
pK18mob-CRISPR::*tet*	pK18mob-CRISPR derivative, CRISPR::*tet*, Km^R, Tc^R	This study
pACYC184	p15A ori, Cm^R, Tc^R	[34]
pJB3kan1	RK2 minimal replicon; Ap^R, Km^R	[35]
pJB3kan1_Rmet2825	Rmet_2825 of CH34 in pJB3kan1; Km^R	This study

Eurogentec: Seraing, Belgium, Km^R: kanamycine resistant, Tc^R: tetracycline resistant, Cm^R: chloramphenicol resistant, Ap^R: ampicillin resistant.

2.2. Growth in the Presence of Metals

Cupriavidus metallidurans CH34, NA1, NA4 and H1130 were cultivated in MM284 at 30 °C up to stationary phase (10^9 CFU/mL) and 10 µL of a ten-fold serial dilution in 10 mM MgSO$_4$ were spotted on MM284 agar plates containing various metal concentrations (Table S1). Colony forming units (CFU) were counted after 4–5 days. Data are presented as log(N)/log(N$_0$) in function of metal concentration, with N and N$_0$ CFUs in the presence and absence (control) of metal, respectively.

2.3. NA4 CRISPR Deletion Construction

The CRISPR region of *C. metallidurans* NA4 was amplified by PCR (Phusion High-Fidelity DNA polymerase) (Fisher Scientific, Merelbeke, Belgium) with primer pairs CRSPR_Fw-Rv (Table S2), providing XbaI/HindIII restriction sites. Afterwards, this PCR product was cloned as a XbaI/HindIII fragment into the mobilizable suicide vector pK18mob. The resulting pK18mob_CRISPR plasmid from an *E. coli* DG1 transformant selected on LB Km50 was further confirmed by sequencing prior to amplifying of the flanking CRISPR sequences by inverse PCR (Phusion High-Fidelity DNA polymerase) with primer pair CRISPR_tet_Fw-Rv (Table S2), providing BcuI/BspTI restriction sites. At the same time, the *tet* gene from pACYC184 (Table 1 [34]) was amplified by PCR (Phusion High-Fidelity DNA polymerase) with primer pair Tet_Fw-Rv (Supplementary Table S1), providing BcuI/BspTI restriction sites. Afterwards, this PCR product was cloned as a BcuI/BspTI fragment into the former inverse PCR product. The resulting pK18mob-CRISPR::*tet* plasmid from an *E. coli* DG1 transformant selected on LB

Tc20 Km50 was further confirmed by sequencing prior to conjugation (with *E. coli* MFDpir as donor host [32]) to *C. metallidurans* NA4. The resulting transformants selected on LB Tc20 were replica plated on LB Tc20 and LB Km1500. NA4 ΔCRISPR::*tet* cells resistant to Tc20 but sensitive to Km1500 were further confirmed by sequencing.

2.4. Construction of Plasmids

PCR amplification of *C. metallidurans* CH34 Rmet_2825 was performed on genomic DNA from *C. metallidurans* CH34 with primer pair Rmet2825_Fw-Rv (Table S2). This amplicon was subsequently cloned into pJB3kan1, which was linearized by PCR amplification with the primers pJB3kan1_Fw-Rv (Table S2), using the GeneArt™ Seamless Cloning and Assembly Enzyme Mix (Fisher Scientific, Merelbeke, Belgium). The resulting pJB3kan1-Rmet2825 plasmid from *E. coli* DG1 transformants selected on LB Km50 was further confirmed by sequencing prior to transformation to *E. coli* MFDpir.

2.5. Conjugation Assay for Testing CRISPR-Cas

Donor (*E. coli* MFDpir pJB3kan1-Rmet2825) and recipient (*C. metallidurans* NA4 or NA4 ΔCRISPR::*tet*) were grown overnight at 37° in LB Km50 DAP, and at 30° in LB, respectively. Fifty μL of donor and recipient were spotted on a 0.45 μm Supor® membrane disc filter (Pall Life Sciences, Hoegaarden, Belgium) that was put on a LB DAP plate. After overnight incubation at 30 °C, cells were resuspended in 1 mL of 10 mM $MgSO_4$ and 10-fold serial diluted on LB Km50 DAP (37 °C), LB (30 °C) and LB Km1500 plates (30 °C) to count CFU of donors, recipients and transconjugants, respectively. Conjugation frequency was measured as the number of transconjugants per donor cell (T/D) and per recipient cell (T/R).

2.6. Plasmid Profiling

The extraction of megaplasmids was based on the method proposed by Andrup et al. [36]. Extracted plasmid DNA was separated by horizontal gel electrophoresis (0.5% Certified Megabase agarose gel (Bio-Rad, Temse, Belgium) in 1X TBE buffer, 100 V, 20 h) in a precooled (4 °C) electrophoresis chamber. After GelRed staining (30 min + overnight destaining at 4 °C in ultrapure water), DNA was visualized and images captured under UV light transillumination (Fusion Fx, Vilber Lourmat, Collégien, France).

2.7. Phenotype Microarray Analysis

Phenotype microarray (PM) analysis was performed using the OmniLog® automated incubator/reader (Biolog Inc., Hayward, CA, USA) following manufacturer's instruction (PM procedures for *E. coli* and other GN Bacteria version 16-Jan-06 with slight modifications). Briefly, cells were suspended in Biolog's inoculation fluid IF-0a (1x) until an optical density (600 nm) of 0.2 was reached. Subsequently, a 1:50 dilution was made in IF-0a (1x) containing dye mix A. Furthermore, 2 mM sodium succinate and 2 μM ferric citrate (Sigma-Aldrich, Overijse, Belgium) were used as carbon sources in PM 3 till 8. All 20 plates (PM-1 through PM-20) inoculated with bacterial cell suspensions, were incubated at 30 °C and cell respiration was measured every 30 min for 144 h. Raw kinetic data were retrieved using the OmniLog—OL_PM_FM/Kin 1.30-: File Management/Kinetic Plot Version software of Biolog. Analysis was carried out with the R-library OPM (version 1.3.64) [37,38]. The area under the curve (AUC) threshold to decide whether a strain is or is not growing in a specific well of the PM, was derived by plotting the AUC values of all PM reactions for each strain, showing in all conditions an almost bimodal distribution. The AUC threshold (one value for all four strains) was determined as the value separating both major peaks (threshold value of 8000) (Figure S1). Negative control wells that contained the inoculated Omnilog™ growth medium without any substrate were measured to normalize differences in inocula and redox dye oxidation between samples.

2.8. Computational Methods

The pan-genome analysis was performed via the MaGe platform [39], which uses MicroScope gene families (MICFAM) that are computed with an algorithm implemented in the SiLiX software [40]. The alignment constraints to compute the MICFAM families were 80% amino-acid identity and 80% amino-acid alignment coverage. The MICFAM is part of the core-genome if associated with at least one gene from every compared genome (see Table S3 for complete data set).

A phylogenetic tree of the genomes was constructed via the MaGe platform from the pairwise genome distances using a neighbor-joining algorithm. The pairwise genome distance was calculated with Mash [41].

The CARD (comprehensive antibiotic resistance database) [42] implementation within the MaGe platform [39] was used to identify known resistance determinants and associated antibiotics. All predictions were strict as defined by CARD, meaning a match above the CARD curated bitscore cut-offs [42–44].

A BLAST search against BacMet (antibacterial biocide and metal resistance genes database) was used to inventory genes predicted to confer resistance to metals and/or antibacterial biocides [45]. The alignment constraints were 35% amino-acid identity and 80% amino-acid alignment coverage.

The different constraints used to compute the MICFAM families and CARD/BacMet BLAST hits can result in minor differences in the number of core genome genes from a particular strain that results in a positive CARD/BacMet hit.

3. Results and Discussion

Four *C. metallidurans* strains were selected: type strain CH34 [3], strain NA1 and NA4 isolated from the drinking water systems onboard the International Space Station that were analyzed previously and had mobilomes divergent from that of CH34 [7], and strain H1130, recently isolated from an invasive human infection [15]. This selection allows comparing the type strain with two strains isolated from a similar environment but with different mobilomes (at least based on elements known in CH34 [7]) and an isolate from a human infection.

3.1. Comparison of General Genome Features

The genome of *C. metallidurans* NA1, NA4 and H1130 was previously sequenced [46,47] and estimated to be 6,833,318 bp, 7,370,364 bp and 7,225,099 bp, respectively (with type strain CH34 being 6,913,352 bp [3]). The G + C content of the genomes are very similar to each other, with 63.76%, 63.27%, 63.50% and 63.82% for NA1, NA4, H1130 and CH34, respectively. NA4 contained the most coding sequences (CDSs) (7467), followed by H1130 (7032), NA1 (6815) and CH34 (6757). All strains contained multiple replicons, namely, one chromosome, one chromid and megaplasmids (>100 kb [19]). Strain NA1 carries two megaplasmids. Strain NA4 carries three megaplasmids and one plasmid. Strain H1130 carries only one megaplasmid (Figure 1).

The core genome contains 4697 MICFAM gene families shared by all four strains, which relates to 70.9%, 70.2%, 65.4% and 69.9% of the total CDSs of CH34, NA1, NA4 and H1130, respectively. This means that roughly 30% to 35% of the CDSs belong to the variable (shared by two or three strains) or strain-specific genome (Figure 2). Strains CH34 and NA4 shared the most gene families (Figure 3). Furthermore, the Mash-distance-based phylogeny (Figure 4) indicated that NA4 and CH34 were the most closely related. In addition, NA4 shared more gene families with H1130 and CH34 than with NA1, which corresponded with the phylogenetic distance. These data indicated that NA1 and NA4 were not the two most similar strains, despite their isolation from the same environment.

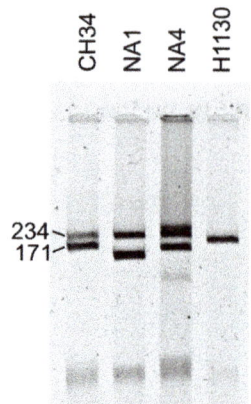

Figure 1. Agarose gel electrophoresis of *Cupriavidus metallidurans* CH34, NA1, NA4 and H1130 (mega)plasmid DNA. The characterized CH34 megaplasmids pMOL30 (234 kb) and pMOL28 (171 kb) serve as reference.

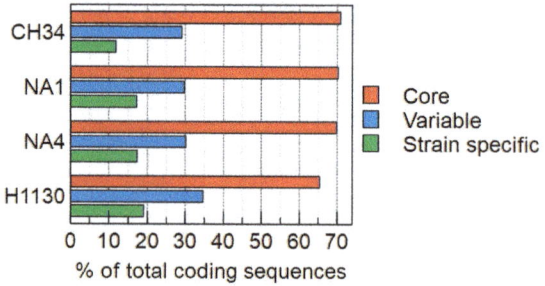

Figure 2. Percentage of coding sequences (CDSs) belonging to the core, variable (shared by two or three strains) and strain-specific genome of *Cupriavidus metallidurans* CH34, NA1, NA4 and H1130.

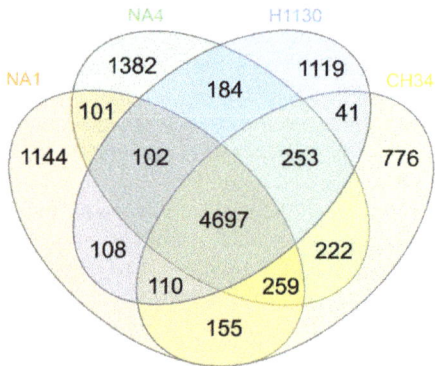

Figure 3. Venn diagram displaying the distribution of shared MicroScope gene families (MICFAM) among *C. metallidurans* CH34, NA1, NA4 and H1130.

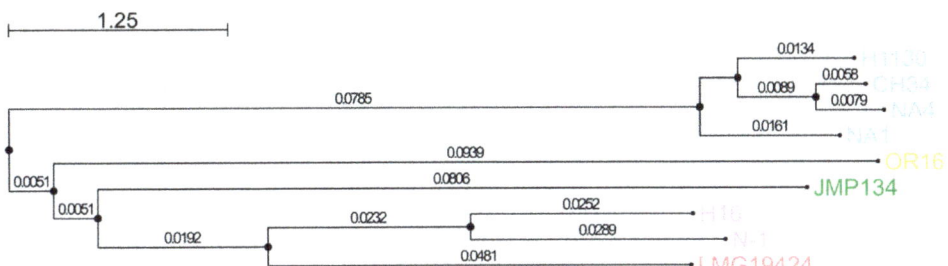

Figure 4. Neighbor-joining phylogenetic tree of *C. metallidurans* CH34, NA1, NA4 and H1130, based on the genome pairwise distance matrix calculated with Mash. *Cupriavidus basilensis*: OR16, *Cupriavidus pinatubonensis*: JMP134, *Cupriavidus necator*: H16 and N-1, and *Cupriavidus taiwanensis*: LMG19424 were included for comparison.

Evidently, with 1827 Microscope gene families shared with *Cupriavidus taiwanensis* LMG19424, *Cupriavidus necator* H16, *Cupriavidus pinatubonensis* JMP134, *Cupriavidus basilensis* OR16 and *Cupriavidus necator* N-1, the *C. metallidurans* strains share more gene families among each other than with strains of different *Cupriavidus* species. Strains CH34, NA1, NA4 and H1130 shared 1977 gene families unique to the *C. metallidurans* species.

The COGnitor module [48] implemented in the MaGe platform was used to compare the CDSs of the core, variable and specific genome assigned to a COG (clusters of orthologous groups) functional category (Figure 5). The latter indicated that for all four strains, COG L (replication, recombination and repair) and U (intracellular trafficking and secretion) are overrepresented on the variable plus specific genome. Other COGs were also significantly overrepresented on the variable plus specific genome for particular strains. For instance, COG D (cell cycle control, division and partitioning) for CH34, NA4 and H1130, and COG V (defense mechanisms) for NA1 and NA4 (see Figure 5 for all significant overrepresentations).

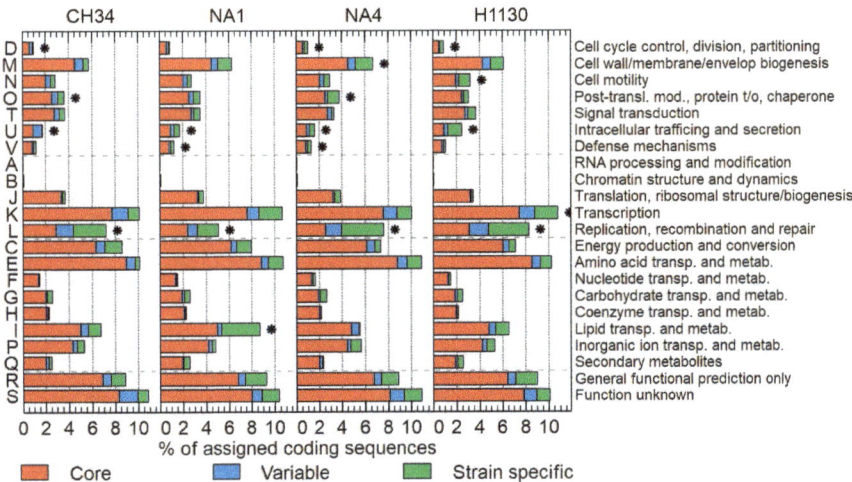

Figure 5. Percentage of CDSs assigned to a COG (clusters of orthologous groups) functional class (general categories: cellular processes and signaling: D, M, N, O, T, U, V; information storage and processing: A, B, J, K, L; metabolism; C, E, F, G, H, I, P, Q; poorly characterized: R, S) belonging to the core, variable of strain-specific genome of *C. metallidurans* CH34, NA1, NA4 and H1130. * Significant ($p < 0.05$; based on hypergeometric distribution) overrepresentation of COG on variable + specific compared to the core genome.

3.2. The Mobilome

Recently, we showed that *C. metallidurans* strains have substantial differences in the diversity and size of their mobile gene pool [7]. However, since this comparison was based on whole-genome hybridization to microarrays containing type strain CH34 oligonucleotide probes, the presence of MGEs other than those in CH34 could not be assessed. Here, the mobilomes of NA1, NA4 and H1130 (including IS elements, transposons, genomic islands and prophages) as well as the presence of CRISPR-Cas systems were scrutinized.

3.2.1. Insertion Sequence Elements and Transposons

ISFinder [49] and ISSaga [50] (+ manual curation) were used to create an inventory of the IS elements, which identified 57, 25, 33 and 91 putative IS elements in CH34 [24], NA1, NA4 and H1130, respectively. It must be noted that this list is based on a draft genome assembly for NA1, NA4 and H1130, which could have an impact on the actual number. Possible identical IS elements present in multiple copies will only be represented as one contig in the genome assembly, as such leading to an underestimation of the number of IS elements in the respective genome [51]. Active IS transposition in CH34 was already observed for IS*Rme1*, IS*Rme3*, IS*Rme5*, IS*Rme15*, IS*1086*, IS*1087B*, IS*1088* and IS*1090* [24,52–58]. Transposition activity of IS*Rme5* > IS*1088* > IS*Rme3* > IS*1087B* > IS*1090* > IS*1086* > IS*Rme15*, at least into the *cnr* target after exposure of AE126, a derivative of CH34 cured from plasmid pMOL30 carrying the main zinc resistance determinant, to 0.8 mM Zn^{2+} [58]. Some of these active IS elements are also carried by NA1 (2 IS*Rme3* copies), NA4 (1 IS*Rme1*, 4 IS*Rme4* and 1 IS*Rme5* copy) and H1130 (16 IS*Rme3* copies) (based on 98% DNA sequence identity cut-off). Next to transposition, IS elements can also cause more extensive/general loss of genetic information by recombination events between identical individual IS copies, e.g., loss of the CH34 genes involved in autotrophy by IS*1071*-mediated excision [24]. Similar observations of IS*1071*-mediated rearrangements affecting the metabolic potential of the host have been described for *Comamonas* sp. strain JS46 [59] and *Cupriavidus pinatubonensis* JMP134 [60]. Thus, these IS elements in CH34, NA1, NA4 and H1130 can play a multifaceted, pivotal role in the adaptation to stress conditions (as shown for CH34) [27,58].

The CH34 genome harbors five distinct transposon families totaling 19 intact transposons. The transposition modules of four transposons are related to those of mercury transposons with Tn*4378*, Tn*4380* and Tn*6050* belonging to the Tn*21*/Tn*501* family, and Tn*6048* to the Tn*5053* family [61]. The transposition module of Tn*6049* could not be categorized. Tn*6048*, Tn*6049* and mercury transposons are also conserved in NA1 (one Tn*6048* copy, one Tn*6049* copy), NA4 (3 mercury transposons, 3 Tn*6049* copies) and H1130 (4 mercury transposons). Tn*6050* appeared to be only present in CH34. No other transposons were identified.

3.2.2. Genomic Islands

The MaGe platform was used to scrutinize the presence of genomic islands (GIs), including those previously identified in CH34. The largest island (109 kb) on the chromosome of CH34 belongs to the large pKLC102/PAGI-2 family of elements that share a core gene set and are integrated downstream of tRNA genes [62,63]. A similar element is present in NA1 (2 copies), NA4 and H1130 as shown by progressive Mauve alignment [64] (Figure S2). The Tn*4371*-family of integrative and conjugative elements CMGI-2, CMGI-3 and CMGI-4 of CH34 were previously designated ICE$_{Tn4371}$6054, ICE$_{Tn4371}$6055 and ΔICE$_{Tn4371}$6056, respectively [65]. CMGI-2 (ICE$_{Tn4371}$6054) and CMGI-3 (ICE$_{Tn4371}$6055) are responsible for CH34's ability to grow on aromatic compounds and to fix carbon dioxide, respectively [7,24]. No Tn*4371*-family genomic island was identified in NA4. One Tn*4731*-family element was identified in NA1, which is highly similar to previously identified elements in *Delftia acidovorans* SPH-1 (DAGI-1; ICE$_{Tn4371}$60370), *Comamonas testosteroni* KF-1 (CTGI-1; ICE$_{Tn4371}$6038) and the partial CMGI-4 (ΔICE$_{Tn4371}$6056) of CH34 [25,65]. The island carries an RND-driven efflux system. In H1130, two Tn*4371*-family genomic islands were identified, one carrying

12 genes (putatively involved in ion transport), while the second could not be correctly defined as the integration/excision and stabilization/maintenance module up to *rlxS* (encoding a relaxase protein) are not located on the same contig as the transfer module (starting from *traR* coding for a transcriptional regulator). Therefore, the accessory genes that are typically located between *rlxS* and *traR* in Tn*4371*-family members could not be properly assessed [65]. All other GIs on CH34's chromosome were not found in the other strains, except CMGI-5 in NA1. CMGI-C and CMGI-E, previously identified on CH34's chromid, are absent in all strains. CMGI-A, -B and -D are conserved in NA4 and H1130, but show limited synteny with NA1. No other genomic islands could be clearly identified in NA1 or NA4. One other genomic island was clearly noticeable in H1130. This 87 kb region, which is absent in CH34, NA1 and NA4, is syntenic with an 80-kb cluster located on the 1.47-Mbp megaplasmid of *Burkholderia xenovorans* LB400. In *B. xenovorans* LB400, this Dit island encodes proteins of abietane diterpenoids metabolism and mediates growth on abietic acid, dehydroabietic acid, palustric acid and 7-oxo-dehydroabietic acid [66] (not included in the phenotypic microarray). Abietane diterpenoids are tricyclic, C-20, carboxylic acid-containing compounds produced by plants and are a key component of the defense systems of coniferous trees [66,67]. This observation also adds evidence to the mobility of this cluster and its distribution among proteobacterial genomes [66]. In addition, two smaller regions (13.6 and 10.3 kb) carrying genes coding for unknown functions and a tyrosine-based site-specific recombinase were identified.

3.2.3. Prophages

The presence of prophages was scrutinized via PHASTER [68] and showed no prophages in type strain CH34 (which was already known) and the presence of intact prophages in NA4 and H1130 as well as incomplete/remnants in H1130, NA1 and NA4 (Table 2). Although mitomycin C exposure did not result in prophage induction (data not shown), a derivative of NA4 exposed to uranium lost the 43.6 kb region predicted as an intact prophage (unpublished data).

Table 2. Prophage detected in *C. metallidurans* NA1, NA4 and H1130.

Strain	Size [a]	Completeness [b]	Score [c]	# [d]	Position	Most Common Phage [e]	GC %
NA1	27.9	questionable	90	32	528,474–556,451	Ralsto_RS138 (NC_029107; 7)	65.35
	17.7	incomplete	20	21	554,542–572,263	Pseudo_NP1 (NC_031058; 3)	64.23
NA4	43.6	intact	100	50	1,706,628–1,750,233	Bordet_BPP_1 (NC_005357; 18)	64.94
	6.1	intact	100	10	1,941,664–1,947,835	Ralsto_PE226 (NC_015297; 6)	60.08
	45.2	intact	150	41	2,145,854–2,191,126	Burkho_Bcep176 (NC_007497; 11)	61.83
	8	incomplete	30	10	2,181,367–2,189,450	Gordon_Nymphadora (NC_031061; 2)	62.44
	120.5	intact	130	125	2,248,504–2,369,042	Salmon_118970_sal3 (NC_031940; 14)	61.84
	44.5	incomplete	30	39	2,545,435–2,589,959	Pseudo_JBD44 (NC_030929; 5)	63.89
H1130	19.3	incomplete	30	21	1,470,543–1,489,908	Burkho_phiE125 (NC_003309; 3)	61.27
	12.7	incomplete	40	19	1,505,965–1,518,710	Bacill_SP_15 (NC_031245; 5)	63.39
	7.9	incomplete	30	9	2,763,200–2,771,108	Entero_phi92 (NC_023693; 4)	58.60
	48.4	intact	110	73	6,748,679–6,797,173	Salmon_SEN34 (NC_028699; 22)	62.09
	16	incomplete	50	29	7,110,429–7,126,525	Clostr_phiCT453B (NC_029004; 4)	61.04

[a] Size in kb; [b] Prediction of whether the region contains an intact or incomplete prophage and [c] score based on PHASTER criteria [68]; [d] number of proteins; [e] the phage with the highest number of proteins most similar to those in the region (between parentheses: accession number; number of proteins).

3.2.4. CRISPR-Cas

The CRISPR-Cas system is an adaptive immunity system that stores memory of past encounters with foreign DNA in spacers that are inserted between direct repeats in CRISPR arrays [69]. CRISPR-Cas systems were detected with CRISPRfinder [70] and CRISPRDetect [71] (default settings). Only positive hits with both were further examined, resulting in the identification of 1 CRISPR-Cas system in NA4. CRISPRTarget [72] identified 5 spacer sequences related to genomic island CMGI-5 of CH34 (which is also present in NA1). CMGI-5 is probably a plasmid remnant and contains besides hypothetical genes, some typical plasmid-related genes such as *repA*, *traY*, *mobA* and *mobB*. To assess if the identified system is active, the conjugation frequency of plasmid pJB3kan1 carrying the CMGI-5 *repA* gene (pJB3kan1_Rmet2825; containing one spacer) was determined for the parental and CRISPR-deleted NA4 strain. CRISPR deletion in NA4 increased the conjugation efficiency 33-fold, indicating an active CRISPR-Cas system in NA4 (Figure 6).

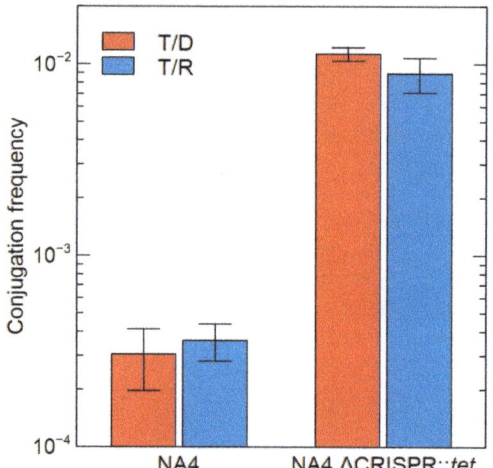

Figure 6. Conjugation frequency of pJB3kan1_Rmet2825 (containing one spacer identified in the NA4 CRISP-Cas system) from donor *E. coli* MFDpir to recipient *C. metallidurans* NA4 and NA4 ΔCRISPR::*tet*, respectively. Median values plus corresponding calculated standard deviations across biological triplicates are shown (T/D = conjugation frequency per donor; T/R = conjugation frequency per recipient).

3.3. The Resistome

3.3.1. Antibiotic Resistance

The CARD [42] implementation within the MaGe platform [39] was used to identify known resistance determinants and associated antibiotics. The latter predicted 33, 36, 33 and 39 proteins involved in antibiotic resistance in CH34, NA1, NA4 and H1130, of which 31, 31, 30 and 39 belonged to the core genome, respectively. No marked difference in tolerance to antibiotics was observed.

3.3.2. Metal Resistance

The antibacterial biocide and metal resistance genes database (BacMet) was used to create an inventory of genes predicted to confer resistance to metals and/or antibacterial biocides [45]. This showed 302, 282, 337 and 302 proteins involved in biocide and metal resistance in CH34, NA1, NA4 and H1130, respectively. Most genes belonged to the core genome (221, 246, 276 and 251 for CH34, NA1, NA4 and H1130, respectively). The compounds (metal and chemical class) to which these genes confer resistance are very similar for all four strains (Figure 7). Genes conferring resistance to nickel, copper, cobalt and the chemical classes acridine and phenanthridine were the most abundant.

For CH34, the predicted genes contained 68 out of the 174 genes that were previously identified to be related to metal resistance (for an overview see [2,3]). Specific analysis of these 174 proteins showed that almost all are conserved in NA1, NA4 and H1130. Exceptions are (*i*) the accessory cluster related to chromate resistance in H1130, (*ii*) the *hmz* cluster in NA4 and H1130, (*iii*) *cdfX* in NA1, NA4 and H1130, and (*iv*) the *dax/gig* cluster in NA1. The latter three are all located on a genomic island. The gene cluster related to chromate resistance on pMOL28 from CH34 contains five additional genes that are strongly induced by chromate in CH34 [73] as well as for the homologous system in *Arthrobacter* sp. FB24 (both at the gene and protein level) [74,75]. The *hmz* cluster is a HME-RND-driven system, belonging to the HME3b (Heavy Metal Efflux) subfamily of the RND superfamily, with no known substrate and transcriptionally silent in *C. metallidurans* CH34 [5,73,76]. The *cdfX* gene of CH34 encodes a putative permease (211 amino acid residues and six predicted transmembrane α-helices) that shares 87% amino-acid identity with PbtF from *Achromobacter xylosoxidans* A8 [5]. Expression of *pbtF* in *A. xylosoxidans* A8 was induced by Pb^{2+}, Cd^{2+} and Zn^{2+}, and although PbtF showed measurable Pb^{2+}-efflux activity, it did not confer increased metal tolerance in *E. coli* GG48 [77]. The *dax* cluster [73], which was renamed *gig* for "gold-induced genes" in Wiesemann et al. [78], is induced by Ag^+ and Au^{3+} [73,79] but not essential for gold resistance [78].

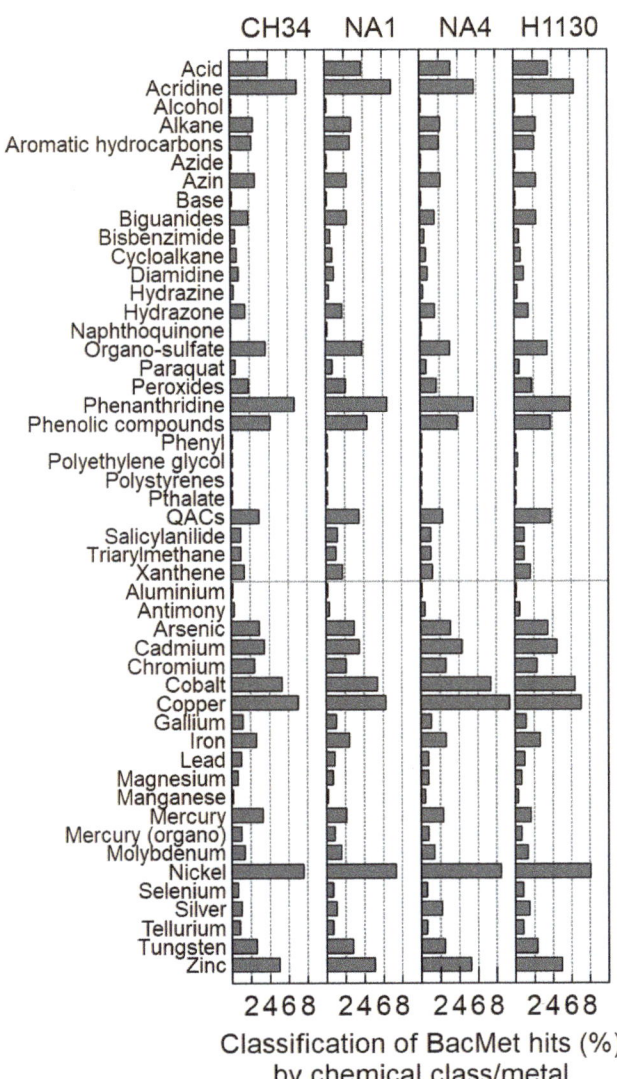

Figure 7. Inventory of *C. metallidurans* CH34, NA1, NA4 and H1130 genes conferring resistance to metals/chemical classes based on the BacMet database (antibacterial biocide and metal resistance genes database).

In agreement with the conservation of these metal resistance determinants, growth in the presence of increasing metal concentrations showed only minor differences between CH34, NA1, NA4 and H1130 (Figure 8). Moreover, the minor strain-dependent differences (see above) did not mediate differences in metal resistance (Figure 8). Essentially, the most noticeable difference in growth was observed in the presence of Ni^{2+}, with higher concentrations tolerated by NA4 and H1130. Initially, the *nccCBA* locus, which is inactivated in CH34 because of a frame shift mutation, was put forward as a possible explanation [12]. However, the frame shift mutation in *nccB* is present in all four strains. However, NA4 and H1130 carry a second *nccYXHCBAN* locus coding for an RND-driven efflux system involved in Ni^{2+} and Co^{2+} resistance. This locus is homologous to that of *C. metallidurans* 31A and KT02, which has been shown to be responsible for resistance to 40 mM Ni^{2+} [80], and is likely responsible for the observed differences. In addition, although the *nimBAC* locus, coding for an RND-driven efflux system putatively involved in Ni^{2+} and Co^{2+} resistance [5], is only inactivated in CH34 (via IS*Rme3* insertion) and not in NA1, NA4 and H1130, growth in the presence of Ni^{2+} is similar for NA1 and CH34. Other observations are the lower resistance of NA1 to Cd^{2+} and to lesser extent Co^{2+} and Ag^+. However, based on the current data, no hypotheses can be put forward to explain these observations.

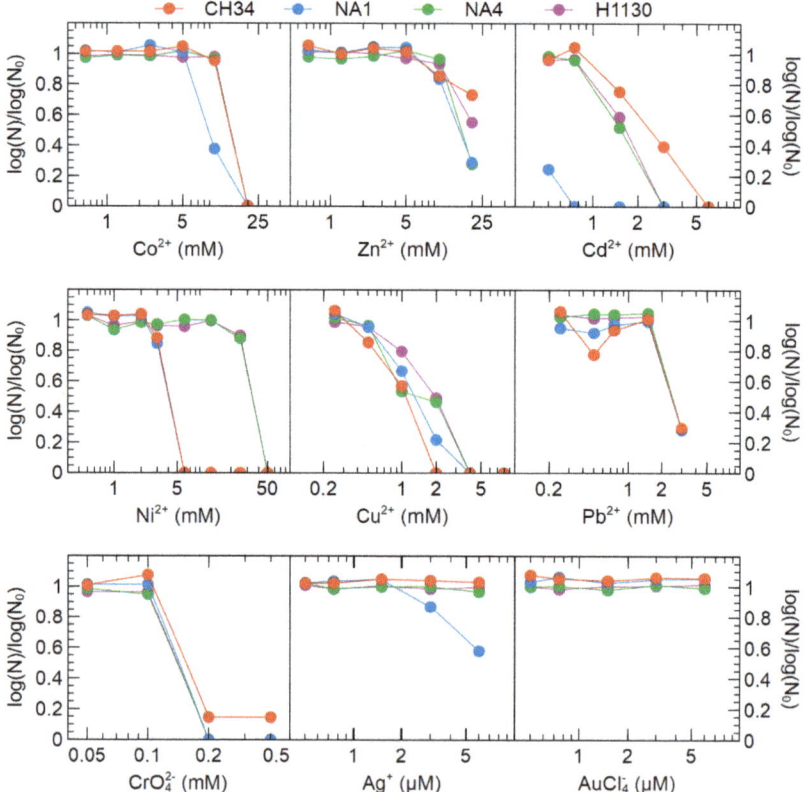

Figure 8. Viable count of *C. metallidurans* CH34, NA1, NA4 and H1130 grown in the presence of various metal concentrations (Table S2). Data are presented as $\log(N)/\log(N_0)$ in function of metal concentration, with N and N_0 the colony forming units (CFUs) in the presence and absence (control) of metal, respectively.

3.4. Phenotypic Microarrays

In order to scrutinize functional differences between the four *C. metallidurans* strains, phenotypic characterization with OmniLog Phenotypic Microarrays (PMs) was conducted. Area under the curve (AUC) values were calculated and a threshold cut-off (8000) was applied to discriminate a positive (growth) from a negative (no-growth) reaction. This revealed an overall phenotypic similarity among the four strains, with 1744 out of the 1920 assays shared (Figures 9–11).

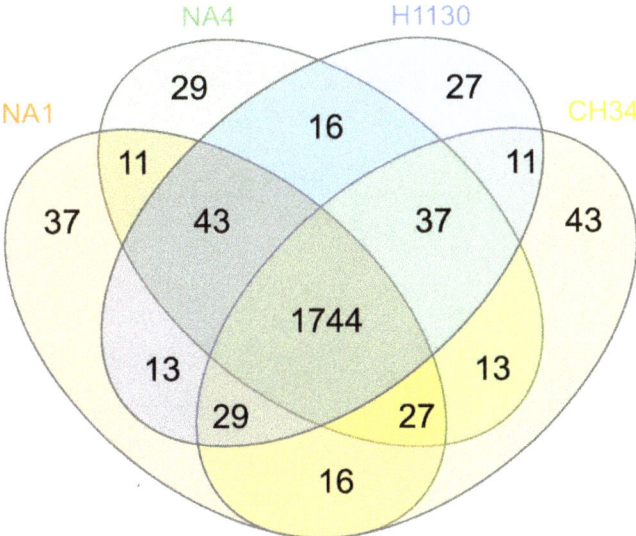

Figure 9. Venn diagram displaying the distribution of OmniLog phenotypic assays shared among *C. metallidurans* CH34, NA1, NA4 and H1130.

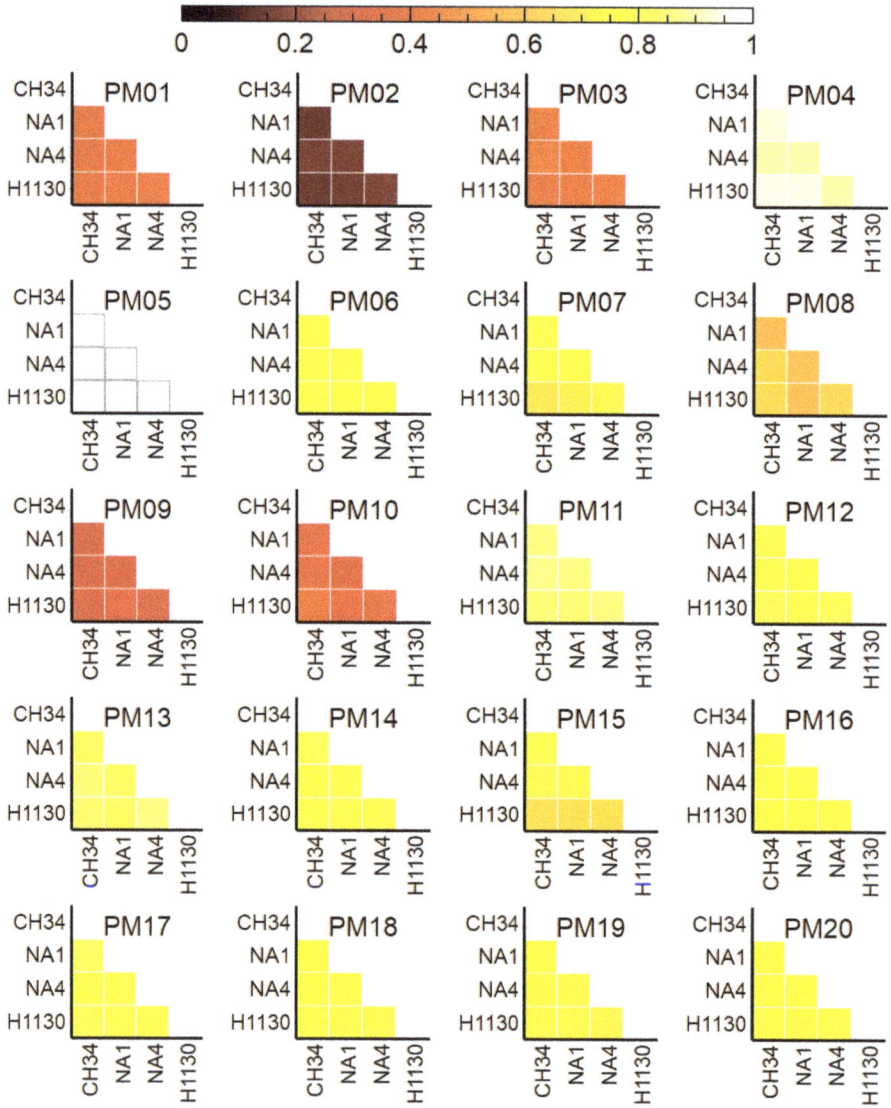

Figure 10. Overview of positive (growth) OmniLog phenotypic assays shared by *C. metallidurans* CH34, NA1, NA4 and H1130 for each PM plate (with 1 being all 96 assays). The assays on each PM plate are detailed in Table S4.

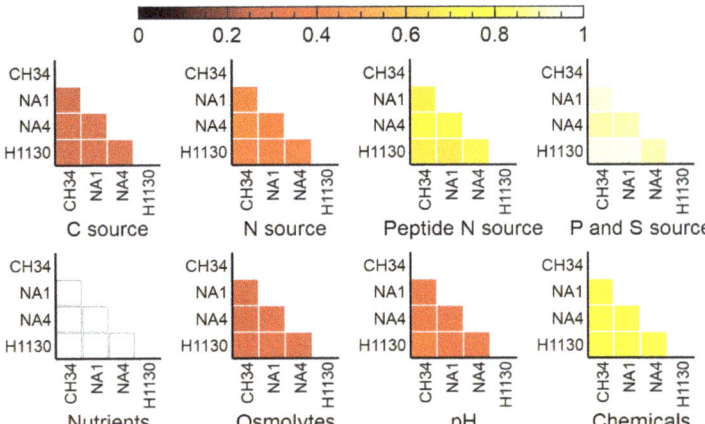

Figure 11. Overview of positive (growth) OmniLog phenotypic assays shared by *C. metallidurans* CH34, NA1, NA4 and H1130 for different metabolic and chemical sensitivity tests (with 1 being all assays shared).

3.4.1. C, N, P and S Sources

Only around 27% to 28% of the C source reactions was positive, which is related to their inability to assimilate sugars and sugar alcohols (Figure 11) [1,3]. All four strains lack a glucose uptake system. The latter is most likely deleted in all four strains as a N-acetylglucosamine-specific phosphotransferase system (PTS)-type transport system essential for glucose uptake (growth) in *Cupriavidus necator* H16 [81,82] is absent in from a large syntenic region (>110 genes) conserved among *C. necator* H16 and *C. metallidurans* CH34, NA1, NA4 and H1130 (data not shown).

A few marked differences were observed for the use of amino acids as N source, in particular for L-leucine, L-tryptophan and L-Valine (Figure 12). Specific for L-tryptophan, growth was observed for NA1, NA4 and H1130 but not for CH34. Aerobic L-tryptophan degradation in *C. metallidurans* most likely occurs via a three-step pathway to anthrilanate requiring tryptophan 2,3-dioxygenase (*kynA*), kynurenine formamidase (*kynB*) and kynureninase (*kynU*). Experimental verification of the anthranilate pathway was achieved by functional expression of the CH34 *kynBAU* operon in *Escherichia coli* after suppressing the stop codon disrupting *kynB* [83]. This amber mutation is not present in NA1, NA4 and H1130, which could explain the observed differences. Similar differences were also observed when growth was scored for dipeptides (N source), as CH34 grew less or not on L-tryptophan-containing dipeptides compared to NA1, NA4 and H1130. Only minor differences were observed for growth on P and S sources (Figure 11, Table S4).

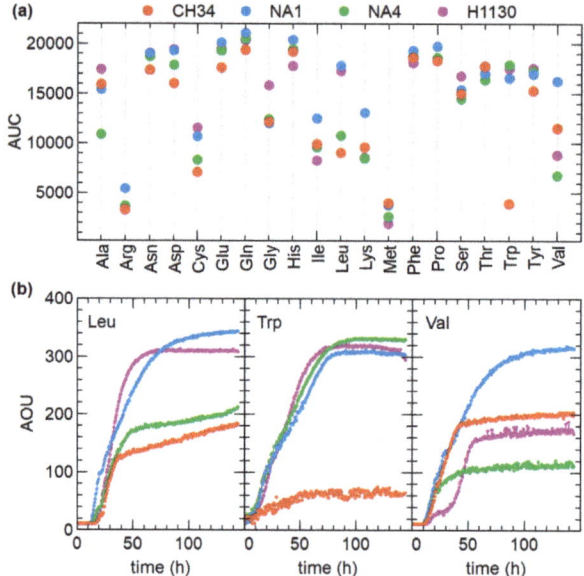

Figure 12. (**a**) Overview of OmniLog phenotypic assays with amino acids as N source (AUC = area under the curve) for *C. metallidurans* CH34, NA1, NA4 and H1130, (**b**) Growth kinetics in the presence of L-leucine, L-tryptophan and L-Valine as N source (AOU = arbitrary OmniLog units).

3.4.2. Osmolytes and pH

The addition of ionic osmolytes had a clear and comparable impact on the growth of strains CH34, NA1, NA4 and H1130, as growth was generally only observed for the lower/lowest concentrations (1% NaCl, 2% Na_2SO_4, 1% sodium formate, 3% urea and 2% sodium lactate). In contrast, addition of up to 20% of the non-ionic osmolyte ethylene glycol had no impact on growth of CH34, NA1, NA4 and H1130.

The effect of pH over the range 3.5 to 10 growth was comparable for CH34, NA1, NA4 and H1130. Growth was inhibited below pH 5 for all strains. Growth at pH 10 was much more pronounced for H1130 than for the other strains (Figure 13).

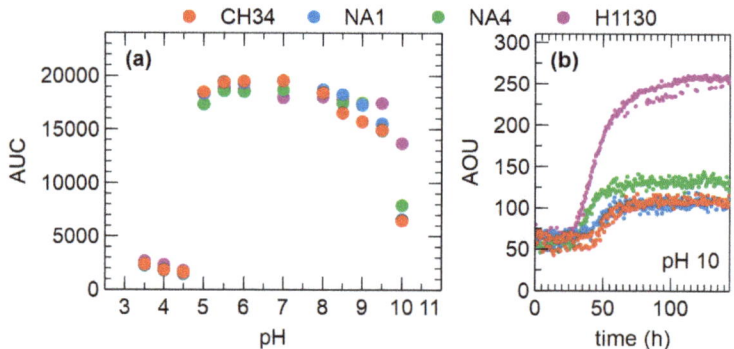

Figure 13. (**a**) Overview of OmniLog phenotypic assays related to pH (AUC = area under the curve) for *C. metallidurans* CH34, NA1, NA4 and H1130, (**b**) Growth kinetics at pH 10 (AOU = arbitrary OmniLog units).

3.4.3. Chemicals

The PM-11 to PM-20 plates carry different chemicals (4 increasing concentrations of each) to test sensitivity, only for eight out of the 240 chemicals tested at least one of the strains was susceptible to the lowest concentration. For more than 50% of the tested chemicals, CH34, NA1, NA4 and H1130 were resistant to the highest concentration included in the phenotypic microarrays (Figure 14).

No growth was observed in the presence of 2,2′-dipyridyl (metal chelator), hydroxyurea (ROS producer) and phenethicillin (a narrow-spectrum, β-lactamase-sensitive penicillin) for all four strains. In contrast to phenethicillin, CH34, NA1, NA4 and H1130 were resistant to (at least one concentration of) all other β-lactam antibiotics tested. Only H1130 grew in the presence of sodium meta- and orthovanadate, and did not grow in the presence of thallium acetate (Figure 15). Strain CH34 and NA4 did not grow in the presence of potassium tellurite (Figure 15). The genetic basis underlying resistance to these metals is poorly understood, therefore, no correlation to the genotype could be established. Strain CH34 and NA1 were susceptible to sodium metaperiodate (oxidizing agent) and tolylfluanid (fungicide), respectively (Figure 15).

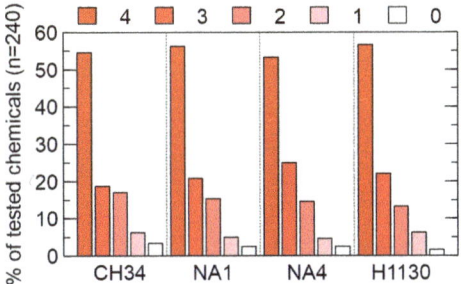

Figure 14. Percentage of tested chemicals (n = 240) to which *C. metallidurans* CH34, NA1, NA4 and H1130 are resistant. Four increasing concentrations are included in the phenotypic microarrays (0: susceptible to the lowest concentration, 1 to 4: resistant to the lowest up to the highest concentration).

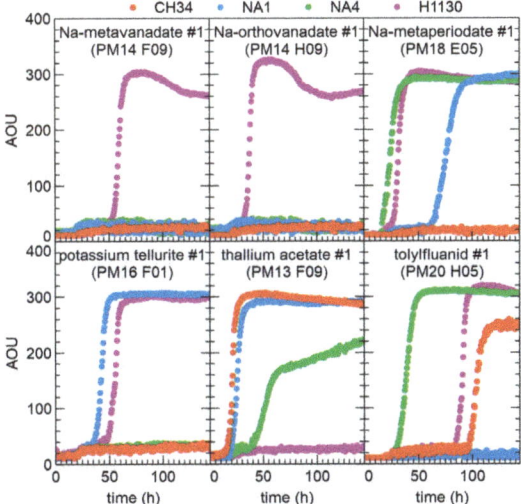

Figure 15. Growth kinetic of *C. metallidurans* CH34, NA1, NA4 and H1130 in the presence of different chemicals (lowest concentration in the phenotypic microarrays is shown) (AOU = arbitrary OmniLog units).

3.4.4. Trait Prediction

Finally, the prediction of Traitar, an automated software framework for the accurate prediction of 67 phenotypes directly from a genome sequence [84], was evaluated by comparison with the generated phenotypic data (OmniLog Phenotypic Microarray data and previous observations/knowledge). Traitar correctly predicted 85% (45 out of 53 analyzed), 81% (38 out of 47), 80% (37 out of 46) and 80% (37 out of 46) of the CH34, NA1, NA4 and H1130 traits, respectively (Figure 16). Although Weimann and colleagues [84] indicated that the phypat classifier assigned more phenotypes at the price of more false-positive predictions, whereas the phypat + PGL classifier assigned fewer phenotypes with fewer false assignments, it appeared that in the case of the *C. metallidurans* strains, phypat + PGL assigned more false-positive predictions.

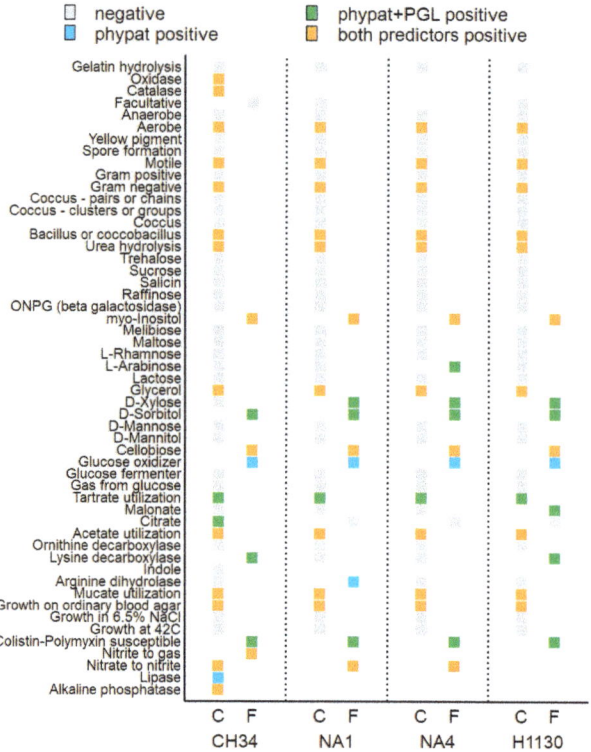

Figure 16. Overview of the correctly (C) and falsely (F) predicted phenotypic traits of *C. metallidurans* CH34, NA1, NA4 and H1130 by Traitar (two classifiers: phypat and phypat + PGL).

4. Conclusions

The comparison of four *C. metallidurans* strains isolated from different environments indicated that metal resistance determinants and properties are maintained in these environments. As most of the metal determinants are on the native megaplasmids, it could be argued that these environments provided a selective pressure for the conservation of these determinants and plasmids. The previously identified differences in the size and diversity of the mobile gene pool were put in perspective by the identification of intact (and remnant) prophages in NA4 and H1130, and a genomic island putatively involved in abietane diterpenoids metabolism in H1130. The latter indicated that mobilome diversity differed (integrative and conjugative elements/genomic islands versus prophages). Furthermore, the mobilome is apparently not directly related to the isolation environment as the NA1 mobilome is shaped more like that of H1130 than that of NA4 isolated from the same environment. In addition, an active CRISPR-Cas system was identified in strain NA4, providing immunity to a plasmid that integrated in CH34 and NA1. Despite the large size of the variable and specific genomes, only minor differences were observed in the global phenomes (as measured by phenotype microarrays) and all four strains were highly resistant to a wide variety of chemicals, much broader than metals. The variable and specific genome were probably acquired through later transfer and perhaps carry functions more essential for survival in challenging and fluctuating environments than general metabolic functions.

Supplementary Materials: The following are available online at http://www.mdpi.com/2073-4425/9/10/507/s1, Figure S1: Density plot of the AUC values of all PM reactions for each strain, Figure S2: Progressive Mauve alignment of CMGI-1 of CH34 with related elements in NA1, NA4 and H1130, Table S1: Metal concentrations used in growth experiments, Table S2: Primers used in this study, Table S3: Pan-genome analysis data set, Table S4: PM area under the curve (AUC) values.

Author Contributions: Conceptualization, R.V.H.; Data curation, R.V.H. and P.M.; Formal analysis, R.V.H. and P.M.; Funding acquisition, R.V.H. and N.L.; Investigation, R.V.H., A.P., A.V.A. and K.M.; Methodology, R.V.H. and P.M.; Project administration, R.V.H.; Software, R.V.H. and P.M.; Supervision, R.V.H., N.L. and B.L.; Validation, R.V.H. and P.M.; Visualization, R.V.H.; Writing—original draft, R.V.H.; Writing—review & editing, R.V.H., A.V.A., K.M. and P.M.

Funding: This work was funded by the European Space Agency (ESA-PRODEX) and the Belgian Science Policy (Belspo) through the COMICS project (C90356).

Conflicts of Interest: The authors declare no conflict of interest.

References

1. Mergeay, M.; Nies, D.; Schlegel, H.G.; Gerits, J.; Charles, P.; Van Gijsegem, F. *Alcaligenes eutrophus* CH34 is a facultative chemolithotroph with plasmid-bound resistance to heavy metals. *J. Bacteriol.* **1985**, *162*, 328–334. [PubMed]
2. Mergeay, M.; Van Houdt, R. *Metal Response in Cupriavidus Metallidurans, Volume I: From Habitats to Genes and Proteins*; Springer International Publishing: Cham, Switzerland, 2015; p. 89.
3. Janssen, P.J.; Van Houdt, R.; Moors, H.; Monsieurs, P.; Morin, N.; Michaux, A.; Benotmane, M.A.; Leys, N.; Vallaeys, T.; Lapidus, A.; et al. The complete genome sequence of *Cupriavidus metallidurans* strain CH34, a master survivalist in harsh and anthropogenic environments. *PLoS ONE* **2010**, *5*, e10433. [CrossRef] [PubMed]
4. Monchy, S.; Benotmane, M.A.; Janssen, P.; Vallaeys, T.; Taghavi, S.; van der Lelie, D.; Mergeay, M. Plasmids pMOL28 and pMOL30 of *Cupriavidus metallidurans* are specialized in the maximal viable response to heavy metals. *J. Bacteriol.* **2007**, *189*, 7417–7425. [CrossRef] [PubMed]
5. Vandenbussche, G.; Mergeay, M.; Van Houdt, R. *Metal Response in Cupriavidus metallidurans, Volume II: Insights into the Structure-Function Relationship of Proteins*; Springer International Publishing: Cham, Switzerland, 2015; p. 70.
6. Diels, L.; Mergeay, M. DNA probe-mediated detection of resistant bacteria from soils highly polluted by heavy metals. *Appl. Environ. Microbiol.* **1990**, *56*, 1485–1491. [PubMed]
7. Van Houdt, R.; Monsieurs, P.; Mijnendonckx, K.; Provoost, A.; Janssen, A.; Mergeay, M.; Leys, N. Variation in genomic islands contribute to genome plasticity in *Cupriavidus metallidurans*. *BMC Genom.* **2012**, *13*, 111. [CrossRef] [PubMed]
8. Kunito, T.; Kusano, T.; Oyaizu, H.; Senoo, K.; Kanazawa, S.; Matsumoto, S. Cloning and sequence analysis of czc genes in *Alcaligenes* sp. strain CT14. *Biosci. Biotechnol. Biochem.* **1996**, *60*, 699–704. [CrossRef] [PubMed]
9. Reith, F.; Rogers, S.L.; McPhail, D.C.; Webb, D. Biomineralization of gold: Biofilms on bacterioform gold. *Science* **2006**, *313*, 233–236. [CrossRef] [PubMed]
10. Schmidt, T.; Schlegel, H.G. Nickel and cobalt resistance of various bacteria isolated from soil and highly polluted domestic and industrial wastes. *FEMS Microbiol. Lett.* **1989**, *62*, 315–328. [CrossRef]
11. Miyake-Nakayama, C.; Masujima, S.; Ikatsu, H.; Miyoshi, S.-I.; Shinoda, S. Isolation and characterization of a new dichloromethane degrading bacterium, *Ralstonia metallidurans*, PD11. *Biocontrol Sci.* **2004**, *9*, 89–93. [CrossRef]

12. Mijnendonckx, K.; Provoost, A.; Ott, C.M.; Venkateswaran, K.; Mahillon, J.; Leys, N.; Van Houdt, R. Characterization of the survival ability of *Cupriavidus metallidurans* and *Ralstonia pickettii* from space-related environments. *Microbol. Ecol.* **2013**, *65*, 347–360. [CrossRef] [PubMed]
13. Mora, M.; Perras, A.; Alekhova, T.A.; Wink, L.; Krause, R.; Aleksandrova, A.; Novozhilova, T.; Moissl-Eichinger, C. Resilient microorganisms in dust samples of the International Space Station-survival of the adaptation specialists. *Microbiome* **2016**, *4*, 65. [CrossRef] [PubMed]
14. Coenye, T.; Spilker, T.; Reik, R.; Vandamme, P.; Lipuma, J.J. Use of PCR analyses to define the distribution of *Ralstonia* species recovered from patients with cystic fibrosis. *J. Clin. Microbiol.* **2005**, *43*, 3463–3466. [CrossRef] [PubMed]
15. Langevin, S.; Vincelette, J.; Bekal, S.; Gaudreau, C. First case of invasive human infection caused by *Cupriavidus metallidurans*. *J. Clin. Microbiol.* **2011**, *49*, 744–745. [CrossRef] [PubMed]
16. D'Inzeo, T.; Santangelo, R.; Fiori, B.; De Angelis, G.; Conte, V.; Giaquinto, A.; Palucci, I.; Scoppettuolo, G.; Di Florio, V.; Giani, T.; et al. Catheter-related bacteremia by *Cupriavidus metallidurans*. *Diagn. Microbiol. Infect. Dis.* **2015**, *81*, 9–12. [CrossRef] [PubMed]
17. Harrison, P.W.; Lower, R.P.; Kim, N.K.; Young, J.P. Introducing the bacterial 'chromid': Not a chromosome, not a plasmid. *Trends Microbiol.* **2010**, *18*, 141–148. [CrossRef] [PubMed]
18. DiCenzo, G.C.; Finan, T.M. The divided bacterial genome: Structure, function, and evolution. *Microbiol. Mol. Biol. Rev.* **2017**, *81*, e00019-17. [CrossRef] [PubMed]
19. Schwartz, E. *Microbial Megaplasmids*; Springer: Berlin/Heidelberg, Germany, 2009; p. 348.
20. Schwartz, E.; Henne, A.; Cramm, R.; Eitinger, T.; Friedrich, B.; Gottschalk, G. Complete nucleotide sequence of pHG1: A *Ralstonia eutropha* H16 megaplasmid encoding key enzymes of H_2-based lithoautotrophy and anaerobiosis. *J. Mol. Biol.* **2003**, *332*, 369–383. [CrossRef]
21. Amadou, C.; Pascal, G.; Mangenot, S.; Glew, M.; Bontemps, C.; Capela, D.; Carrere, S.; Cruveiller, S.; Dossat, C.; Lajus, A.; et al. Genome sequence of the β-rhizobium *Cupriavidus taiwanensis* and comparative genomics of rhizobia. *Genom. Res.* **2008**, *18*, 1472–1483. [CrossRef] [PubMed]
22. Trefault, N.; De la Iglesia, R.; Molina, A.M.; Manzano, M.; Ledger, T.; Perez-Pantoja, D.; Sanchez, M.A.; Stuardo, M.; Gonzalez, B. Genetic organization of the catabolic plasmid pJP4 from *Ralstonia eutropha* JMP134 (pJP4) reveals mechanisms of adaptation to chloroaromatic pollutants and evolution of specialized chloroaromatic degradation pathways. *Environ. Microbiol.* **2004**, *6*, 655–668. [CrossRef] [PubMed]
23. Van Houdt, R.; Mergeay, M. Genomic context of metal response genes in *Cupriavidus metallidurans* with a focus on strain CH34. In *Metal Response in Cupriavidus Metallidurans, Volume I: From Habitats to Genes and Proteins*; Mergeay, M., Van Houdt, R., Eds.; Springer International Publishing: Cham, Switzerland, 2015; pp. 21–44.
24. Mijnendonckx, K.; Provoost, A.; Monsieurs, P.; Leys, N.; Mergeay, M.; Mahillon, J.; Van Houdt, R. Insertion sequence elements in *Cupriavidus metallidurans* CH34: Distribution and role in adaptation. *Plasmid* **2011**, *65*, 193–203. [CrossRef] [PubMed]
25. Van Houdt, R.; Monchy, S.; Leys, N.; Mergeay, M. New mobile genetic elements in *Cupriavidus metallidurans* CH34, their possible roles and occurrence in other bacteria. *Antonie Leeuwenhoek* **2009**, *96*, 205–226. [CrossRef] [PubMed]
26. Juhas, M.; van der Meer, J.R.; Gaillard, M.; Harding, R.M.; Hood, D.W.; Crook, D.W. Genomic islands: Tools of bacterial horizontal gene transfer and evolution. *FEMS Microbiol. Rev.* **2009**, *33*, 376–393. [CrossRef] [PubMed]
27. Vandecraen, J.; Chandler, M.; Aertsen, A.; Van Houdt, R. The impact of insertion sequences on bacterial genome plasticity and adaptability. *Crit. Rev. Microbiol.* **2017**, *43*, 709–730. [CrossRef] [PubMed]
28. Obeng, N.; Pratama, A.A.; Elsas, J.D.V. The significance of mutualistic phages for bacterial ecology and evolution. *Trends Microbiol.* **2016**, *24*, 440–449. [CrossRef] [PubMed]
29. Bondy-Denomy, J.; Davidson, A.R. When a virus is not a parasite: The beneficial effects of prophages on bacterial fitness. *J. Microbiol.* **2014**, *52*, 235–242. [CrossRef] [PubMed]
30. Bochner, B.R.; Gadzinski, P.; Panomitros, E. Phenotype microarrays for high-throughput phenotypic testing and assay of gene function. *Genom. Res.* **2001**, *11*, 1246–1255. [CrossRef] [PubMed]
31. Mergeay, M.; Houba, C.; Gerits, J. Extrachromosomal inheritance controlling resistance to cadmium, cobalt, copper and zinc ions: Evidence from curing a *Pseudomonas*. *Arch. Int. Physiol. Biochim. Biophys.* **1978**, *86*, 440–442.

32. Ferrieres, L.; Hemery, G.; Nham, T.; Guerout, A.M.; Mazel, D.; Beloin, C.; Ghigo, J.M. Silent mischief: Bacteriophage Mu insertions contaminate products of *Escherichia coli* random mutagenesis performed using suicidal transposon delivery plasmids mobilized by broad-host-range RP4 conjugative machinery. *J. Bacteriol.* **2010**, *192*, 6418–6427. [CrossRef] [PubMed]
33. Schafer, A.; Tauch, A.; Jager, W.; Kalinowski, J.; Thierbach, G.; Puhler, A. Small mobilizable multi-purpose cloning vectors derived from the *Escherichia coli* plasmids pK18 and pK19: Selection of defined deletions in the chromosome of *Corynebacterium glutamicum*. *Gene* **1994**, *145*, 69–73. [CrossRef]
34. Chang, A.C.; Cohen, S.N. Construction and characterization of amplifiable multicopy DNA cloning vehicles derived from the p15A cryptic miniplasmid. *J. Bacteriol.* **1978**, *134*, 1141–1156. [PubMed]
35. Blatny, J.M.; Brautaset, T.; Winther-Larsen, H.C.; Haugan, K.; Valla, S. Construction and use of a versatile set of broad-host-range cloning and expression vectors based on the RK2 replicon. *Appl. Environ. Microbiol.* **1997**, *63*, 370–379. [PubMed]
36. Andrup, L.; Barfod, K.K.; Jensen, G.B.; Smidt, L. Detection of large plasmids from the *Bacillus cereus* group. *Plasmid* **2008**, *59*, 139–143. [CrossRef] [PubMed]
37. Vaas, L.A.; Sikorski, J.; Hofner, B.; Fiebig, A.; Buddruhs, N.; Klenk, H.P.; Goker, M. opm: An R package for analysing OmniLog® phenotype microarray data. *Bioinformatics* **2013**, *29*, 1823–1824. [CrossRef] [PubMed]
38. Vaas, L.A.; Sikorski, J.; Michael, V.; Goker, M.; Klenk, H.P. Visualization and curve-parameter estimation strategies for efficient exploration of phenotype microarray kinetics. *PLoS ONE* **2012**, *7*, e34846. [CrossRef] [PubMed]
39. Vallenet, D.; Belda, E.; Calteau, A.; Cruveiller, S.; Engelen, S.; Lajus, A.; Le Fevre, F.; Longin, C.; Mornico, D.; Roche, D.; et al. MicroScope—An integrated microbial resource for the curation and comparative analysis of genomic and metabolic data. *Nucleic Acids Res.* **2013**, *41*, D636–647. [CrossRef] [PubMed]
40. Miele, V.; Penel, S.; Duret, L. Ultra-fast sequence clustering from similarity networks with SiLiX. *BMC Bioinf.* **2011**, *12*, 116. [CrossRef] [PubMed]
41. Ondov, B.D.; Treangen, T.J.; Melsted, P.; Mallonee, A.B.; Bergman, N.H.; Koren, S.; Phillippy, A.M. Mash: Fast genome and metagenome distance estimation using MinHash. *Genom. Biol.* **2016**, *17*, 132. [CrossRef] [PubMed]
42. McArthur, A.G.; Waglechner, N.; Nizam, F.; Yan, A.; Azad, M.A.; Baylay, A.J.; Bhullar, K.; Canova, M.J.; De Pascale, G.; Ejim, L.; et al. The comprehensive antibiotic resistance database. *Antimicrob. Agents Chemother.* **2013**, *57*, 3348–3357. [CrossRef] [PubMed]
43. McArthur, A.G.; Wright, G.D. Bioinformatics of antimicrobial resistance in the age of molecular epidemiology. *Curr. Opin. Microbiol.* **2015**, *27*, 45–50. [CrossRef] [PubMed]
44. Jia, B.; Raphenya, A.R.; Alcock, B.; Waglechner, N.; Guo, P.; Tsang, K.K.; Lago, B.A.; Dave, B.M.; Pereira, S.; Sharma, A.N.; et al. CARD 2017: Expansion and model-centric curation of the comprehensive antibiotic resistance database. *Nucleic Acids Res.* **2017**, *45*, D566–D573. [CrossRef] [PubMed]
45. Pal, C.; Bengtsson-Palme, J.; Rensing, C.; Kristiansson, E.; Larsson, D.G. BacMet: Antibacterial biocide and metal resistance genes database. *Nucleic Acids Res.* **2014**, *42*, D737–D743. [CrossRef] [PubMed]
46. Monsieurs, P.; Provoost, A.; Mijnendonckx, K.; Leys, N.; Gaudreau, C.; Van Houdt, R. Genome sequence of *Cupriavidus metallidurans* Strain H1130, isolated from an invasive human infection. *Genom. Announc.* **2013**, *1*, e01051-13. [CrossRef] [PubMed]
47. Monsieurs, P.; Mijnendonckx, K.; Provoost, A.; Venkateswaran, K.; Ott, C.M.; Leys, N.; Van Houdt, R. Genome sequences of *Cupriavidus metallidurans* strains NA1, NA4, and NE12, isolated from space equipment. *Genom. Announc.* **2014**, *2*, e00719-14. [CrossRef] [PubMed]
48. Tatusov, R.L.; Galperin, M.Y.; Natale, D.A.; Koonin, E.V. The COG database: A tool for genome-scale analysis of protein functions and evolution. *Nucleic Acids Res.* **2000**, *28*, 33–36. [CrossRef] [PubMed]
49. Siguier, P.; Perochon, J.; Lestrade, L.; Mahillon, J.; Chandler, M. ISfinder: The reference centre for bacterial insertion sequences. *Nucleic Acids Res.* **2006**, *34*, D32–D36. [CrossRef] [PubMed]
50. Varani, A.M.; Siguier, P.; Gourbeyre, E.; Charneau, V.; Chandler, M. ISsaga is an ensemble of web-based methods for high throughput identification and semi-automatic annotation of insertion sequences in prokaryotic genomes. *Genom. Biol.* **2011**, *12*, R30. [CrossRef] [PubMed]
51. Ricker, N.; Qian, H.; Fulthorpe, R.R. The limitations of draft assemblies for understanding prokaryotic adaptation and evolution. *Genomics* **2012**, *100*, 167–175. [CrossRef] [PubMed]

52. Dong, Q.; Sadouk, A.; van der Lelie, D.; Taghavi, S.; Ferhat, A.; Nuyten, J.M.; Borremans, B.; Mergeay, M.; Toussaint, A. Cloning and sequencing of IS*1086*, an *Alcaligenes eutrophus* insertion element related to IS*30* and IS*4351*. *J. Bacteriol.* **1992**, *174*, 8133–8138. [CrossRef] [PubMed]
53. Collard, J.M.; Provoost, A.; Taghavi, S.; Mergeay, M. A new type of *Alcaligenes eutrophus* CH34 zinc resistance generated by mutations affecting regulation of the *cnr* cobalt-nickel resistance system. *J. Bacteriol.* **1993**, *175*, 779–784. [CrossRef] [PubMed]
54. Grass, G.; Grosse, C.; Nies, D.H. Regulation of the *cnr* cobalt and nickel resistance determinant from *Ralstonia* sp. strain CH34. *J. Bacteriol.* **2000**, *182*, 1390–1398. [CrossRef] [PubMed]
55. Talat, M.-E. Genetic Mechanism of Heavy Metal Resistance of *Pseudomonas aeruginosa* CMG103. Ph.D. Thesis, University of Karachi, Karachi, Pakistan, 2000.
56. Schneider, D.; Faure, D.; Noirclerc-Savoye, M.; Barriere, A.C.; Coursange, E.; Blot, M. A broad-host-range plasmid for isolating mobile genetic elements in gram-negative bacteria. *Plasmid* **2000**, *44*, 201–207. [CrossRef] [PubMed]
57. Tibazarwa, C.; Wuertz, S.; Mergeay, M.; Wyns, L.; van der Lelie, D. Regulation of the *cnr* cobalt and nickel resistance determinant of *Ralstonia eutropha* (*Alcaligenes eutrophus*) CH34. *J. Bacteriol.* **2000**, *182*, 1399–1409. [CrossRef] [PubMed]
58. Vandecraen, J.; Monsieurs, P.; Mergeay, M.; Leys, N.; Aertsen, A.; Van Houdt, R. Zinc-induced transposition of insertion sequence elements contributes to increased adaptability of *Cupriavidus metallidurans*. *Front. Microbiol.* **2016**, *7*, 359. [CrossRef] [PubMed]
59. Providenti, M.A.; Shaye, R.E.; Lynes, K.D.; McKenna, N.T.; O'Brien, J.M.; Rosolen, S.; Wyndham, R.C.; Lambert, I.B. The locus coding for the 3-nitrobenzoate dioxygenase of *Comamonas* sp. strain JS46 is flanked by IS*1071* elements and is subject to deletion and inversion events. *Appl. Environ. Microbiol.* **2006**, *72*, 2651–2660. [CrossRef] [PubMed]
60. Clement, P.; Pieper, D.H.; Gonzalez, B. Molecular characterization of a deletion/duplication rearrangement in *tfd* genes from *Ralstonia eutropha* JMP134(pJP4) that improves growth on 3-chlorobenzoic acid but abolishes growth on 2,4-dichlorophenoxyacetic acid. *Microbiology* **2001**, *147*, 2141–2148. [CrossRef] [PubMed]
61. Mindlin, S.; Petrova, M. Mercury resistance transposons. In *Bacterial Integrative Mobile Genetic Elements*; Roberts, A.P., Mullany, P., Eds.; Landes Biosciences: Austin, TX, USA, 2013; pp. 33–52.
62. Miyazaki, R.; Minoia, M.; Pradervand, N.; Sentchilo, V.; Sulser, S.; Reinhard, F.; van der Meer, J.R. The *clc* Element and Related Genomic Islands in *Proteobacteria*. In *Bacterial Integrative Mobile Genetic Elements*; Roberts, A.P., Mullany, P., Eds.; Landes Bioscience: Austin, TX, USA, 2013; pp. 261–272.
63. Klockgether, J.; Wurdemann, D.; Reva, O.; Wiehlmann, L.; Tummler, B. Diversity of the abundant pKLC102/PAGI-2 family of genomic islands in *Pseudomonas aeruginosa*. *J. Bacteriol.* **2007**, *189*, 2443–2459. [CrossRef] [PubMed]
64. Darling, A.E.; Mau, B.; Perna, N.T. progressiveMauve: Multiple genome alignment with gene gain, loss and rearrangement. *PLoS ONE* **2010**, *5*, e11147. [CrossRef] [PubMed]
65. Van Houdt, R.; Toussaint, A.; Ryan, M.P.; Pembroke, J.T.; Mergeay, M.; Adley, C.C. The Tn*4371* ICE Family of Bacterial Mobile Genetic Elements. In *Bacterial Integrative Mobile Genetic Elements*; Roberts, A.P., Mullany, P., Eds.; Landes Bioscience: Austin, TX, USA, 2013; pp. 179–200.
66. Smith, D.J.; Park, J.; Tiedje, J.M.; Mohn, W.W. A large gene cluster in *Burkholderia xenovorans* encoding abietane diterpenoid catabolism. *J. Bacteriol.* **2007**, *189*, 6195–6204. [CrossRef] [PubMed]
67. Byun-McKay, A.; Godard, K.A.; Toudefallah, M.; Martin, D.M.; Alfaro, R.; King, J.; Bohlmann, J.; Plant, A.L. Wound-induced terpene synthase gene expression in Sitka spruce that exhibit resistance or susceptibility to attack by the white pine weevil. *Plant Physiol.* **2006**, *140*, 1009–1021. [CrossRef] [PubMed]
68. Arndt, D.; Grant, J.R.; Marcu, A.; Sajed, T.; Pon, A.; Liang, Y.; Wishart, D.S. PHASTER: A better, faster version of the PHAST phage search tool. *Nucleic Acids Res.* **2016**, *44*, W16–W21. [CrossRef] [PubMed]
69. Sorek, R.; Lawrence, C.M.; Wiedenheft, B. CRISPR-mediated adaptive immune systems in bacteria and archaea. *Annu. Rev. Biochem.* **2013**, *82*, 237–266. [CrossRef] [PubMed]
70. Grissa, I.; Vergnaud, G.; Pourcel, C. CRISPRFinder: A web tool to identify clustered regularly interspaced short palindromic repeats. *Nucleic Acids Res.* **2007**, *35*, W52–W57. [CrossRef] [PubMed]
71. Biswas, A.; Staals, R.H.; Morales, S.E.; Fineran, P.C.; Brown, C.M. CRISPRDetect: A flexible algorithm to define CRISPR arrays. *BMC Genom.* **2016**, *17*, 356. [CrossRef] [PubMed]

72. Biswas, A.; Fineran, P.C.; Brown, C.M. Computational Detection of CRISPR/crRNA Targets. *Methods Mol. Biol.* **2015**, *1311*, 77–89. [PubMed]
73. Monsieurs, P.; Moors, H.; Van Houdt, R.; Janssen, P.J.; Janssen, A.; Coninx, I.; Mergeay, M.; Leys, N. Heavy metal resistance in *Cupriavidus metallidurans* CH34 is governed by an intricate transcriptional network. *BioMetals* **2011**, *24*, 1133–1151. [CrossRef] [PubMed]
74. Henne, K.L.; Nakatsu, C.H.; Thompson, D.K.; Konopka, A.E. High-level chromate resistance in *Arthrobacter* sp. strain FB24 requires previously uncharacterized accessory genes. *BMC Microbiol.* **2009**, *9*, 199. [CrossRef] [PubMed]
75. Henne, K.L.; Turse, J.E.; Nicora, C.D.; Lipton, M.S.; Tollaksen, S.L.; Lindberg, C.; Babnigg, G.; Giometti, C.S.; Nakatsu, C.H.; Thompson, D.K.; et al. Global proteomic analysis of the chromate response in *Arthrobacter* sp. strain FB24. *J. Proteome Res.* **2009**, *8*, 1704–1716. [CrossRef] [PubMed]
76. Nies, D.H.; Rehbein, G.; Hoffmann, T.; Baumann, C.; Grosse, C. Paralogs of genes encoding metal resistance proteins in *Cupriavidus metallidurans* strain CH34. *J. Mol. Microbiol. Biotechnol.* **2006**, *11*, 82–93. [CrossRef] [PubMed]
77. Hlozkova, K.; Suman, J.; Strnad, H.; Ruml, T.; Paces, V.; Kotrba, P. Characterization of *pbt* genes conferring increased Pb^{2+} and Cd^{2+} tolerance upon *Achromobacter xylosoxidans* A8. *Res. Microbiol.* **2013**, *164*, 1009–1018. [CrossRef] [PubMed]
78. Wiesemann, N.; Mohr, J.; Grosse, C.; Herzberg, M.; Hause, G.; Reith, F.; Nies, D.H. Influence of copper resistance determinants on gold transformation by *Cupriavidus metallidurans* strain CH34. *J. Bacteriol.* **2013**, *195*, 2298–2308. [CrossRef] [PubMed]
79. Reith, F.; Etschmann, B.; Grosse, C.; Moors, H.; Benotmane, M.A.; Monsieurs, P.; Grass, G.; Doonan, C.; Vogt, S.; Lai, B.; et al. Mechanisms of gold biomineralization in the bacterium *Cupriavidus metallidurans*. *Proc. Natl. Acad. Sci. USA* **2009**, *106*, 17757–17762. [CrossRef] [PubMed]
80. Schmidt, T.; Schlegel, H.G. Combined nickel-cobalt-cadmium resistance encoded by the *ncc* locus of *Alcaligenes xylosoxidans* 31A. *J. Bacteriol.* **1994**, *176*, 7045–7054. [CrossRef] [PubMed]
81. Orita, I.; Iwazawa, R.; Nakamura, S.; Fukui, T. Identification of mutation points in *Cupriavidus necator* NCIMB 11599 and genetic reconstitution of glucose-utilization ability in wild strain H16 for polyhydroxyalkanoate production. *J. Biosci. Bioeng.* **2012**, *113*, 63–69. [CrossRef] [PubMed]
82. Raberg, M.; Peplinski, K.; Heiss, S.; Ehrenreich, A.; Voigt, B.; Doring, C.; Bomeke, M.; Hecker, M.; Steinbuchel, A. Proteomic and transcriptomic elucidation of the mutant *Ralstonia eutropha* G+1 with regard to glucose utilization. *Appl. Environ. Microbiol.* **2011**, *77*, 2058–2070. [CrossRef] [PubMed]
83. Kurnasov, O.; Jablonski, L.; Polanuyer, B.; Dorrestein, P.; Begley, T.; Osterman, A. Aerobic tryptophan degradation pathway in bacteria: Novel kynurenine formamidase. *FEMS Microbiol. Lett.* **2003**, *227*, 219–227. [CrossRef]
84. Weimann, A.; Mooren, K.; Frank, J.; Pope, P.B.; Bremges, A.; McHardy, A.C. From genomes to phenotypes: Traitar, the microbial trait analyzer. *mSystems* **2016**, *1*, e00101-16. [CrossRef] [PubMed]

© 2018 by the authors. Licensee MDPI, Basel, Switzerland. This article is an open access article distributed under the terms and conditions of the Creative Commons Attribution (CC BY) license (http://creativecommons.org/licenses/by/4.0/).

Article

Unintentional Genomic Changes Endow *Cupriavidus metallidurans* with an Augmented Heavy-Metal Resistance

Felipe A. Millacura [1], Paul J. Janssen [2], Pieter Monsieurs [2], Ann Janssen [2], Ann Provoost [2], Rob Van Houdt [2] and Luis A. Rojas [3],*

[1] School of Biological Sciences, University of Edinburgh, Edinburgh EH9 3JQ, UK; s1647595@sms.ed.ac.uk
[2] Interdisciplinary Biosciences, Belgian Nuclear Research Centre, SCK•CEN, 2400 Mol, Belgium; pjanssen@sckcen.be (P.J.J.); pieter.monsieurs@sckcen.be (P.M.); ann.janssen@sckcen.be (A.J); ann.provoost@sckcen.be (A.P.); rvhoudto@sckcen.be (R.V.H.)
[3] Chemistry Department, Faculty of Sciences, Universidad Católica del Norte, UCN, Antofagasta 1240000, Chile
* Correspondence: luis.rojas02@ucn.cl; Tel.: +56-55-235-5629

Received: 6 October 2018; Accepted: 8 November 2018; Published: 13 November 2018

Abstract: For the past three decades, *Cupriavidus metallidurans* has been one of the major model organisms for bacterial tolerance to heavy metals. Its type strain CH34 contains at least 24 gene clusters distributed over four replicons, allowing for intricate and multilayered metal responses. To gain organic mercury resistance in CH34, broad-spectrum *mer* genes were introduced in a previous work via conjugation of the IncP-1β plasmid pTP6. However, we recently noted that this CH34-derived strain, MSR33, unexpectedly showed an increased resistance to other metals (i.e., Co^{2+}, Ni^{2+}, and Cd^{2+}). To thoroughly investigate this phenomenon, we resequenced the entire genome of MSR33 and compared its DNA sequence and basal gene expression profile to those of its parental strain CH34. Genome comparison identified 11 insertions or deletions (INDELs) and nine single nucleotide polymorphisms (SNPs), whereas transcriptomic analysis displayed 107 differentially expressed genes. Sequence data implicated the transposition of IS*1088* in higher Co^{2+} and Ni^{2+} resistances and altered gene expression, although the precise mechanisms of the augmented Cd^{2+} resistance in MSR33 remains elusive. Our work indicates that conjugation procedures involving large complex genomes and extensive mobilomes may pose a considerable risk toward the introduction of unwanted, undocumented genetic changes. Special efforts are needed for the applied use and further development of small nonconjugative broad-host plasmid vectors, ideally involving CRISPR-related and advanced biosynthetic technologies.

Keywords: *Cupriavidus*; heavy metals; genomic islands; genomic rearrangements; metal resistance genes

1. Introduction

Since life appeared on Earth some 3.7 billion years ago, microorganisms have undergone molecular changes to adapt (i.e., respond to selection) to the harsh conditions of their natural habitats, including extreme temperatures, pH, salinity, UV and ionizing radiation, and heavy metals [1]. In addition, global fluctuations in the composition of the atmosphere, the oceans, and the earth crust have elicited genomic changes in microbes over eons of time and hence contributed to microbial diversity [2]. Certain microorganisms had been adapted to heavy metals and radionuclides prior to human appearance. However, during the past few hundred years, anthropogenic influences have forced microbes to also adapt to pollutants previously nonexistent (xenobiotics) [3–6], as well as to increased concentrations of heavy metals and radionuclides [7–9].

Members of the beta-proteobacterial genus *Cupriavidus* are prime examples of microbial endurance, possessing a variety of genomic islands involved in the resistance to heavy metals or the degradation of aromatics or xenobiotics [10–20]. They all typically display a bipartite chromosomal structure with one chromosomal replicon bearing the marks of a plasmid-type maintenance and replication system henceforth called "chromid". In addition, most *Cupriavidus* strains carry one or two dispensable megaplasmids with a size of 100 kb or more. The model organism for heavy metal resistance, *Cupriavidus metallidurans* strain CH34, carries two megaplasmids, pMOL28 and pMOL30. Together, the four CH34 replicons encode resistance markers for a plethora of heavy metals including copper, nickel, zinc, cobalt, cadmium, chrome, lead, silver, gold, mercury, caesium, selenium, strontium, and uranium [12,21–25]. These resistances are mainly related to a variety of metal reduction and efflux systems [26–28]. Aromatics degradation, on the other hand, is carried out by various bacterial multicomponent mono- and di-oxygenases solely encoded by genes on the chromosome and chromid [29].

In an effort to improve the inorganic and organic mercury resistance of *C. metallidurans* CH34 and thus improve its utility in cleaning up mercury-contaminated environments, the IncP-1β plasmid pTP6, providing additional *mer* genes [30], was introduced in CH34 by biparental mating, leading to strain MSR33 [31]. These extra *mer* genes are part of a transposon, Tn*50580*, which is necessary for broad-spectrum (organomercury) resistance, including two genes, *merG* and *merB*, not native to CH34 (i.e., CH34 only contains two narrow-spectrum mercury resistance *merRTPADE* operons, one on each megaplasmid, conferring resistance to inorganic mercury). MerB plays a key role in methylmercury degradation through its unique ability to cleave the carbon-mercury bond in methylmercury and the subsequent shuttling of ionic mercury to MerA to reduce it to the less harmful elemental mercury.

In comparison to its parental strain CH34, strain MSR33 became 240% more resistant to inorganic mercury and gained resistance to methylmercury by incorporating the previously nonpresent *merBG* genes. Other metal resistances (as tested for chrome and copper) were seemingly unaffected, and the pTP6 plasmid was stably maintained for over 70 generations under nonselective pressure [31]. However, when we recently tested the resistances for both strains to additional metals (i.e., cadmium, cobalt, and nickel), we noted a significant increase of metal resistance for strain MSR33 compared to CH34 (this study). As this was fully unexpected, since the only difference between the two strains should be an additional *mer* gene dosage, implicating only an improved mercury resistance, we set out to investigate the reasons for this phenomenon. Considering the genome plasticity [32–34] and the intricate relationships between metal resistance loci [24] in *C. metallidurans* CH34, we decided to sequence the full genome of strain MSR33 and compare the sequence of its replicons with the corresponding replicons of the parental strain CH34 and with plasmid pTP6. We also performed microarray-based expression analysis on both CH34 and MSR33 gene sets to determine whether genomic differences could be correlated with differences in the expression of individual genes.

2. Materials and Methods

2.1. Strains and Culture Conditions

Cupriavidus metallidurans strains CH34 and MSR33, obtained from respectively the SCK•CEN (Mol, Belgium) and the Univesidad Católica del Norte (UCN) (Antofagasta, Chile) culture collections, were cultivated at 30 °C and 200 revolutions per minute (rpm) on a shaker in dark, aerobic conditions in a Tris-buffered mineral medium (MM284) [35] with 0.4% (w/v) succinate as the sole carbon source. *Escherichia coli* JM109 pTP6 (also obtained from the UCN culture collection) was cultured at 37 °C on an M9 minimal medium [36] supplemented with 0.4% (w/v) glucose.

2.2. Synthetic Construct Generation

The *merTPAGB*$_1$ gene cluster of pTP6 was amplified by PCR using Phusion High-Fidelity DNA polymerase (ThermoFisher, Aalst, Belgium) with primer pair pTP6mer_Fw-Rv (Table S1). In tandem, the broad-host-range cloning vector pBBR1MCS-2 [37] was linearized by PCR with the same DNA polymerase and primer pair pBBR1MCS-2_GA_Fw-Rv (Table S1), providing homologous ends with the amplified $merTPAGB_1$ locus. These compatible PCR products were end-ligated using the Invitrogen GeneArt® Seamless Cloning and Assembly Enzyme Mix (ThermoFisher). After transforming *E. coli* DG1 with the ligation mix and selection on Lysogeny Broth (LB) with 50 µg/mL kanamycin (Km), four randomly chosen transformants were tested by DNA digestion and fragment sizing for correct plasmid construction holding the $merTBAGB_1$ genes. The plasmid gene construct of one transformant was further verified by sequencing its insert using forward and reverse cloning primers (Table S1) prior to electroporation into *C. metallidurans* CH34 on an Eppendorf 2510 Electroporator (Eppendorf, Aarschot, Belgium) using conditions as described previously [38].

2.3. Estimation of Bacterial Tolerance to Metals

Strains CH34 and MSR33 were grown overnight in MM284 liquid media with 0.4% (w/v) succinate as the sole carbon source and thereafter used as a pre-inoculum (1% v/v) for a freshly prepared 200 µL culture supplemented with increased concentrations of Hg^{2+} (from 0.0625 mM to 8 mM), Cd^{2+} (from 0.25 mM to 8 mM), and increasing steps (1 mM) of Co^{2+} or Ni^{2+} (from 5 mM to 20 mM). Cultures for metal contact were grown in microtiter plates at 30 °C on a rotary shaker at 120 rpm. Metal ion solutions were prepared from soluble salts of analytical grade ($CdCl_2$, $HgCl_2$, $CoCl_2 \cdot 6H_2O$, and $NiCl_2 \cdot 6H_2O$) in double-deionized water and were filter-sterilized before use. The lowest metal concentration that prevented growth after 48 h (i.e., showing no growth as measured at OD_{600} by a CLARIOstar microplate reader (BMG LabTech, Offenburg, Germany)), was considered the minimum inhibitory concentration (MIC) (Table 1 and Table S2). All MIC analyses were performed using biological triplicates.

Table 1. Minimal inhibitory concentration (MIC) of heavy metals for *Cupriavidus metallidurans* strains MSR33 and CH34.

Strain/Metal in mM	Hg^{2+}	Cd^{2+}	Ni^{2+}	Co^{2+}
C. metallidurans CH34	0.01	2.0	10.0	11.0
C. metallidurans MSR33	0.10	4.0	12.0	20.0
C. metallidurans CH34 (pBBR::$merTPAGB_1$)	0.10	2.0	ND [1]	ND [1]

[1] ND: not determined.

2.4. Plasmid Copy Number Determination

Single-copy (i.e., "replicon-unique") genes were taken as representatives of the chromosome (*cadA*), chromid (*zniA*), and plasmids pMOL30 (*nccA*), pMOL28 (*cnrA*), and pTP6 (*merG*). Primer pairs were designed to amplify 150 bp of each gene (Table S1). Real-time PCR was performed on a 7500 Applied Biosystems Fast Real-Time PCR System (ThermoFisher) using QiaGen RT^2 Sybr Green Rox qPCR Mastermix (ThermoFisher), 20 ng of MSR33 or CH34 genomic DNA as a template, and 0.2 µM of each primer. To reduce nonspecific amplification, we incubated this mixture at 95 °C for 10 min as part of a hot start PCR setup. Next, a 40-cycle amplification and quantification protocol (15 s at 95 °C, 15 s at 58 °C, and 30 s at 60 °C) was performed with a single fluorescence measurement for each cycle. Finally, a melting curve program (15 s at 95 °C, 60 s at 60 °C, 30 s at 95 °C, and 15 s at 60 °C) was carried out. Plasmid copy numbers for both strains were determined using the absolute method (allowing estimates of both the absolute and relative number of plasmids per cell), following earlier described protocols [39]. Standard curves were created with 7500 Fast Software v2.3 of Applied Biosystems (Foster City, CA, USA), using serial (10-fold) dilutions of genomic DNA in a linear range from 20 ng to 0.2 pg. The qPCR efficiencies were calculated from slopes of the log-linear portion of the calibration curves from the equation $E = 10^{(1/slope)}$. Using the linear equation obtained from each calibration curve, log DNA copy numbers were derived by intersecting the obtained C_t values. All analyses were done in triplicate.

2.5. Illumina Sequencing and Assembly

MSR33 cells were grown overnight in MM284 liquid medium, and genomic DNA was extracted by using the Qiagen QIAamp DNA Mini Kit (ThermoFisher), following the instructions of the manufacturer. DNA quantity and quality were measured using a DropSense (Trinean, Piscataway, NJ, USA). Genome sequencing was performed on a HiSeq 2500 apparatus (Illumina, San Diego, CA, USA) using 2 × 250 bp paired-end reads. The reads were trimmed using the Trimmomatic tool [40] and their quality assessed using in-house Perl and shell scripts in combination with SAMtools [41], BEDTools [42], and a Burrows–Wheeler aligner with maximum exact matches (bwa-mem) [43].

The entire genome of *C. metallidurans* CH34 was sequenced and largely annotated [12,22]. Sequences for a chromosome (NC_007973.1), a chromid (NC_007974.2), and the large plasmids pMOL28 (NC_007972.2) and pMOL30 (NC_007971.2) were all obtained from GenBank and used as a reference for the assembly and annotation of trimmed sequences of the MSR33 genome. The DNA sequence of pTP6 plasmid, also obtained from Genbank (AM048832), was used for sequence comparisons between the CH34 and MSR33 sequence data sets. The full *C. metallidurans* MSR33 genome sequence (this study) is available from the NCBI Sequence Read Archive (SRA) under accession number PRJNA493617.

2.6. Total RNA Isolation and Microarray

Cupriavidus metallidurans MSR33 and CH34 cells were both cultivated in triplicate on MM284 medium supplemented with 0.4% (w/v) succinate. Cell samples were taken at the middle exponential phase (OD_{600} 0.6–0.7) and centrifuged at 16,000 × g and 4 °C for 5 min. Pellets were quick-frozen with liquid nitrogen and kept at −80 °C for further analysis. Total RNA extraction was performed as described previously [24] by using an SV Total RNA Isolation System kit (Promega Benelux, Leiden, the Netherlands) according to the manufacturer's recommendations. Samples were cleaned and concentrated using a Nucleospin RNA cleanup XS kit (Macherey-Nagel, Düren, Germany). Concentrated RNA samples (10–20 µg) were retrotranscribed using the Invitrogen Superscript™ Direct cDNA Labeling System (ThermoFisher) and labeled by incorporation of Cy3-dCTP (ref PA53021, control condition) and Cy-5dCTP (ref PA55021, experimental condition) by Pronto!™ Long Oligo/cDNA Hybridization Solution (supplied with the Corning® Pronto! Universal Microarray Hybridization kit from Merck/Sigma-Aldrich, Overijse, Belgium), following the manufacturer's instructions. The microarrays we used were designed with 60-*mer* probes for 6205 Open Reading Frames that were spotted in triplicate onto glass slides (UltraGPS, Corning, NY, USA) using a MicroGrid system (BioRobotics, Cambridge, UK) at the microarray platform at SCK•CEN (Mol, Belgium). The spotted slides were cross-linked and placed in the presoaking solutions from the Pronto Kit (Promega, Madison, WI, USA). Analyses were performed on RNAs retrieved from CH34 and MSR33 cells, using respectively Cy3-dCTP and Cy5-dCTP incorporation and determination of Cy3/Cy5 signal intensity ratios. Labeled cDNA was resuspended in the universal hybridization buffer (Pronto kit), mixed, and added to the spotted slide for overnight hybridization at 42 °C in a Tecan HS4800 Pro hybridization station (Tecan Group Ltd., Männedorf, Switzerland). Afterwards, the slide was washed according to Pronto kit's protocol. Slides were scanned (at 532 and 635 nm) using the GenePix Personal 4100A microarray scanner (Molecular Devices, San Jose, CA, USA). All post-hybridization analyses were performed as described before [24]. In brief, spot signals were qualified using GenePix Pro v.6.0.1 software, and raw median density data were imported into R version 3.3.2 (https://cran.rstudio.com/) for statistical analysis using the LIMMA package version 2.15.15 (http://bioinf.wehi.edu.au/limma/), as available from Bioconductor (https://bioconductor.org). Background correction, normalizations, t-statistics, and p-value corrections were done as before [24]. Only log-transformed expression results with a p-value > 0.05 were considered for data interpretation (Table S2).

3. Results and Discussion

3.1. The Influence of Plasmid pTP6 on Increased Heavy Metal Resistance in MSR33

The *C. metallidurans* strains CH34 and MSR33 have the same genetic background, but MSR33 has, compared to its parental strain CH34, an extra plasmid 54 kb in size [31]. This plasmid, pTP6, is a broad-host-range IncP-1β plasmid originally isolated from mercury-polluted sediments [30]. It carries a transposon with *mer* genes that are not native to CH34 (i.e., *merG*) that encode an organomercurial transporter, and a pair of duplicate *merB* genes that encode periplasmic organomercurial lyases (Table S3). These additional *mer* genes in strain MSR33, via plasmid pTP6, grant this strain a 2.4-fold increased resistance for Hg^{2+} and a 16-fold increased resistance for CH_3Hg^+ [31]. In our hands, we noted a much-improved Hg^{2+} resistance for strain MSR33, with a 10-fold increase in comparison to its parental strain CH34 (Table 1). As an added note, in contrast to Rojas et al. [31], who performed MIC analyses on solid media, we performed our MIC analyses in a liquid medium, increasing metal bioavailability. Hence, sensitivity (i.e., the MIC (Hg) for CH34) was 0.05 mM in Rojas et al.'s study [31] but 0.01 mM in our study]. Surprisingly, when we tested MICs for the metals cadmium, nickel, and cobalt, we found a 2-fold increased resistance in MSR33 to Cd^{2+} and Co^{2+} and a 1.2-fold increased resistance to Ni^{2+} (Table 1).

The only genes on pTP6 with relevance to metal resistance are mercury resistance genes situated on transposon Tn*50580* as the two clusters $merR_1TPAGB_1$ and $merR_2B_2D_2E$ [30]. All other genes are typical for the backbones of self-transmissible and promiscuous IncP-1 plasmids from subgroups α and β and are involved in replication (*trfA*, *ssb*), plasmid maintenance, partitioning, control (*kfrABC*, *incC*, *korABC*, *kluAB*, *klcAB*, *kleABEF*, and a remnant of the resolvase gene *parA*), conjugal transfer (*traCDEFGHIJKL*), mating pair formation (*trbABCDEFGHIJKLMNO*), and transposition (*tniABQR* as part of Tn*50580*). Two more genes, *upf30.5* and *upf31.0*, are located downstream of *trbP* and encode, respectively, a putative outer membrane protein and a site-specific methylase (Figure S1). Except for the *upf* genes and the plasmid maintenance, partitioning, and control genes, pTP6 genes have at least one counterpart on one of the replicons of CH34. Taken together, we did not expect the pTP6 genes, other than the above-mentioned *mer* genes, to play any significant role in the augmented metal resistance of strain MSR33. Nonetheless, we decided to generate a new synthetic construct by cloning the $merTPAGB_1$ gene cluster in the low copy number broad-host-range cloning vector pBBR1MCS-2 [37]. This small plasmid only contained two genes, *rep* and *mob*, involved in, respectively, plasmid replication and mobilization, as well as a kanamycin resistance marker (Km^R). Strain CH34 transformed with this new construct reached the same level of inorganic mercury resistance as the MSR33 strain but did not show an increase in cadmium resistance (Table 1). From this we deduced that the *mer* genes of pTP6 exerted a positive effect on host resistance to mercury. However, neither the concomitant increase of cadmium resistance in MSR33 nor its higher resistance to nickel and cobalt could be readily explained by the presence of the auxiliary, pTP6-associated mercury resistance genes in this strain.

We also determined the effect of various mixtures containing both mercury and cadmium on the growth of strains CH34 and MSR33. In general, the combination of the two metals was expected to be more toxic to cells than the corresponding metals alone. Indeed, strain CH34 was capable of growing in up to 1 mM of Cd^{2+} when combined with 6.25 μM of Hg^{2+}, and cellular growth was diminished beyond these threshold metal concentrations (Table S4). Instead, strain MSR33 was capable of growing in up to 4 mM of Cd^{2+} when combined with 6.25 μM of Hg^{2+} (Table S4). However, even in combination with mercury, cadmium resistance in MSR33 was still twice as high as the cadmium resistance in CH34. Moreover, strain MSR33 showed much higher tolerance to mercury in combination with cadmium, and growth was only affected at the threshold metal concentrations of 125 μM Cd^{2+} and 100 μM Hg^{2+} (Table S4).

To exclude the possibility that the increased resistance in strain MSR33 to Hg^{2+} and Cd^{2+}, either alone or in combination, was a mere effect of gene dosage, we also determined the plasmid copy number for all replicons in strain MSR33 and strain CH34 using quantitative PCR (Table S5). The presence of the pTP6 plasmid in strain MSR33 (plasmid copy number (PCN) = 1.8) did not alter the relative copy numbers for the chromid and plasmids pMOL28 and pMOL30 with respect to the calculated chromosomal copy number taken as a reference. In addition, the Tn*50580* transposon carrying the broad spectrum *mer* gene cluster on plasmid pTP6 did not transpose (verified by genome sequence analysis present in Section 3.2).

The CH34 strain carries on its genome a total of four *mer* gene clusters: *merRTPA* on the chromosome, one complete *merRTPADE* on both plasmids pMOL28 and pMOL30, and one truncated *merRT∆P*, also on pMOL30 [12]. One could argue that the presence of a single copy of pTP6 in strain MSR33 raises the number of *mer* genes in strain MSR33, with one unit for genes *T*, *P*, *A*, *D*, and *E*, and two units for gene *R* (a second *merR* gene is located on the right-hand part of Tn*50580* on pTP6 [30] (Table S6), but the *merG* or *merB* genes of pTP6 were not considered here as they only play a role in organomercurial resistance). Considering the calculated PCN values of all replicons, the theoretical abundance of these genes increased by roughly 50–78% (Table S6). MerR and *merD* are transcriptional regulators that compete for the same operator sequence in the *merR–merT* intergenic region. The *merT* product is an inner membrane protein involved in the transport of Hg^{2+} ions into the cell cytoplasm. The *merP* product is a small periplasmic Hg^{2+}-sequestering protein that shuttles Hg^{2+} to the mercurial reductase MerA, which converts it into the significantly less toxic Hg(0) that then is allowed to leave the cell by passive diffusion. The *merE* product, finally, is another inner membrane protein and may play a role in the uptake of both CH_3-Hg^+ and Hg^{2+}. All these genes appeared to be intact, and there was no reason for us to assume that any of the multiple-copy genes would be dysfunctional or, with respect to each other (from gene to gene or copy to copy), would be differently transcribed or expressed (owing to limitations of gene-specific primer or probe design for multiple copies of these genes in qPCR or hybridization-based microarray procedures, no *mer* gene-specific expression data are available). A plausible explanation for the positive effect of the pTP6 *mer* genes on host mercury resistance could thus lay in the stoichiometry of *mer* gene products, particularly those involved in Hg^{2+} sequestration (*merP*) and transport (*merT/merE*). Nevertheless, the increased resistance to Cd^{2+} in MSR33 cannot be readily explained by a stoichiometric change in *mer* gene products. Also, when a single *merTPAGB* was introduced on a small plasmid into strain CH34, the increased Hg^{2+} resistance was still there, but the increased Cd^{2+} resistance was no longer seen (Table 1). From this we had to conclude that the MSR33 genetic background, besides the extra plasmid pTP6, may actually have differed from the genetic background of its parental strain CH34. In other words, the MSR33 genome had undergone genetic changes leading to an improved resistance to cadmium and possibly also to other metals. Such a genomic adaptation appears to be common to IncP-1 plasmid backbones [44]. In order to get to the bottom of this we decided to determine the DNA sequence of the entire genome of strain MSR33 and register in detail which genetic changes occurred with respect to the reference genomes of strain CH34 and plasmid pTP6.

3.2. The C. metallidurans MSR33 Genome Showed Multiple Insertions or Deletions and Single Nucleotide Polymorphisms

The whole-genome resequencing of MSR33 (since the known genome sequences of strain CH34 and plasmid pTP6 served as references, we considered this effort as a resequencing project) revealed a total of eight insertions and three deletions (Table 2), and nine single nucleotide polymorphisms (Table 3), all changes being located predominantly across the four replicons of the CH34 backbone, with only one genomic change occurring in plasmid pTP6 (Table 2, Figure 1). Most of the insertions (six out of eight) were found to be related to IS*1088*, an insertion element belonging to the IS*30* family with a typical size range of 1000–1250 bp [45]. It should be noted at this point that the CH34 genome indigenously harboured nine copies of IS*1088*, distributed on its chromosome and chromid but not on its megaplasmids [12], bringing the total of IS*1088* copies in the MSR33 genome to 15.

The majority of the new IS*1088* copies in the MSR33 genome were located on the chromosome and chromid, where all IS*1088* copies indigenous to CH34 resided, but one IS*1088* copy transposed into the *cnrY* gene (Rmet_6205) of pMOL28 (Table 2). This gene was part of the *cnrYXHCBAT* locus involved in the inducible cobalt and nickel resistance in strain CH34, and encoded the anti-sigma factor CnrY that tethered, in conjunction with the sensor protein CnrX, the sigma factor CnrH, but released it in the presence of Ni^{2+} or Co^{2+} [46,47]. The sigma factor CnrH promoted transcription of its own locus *cnrYXH*, but also of the structural locus *cnrCBA*, encoding a resistance nodulation division (RND)-driven efflux system [12,48]. The inactivation of the *cnrY* gene in MSR33 by IS*1088* inevitably led to the constitutive derepression of *cnrCBAT* transcription and explained the increased cobalt and nickel resistance we observed for MSR33 (Table 1) (see also gene expression results in Section 3.3). A similar phenomenon was previously seen for spontaneous mutants of a pMOL30-less CH34 derivative, strain AE126 [35], which showed a significantly increased resistance to cobalt and nickel [49] and which was later acknowledged as being an IS- and frameshift-mediated inactivation of *cnrY* and *cnrX* [50].

The other five genes affected by the insertion of an IS*1088* element were Rmet_0312 (*nptA*) and Rmet_2860 (*tauB*), lying on the chromosome; and Rmet_4160 (*pelF*), Rmet_4867 (*acrA*), and Rmet_5682 (*nimB*), lying on the chromid (Table 2). The first three genes encode proteins with general cellular functions, and their inactivation is very unlikely to affect heavy metal resistance in strain MSR33. The fourth gene, *acrA*, encodes a membrane fusion protein and is part of an intact *acrABC* operon whose gene products form a tripartite multidrug efflux system. The last gene, *nimB*, is involved in efflux-mediated heavy metal resistance, encoding also a membrane fusion protein resembling other membrane metal-binding fusion proteins in structure and function (e.g., CzcB, CnrB, CusB, and ZneB) by forming a periplasmic bridge between the cytoplasmic porter and the outer membrane channel [48]. Nonetheless, taking also into account that the *nimA* gene is already inactivated in strain CH34 (and MSR33) by the presence of the insertion sequence element IS*Rme3* [12], it is hard to see how the inactivation of *nimB* would result in the increased metal resistance we observed in strain MSR33. Two insertions were not attributed to IS*1088*. One appeared to be the result of a Tn*3*-related transposition event affecting gene Rmet_5388, encoding a tentative ApbE-like lipoprotein, while the other concerned an unknown mutational event in gene Rmet_5508 resulting in the insertion of a nucleotide triplet (+CTT) (Table 2). This gene encodes a long-chain fatty-acid CoA-ligase and also underwent a triplet deletion (-CGG) just a few nucleotides downstream of the triplet insert. As a combined result, the actual change at the protein level remained perfectly in-frame and gave a protein of the same length, but led to an altered peptide sequence at positions 149–153 (i.e., Xxx-Leu-**Arg**-**Phe**-**Ala**-Gln-Xxx in CH34 to Xxx-Leu-**Phe**-**Ala**-**Lys**-Gln-Xxx in MSR33 (amino acidic sequence change from **Arg**-**Phe**-**Ala** to **Phe**-**Ala**-**Lys**). The third deletion in MSR33 occurred in plasmid pTP6, effectively destroying the genes *upf30.5*, *upf31.0*, and *parA*, immediately preceding Tn*50580*. Except for *cnrY* and *nimB*, none of the aforementioned genes are in any way associated with metal resistance.

Table 2. Insertions (INs) and deletions (DELs) present in the genome of *C. metallidurans* MSR33.

Type	Replicon (#)	Start	End	Size (bp)	Description	Targeted Gene(s)	Function (MaGe Annotation) (*)
IN	CHR1	328395	328395	1102	IS*1088*	Rmet_0312	putative transporter
IN	CHR1	3106728	3106728	1102	IS*1088*	Rmet_2857 (*tauB*)	taurine ABC transporter ATP-binding protein
DEL	CHR2	602035	602490	455		Rmet_4033	LysR family transcriptional regulator
IN	CHR2	741818	741818	1102	IS*1088*	Rmet_4160 (*pelF*)	EPS biosynthesis, biofilm formation
IN	CHR2	1529231	1529231	1102	IS*1088*	Rmet_4867 (*acrA*)	membrane fusion protein, multidrug efflux
IN	CHR2	2113815	2113815	256	*tniA*/Tn*3*	Rmet_5388 (*apbE*)	ApbE-like lipoprotein
IN	CHR2	2253560	2253560	3	+CTT	Rmet_5508	long-chain-fatty-acid-CoA ligase
DEL	CHR2	2253566	2253568	3	−CGG	Rmet_5508	long-chain-fatty-acid-CoA ligase
IN	CHR2	2440975	2440975	1104	IS*1088*	Rmet_5682 (*nimB*)	membrane fusion protein, heavy metal transport
IN	pMOL28	53484	53484	1104	IS*1088*	Rmet_6205 (*cnrY*)	antisigma factor
DEL	pTP6	26155	27385	1230		*upf30.5*, *upf31.0*, *parA*	respectively an outer membrane protein, a DNA methylase, and a plasmid partition protein

Chromosome: CHR1; chromid: CHR2. * Genomic and Metabolics analysis software (www.genoscope.cns.fr).

Table 3. Single nucleotide polymorphisms (SNPs) detected in the genome of *C. metallidurans* MSR33.

Replicon (#)	Position	SNP	SNP Type	Affected Gene	Gene Description *
CHR1	333850	A→G	intergenic (+201/−75)	Rmet_0314 →/→ ssb1	putative transporter, major facilitator family/single-stranded DNA-binding protein (helix-destabilizing protein)
CHR1	645608	A→G	A23A (GCA→GCG)	Rmet_0598 →	Ser/Thr protein phosphatase family protein
CHR1	2400725	G→A	intergenic (−77/+116)	greA ←/← carB	transcription elongation factor/carbamoyl-phosphate synthase large subunit
CHR1	3418147	A→G	V295A (GTC→GCC)	dppF ←	dipeptide transporter; ATP-binding component of ABC superfamily
CHR1	3444412	A→G	V17A (GTC→GCC)	NirJ ←	heme d_1 biosynthesis protein
CHR1	3456113	A→G	V76A (GTG→GCG)	acyP ←	acylphosphatase
CHR2	2253543	A→T	V158E (GTG→GAG)	Rmet_5508 ←	long-chain-fatty-acid-CoA ligase
CHR2	2253553	G→T	P155T (CCG→ACG)	Rmet_5508 ←	long-chain-fatty-acid-CoA ligase
CHR2	2529357	T→C	G264G (GGA→GGG)	Rmet_5769 ←	esterase

#Chromosome: *CHR1*; chromid: *CHR2*; arrows show positive → or negative ← gene orientation; * www.genoscope.cns.fr.

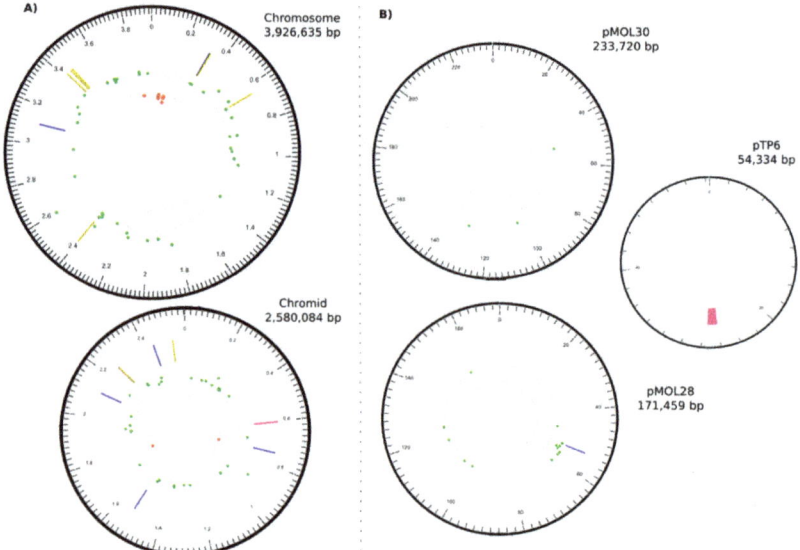

Figure 1. Correlation map between Insertions or Deletions (INDELs), Single nucleotide polymorphisms (SNPs), and transcriptional changes found for *C. metallidurans* MSR33. Concentric circles (ring) displayed from the outside inwards: (**A**) In ring 1, chromosome and chromid size scale in megabases (Mb) using a 20 kilobase (kb) window; (**B**) plasmids size scale in kb using a 2 kb window. In ring 2, position of insertion (blue), deletion (pink), and SNPs (mustard). INDELs and SNPs listed in Tables 2 and 3. In ring 3, each dot represents a single gene basal expression, overexpression ($\log_2$ *ratio* > 1, green), or repression ($\log_2$ *ratio* < −1, red), with a *p*-value < 0.05. Plotted genes listed by function in Table S7. Circos plot created with Circa (http://omgenomics.com/circa).

In addition to the eight insertions and three deletions in the MSR33 genome, sequence analysis revealed the presence of nine single nucleotide polymorphisms (Table 3). Two of those occurred in intergenic regions on the chromosome, without apparent disruption of gene regulatory elements, while the other seven occurred in protein-encoding genes (four on the chromosome and three on the chromid). Except for the "silent" mutation (i.e., no aa change) at position 645608, these Single nucleotide polymorphisms (SNPs) caused aa substitutions in the corresponding gene products (Table 3). No SNPs were detected in any of the plasmids. Remarkably, gene Rmet_5508 lying on the chromid (CHR2) once again was a target for mutation, displaying two SNPs in the immediate vicinity of the aforementioned triplet insertion and deletion in this gene (Table 3), bringing the full change of this region from **RFA**Q**KPAY**V to **FAK**Q**KT**AY**E** (changes are in bold and underlined). It is uncertain whether these protein changes would have any effect on the cellular and metabolic functions in MSR33.

Taken together, of the 11 Insertions or Deletions (INDELs) and nine SNPs identified by the whole genome resequencing of the *C. metallidurans* strain MSR33, only the *cnrY* inactivation by IS*1088* and the concomitant derepression of *cnrB* (see above) may be directly linked to the observed augmented heavy metal resistance in this strain, at least for Co^{2+} and Ni^{2+} (see above). None of the other genomic changes seemed to play a role in this augmentation. The augmented resistance for Cd^{2+} in strain MSR33 (Table 1), however, remains a puzzle. The fact that such augmentation for Cd^{2+} was only noted for MSR33 with an altered genome (i.e., with 17 INDELs and six SNPs), but not in CH34 transformed with plasmid pBBR::*merTPAGB*$_1$ (Table 1) strongly indicates that the MSR33 genetic background was at play. In basic terms, bacterial resistance to toxic metals depends on two cellular processes, metal binding and metal transport, with the former generally being an intrinsic part of the latter. It is well established that many proteins or peptides that mediate the transport, buffering, or detoxification of metal ions in living cells have metal-binding domains (MBDs) in which certain amino acid residues (e.g., cysteine), as well as their structural layout and relative position to each other, play a key role in metal selectivity and specificity [51,52]. While some of these proteins might be highly metal-specific, other proteins follow a more relaxed, nonspecific mode of metal binding. For instance, divalent metal uptake in *C. metallidurans* is governed by a battery of redundant transporters that display a minimal degree of metal cation selectivity [53]. Depending on the environment, this may lead to a cytoplasmic pool of unsolicited metal ions that at some point, particularly when reaching a toxic threshold, need to be removed by the cell. In *C. metallidurans* this was done by one of three efflux systems: Cation diffusion facilitators (CDF), P-type ATPases, and the earlier mentioned RND-driven transenvelope transporters (HME-RND). Their main task in *C. metallidurans*, because of this bacterium's adaptation to metal-rich environments, was to balance the cytoplasmic and periplasmic concentrations of unwanted transition metals by entering the cellular arena and going into competition for metal cations with the "frivolous" metal uptake systems. Interestingly, all three types of metal efflux systems seemed to possess some degree of frivolity toward metal ions as well, albeit perhaps not as outspoken as for the metal uptake systems. The *C. metallidurans* CzcD exporter (Rmet_5979), for instance, allowed as a CDF protein Zn^{2+}, Cd^{2+}, or Co^{2+} as a substrate [54], whereas the DmeF and FieF exporters (Rmet_0198 and Rmet_3406) displayed as CDF family members broad metal specificity for Zn^{2+}, Cd^{2+}, Co^{2+}, and Ni^{2+} [55]. It is worth mentioning that disruption of the *dmeF* gene in strain CH34 dramatically lowered the resistance for Co^{2+} (but not for Zn^{2+}, Cd^{2+}, and Ni^{2+}), indicating a complex interplay between the DmeF exporter and the CzcCBA and CnrCBA efflux pumps (possibly partially obscured by the action of other metal resistance systems) [55]. Moreover, CDF proteins can play diverse roles and may possess different metal ion selectivity depending on the environmental conditions (i.e., by adjusted K_d values for certain metals) [56]. In addition, the eight metal resistance-related P_{1B}-type ATPases currently identified in strain CH34 can be subdivided into two groups according to their substrate profile [28]: Those that extrude Cu^+ and Ag^+ (CupA and CupF) and those that extrude Zn^{2+}, Cd^{2+}, Co^{2+}, or Pb^{2+} (ZntA, CadA, PbrA, and CzcP). These exporters mainly differ in the presence of unique amino acid sequences in their transmembrane MBDs, hence defining their metal specificity. But even within a subgroup, differences may exist in terms of metal affinity. For example, CzcP encoded by plasmid pMOL30 is unable to mediate Zn resistance on its own but rather augments the metal exportability of the ZntA, CadA, and PbrA exporters [57]. In a similar fashion, the five active HME-RND efflux systems in strain CH34 displayed a limited substrate spectrum, pumping out either the monovalent metal cations Cu^+ and Ag^+ (CusA, SilA) or the divalent metal cations Zn^{2+}, Ni^{2+}, and Co^{2+}, with occasionally also Cd^{2+} (ZniA, CnrA, CzcA) [28] (Figure 2). In such HME-RND systems, two steps of heavy-metal extrusion were discerned, the periplasmic and the transenvelope efflux (Figure 2). Each step involved the interaction of metals with MBDs within the Membrane Fusion Protein (MFP) and RND proteins. Sometimes, the delivery of periplasmic metal ions to the typical $C_3B_6A_3$-complex is facilitated by a small periplasmic metallochaperone, as is the case for the *E. coli* CusCBFA system [58] (and likely, based on CusF aa sequence similarities, also the CusCBAF complex of strain CH34). Little is known about the substrate specificity of the metal-binding proteins of HME-RND efflux complexes. Apparently, metal-induced conformational changes in the $C_3B_6A_3$-complex are required in order to create a proper metal-guiding $C_3B_6A_3$ channel for metal export to take place [48,59–61].

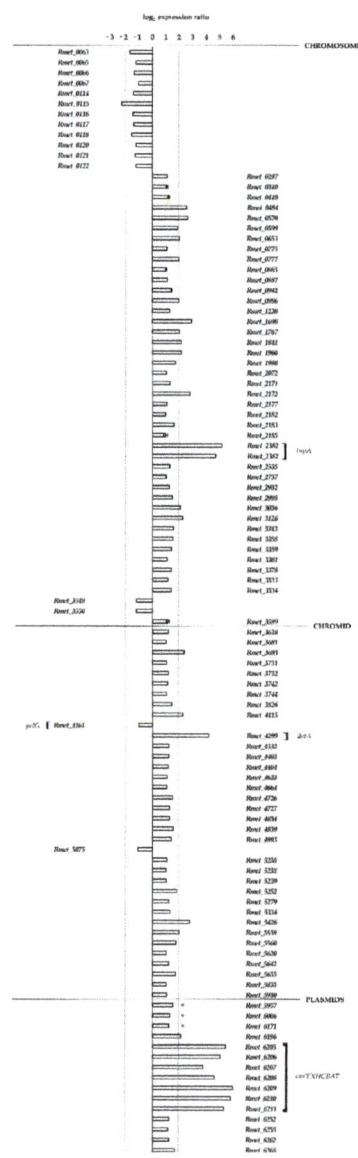

Figure 2. Transcriptional changes in *C. metallidurans* MSR33 with respect to CH34, with both strains grown under equal and nonselective conditions (see methods). Bar graphs show the significantly (p-value < 0.05) higher expression ($\log_2$ ratio > +1) and lower expression ($\log_2$ ratio < −1) of MSR33 genes (with CH34 gene expression levels as reference). Transcriptional changes from genes pertaining to all replicons are depicted. Genes indicated with an * are part of pMOL30.

As mentioned, it is not inconceivable that the genetic changes in MSR33 instigated cellular conditions or pleiotropic effects that were generally favourable for Cd^{2+} detoxification and hence led to the observed improvement in Cd^{2+} resistance. Possibly, this involved the temporal recruitment of one or more metal binding export proteins, from known metal resistant systems or from hitherto unknown export systems, able to bind Cd^{2+}. The transition metals cadmium and mercury belong to Group 12 of chemical elements in the periodic table, together with zinc and copernicium. Although these four metals differ in significant respects, they also have common properties. Particularly, Cd and Hg are similar in their outer shell electron configuration ($d^{10}s^2$) and atomic radius (ca. 150 pm), and their cations both have a high affinity for sulfhydryl groups in cellular compounds and proteins (i.e., in methionine and cysteine residues). From this perspective, competition between Cd^{2+} and Hg^{2+} for certain MBDs cannot be excluded. Bacterial evolution and adaptation to new or rapidly changing environments implies a delicate balance between the safeguarding of genome integrity and the tolerance for genome instability. A too-rigid genome inevitably will lead to the demise of innovative power and hence adaptability of the host, whereas a too plastic or "fluid" genome may lead to disadvantageous mutations and cell growth arrest, or even cell death. This balance between beneficiary and perilous change in a bacterial genome also relates to the general fitness and the energy household of its host. Members of the genus *Cupriavidus*, and in particular *C. metallidurans*, appear to be masters in adaptation as they are home to a wide variety of habitats, often in extreme conditions [62–65]. The introduction of the 54 kb plasmid pTP6 into a strain already carrying two large replicons of 3.9 and 2.6 Mb (chromosome and chromid, respectively) and two megasized plasmids of 171 and 234 kb (pMOL28 and pMOL30, respectively) could be seen as a serious additional burden to the host regardless of whether or not phenotypical or physiological changes occur.

Because the charting of genomic changes in MSR33 with respect to its parental strain CH34 did not provide us with any clues or direct evidence on the involvement of certain genetic loci or of any of the known metal resistance determinants on the observed augmented Cd^{2+} resistance, we decided to compare the basal gene expression data for strains CH34 and MSR33 using RNA microarray technology in an attempt to associate their gene expression profiles with strain-specific physiological behaviour, with a focus on differentially expressed (DE) genes that might be involved in the cellular detoxification of heavy metals such as Cd^{2+}.

3.3. Transcriptional Analysis of Strain MSR33

Strains CH34 and MSR33, which were equally grown in nonselective conditions without any metal-related stress, were investigated for basal gene expression levels, and their expression profiles were compared. A total of 107 DE genes showed statistically significant changes in their expression (Figure 2). Affected genes pertained to the main chromosome (55 genes), the chromid (36 genes), and the plasmids pMOL30 (3 genes) and pMOL28 (13 genes) (Figure 2). In general terms, the products of these 107 genes could be grouped according to their predicted annotated function [12]: Catalytic function (35 genes), transport (24 genes), transcriptional regulation (11 genes), recombination (6 genes), movement and chemotaxis (9 genes), and miscellaneous (22 genes) (Table S7). Most of these genes had a higher expression in strain MSR33, with only 16 genes in strain MSR33 showing a lower expression level. What is immediately striking is that all genes of the *cnrYXHCBAT* operon on plasmid pMOL28 had a significantly higher expression (Figure 2), with $\log_2$ fold changes ranging from four to six. As we know from MSR33 whole genome sequence analysis, this high expression of the *cnr* locus in strain MSR33 was the direct result of IS*1088*-mediated *cnrY* inactivation and hence derepression of the *cnr* locus, explaining the increased resistance we observed for strain MSR33 to Co^{2+} and Ni^{2+} (see previous sections). Equally noticeable is the complete absence of an altered expression of the *czc* locus on pMOL30, strongly suggesting that the augmented Cd^{+2} resistance we see for strain MSR33 was independent of this locus. This, in fact, corroborates earlier findings about the pMOL30-less CH34 derivative AE126 (which, like MSR33, also has an IS*1088*-mediated inactivated *cnrY* gene on the remaining plasmid pMOL28): Vandecraen et al. [50] showed, next to a heightened resistance to Zn^{2+}, Co^{2+}, and Ni^{2+}, a 2-fold increased resistance to Cd^{2+}. Adding another level of complexity, when strains AE126 and AE104 (a CH34 derivative lacking both pMOL28 and pMOL30 plasmids) [35] were transformed with pTP6, these strains, like MSR33, gained an improvement in Cd^{2+} resistance, albeit to a lesser extent (Rojas LA, personal communication).

This would indicate that the augmentation of Cd^{2+} resistance in MSR33 by pTP6 conjugation should be seen as a layered process brought about by multiple factors and possibly diverse mechanisms supporting each other. We cannot say at this point what these mechanisms precisely are and how and when they are triggered, as we have no information about the genomic changes in pTP6 conjugants of AE104 and AE126 (as pTP6 conjugation in CH34 causes genomic changes, this would most likely also be the case for pTP6 conjugation in strains AE104 and AE126, but not necessarily involving the same genomic changes). Clearly, further studies are needed to understand the augmented metal resistance in pTP6 conjugants of CH34 and its derivatives, including (1) the resequencing of pTP6-conjugated AE104 and AE126 strains and (2) the extensive RNAseq-based genetic response analyses for a wider range of heavy metals in all three pTP6-conjugated strains. The additional possibility that some of the observed genetic changes were already introduced to the recipient CH34 strain prior to conjugation with pTP6 cannot be entirely excluded. Lastly, the plasmid-curing procedures used to obtain strains AE104 and AE126 (i.e., applying mitomycin C, nalidixic acid, or hydroxyurea to growing CH34 cells [35]) may also have had mutagenic effects or may have induced transposition activity. In this respect, it would be best, in the frame of future studies, to resequence these strains as well.

A very high log_2 fold difference in the expression of >4 was also noted in strain MSR33 for the Rmet_4229 gene, a *dctA* paralogue whose product was functionally annotated as a C4-dicarboxylate transporter and which is unlikely to have any connection to metal detoxification or resistance, and gene Rmet_2382, originally identified in CH34 as a transposase-encoding *tnpA* gene (IS*1088*) (Figure 2). Intermediate high log_2 fold changes of >2 were seen in strain MSR33 for another 17 genes (Table S7), whereas the remaining 72 genes showed a log_2 fold change between one and two. None of these genes is thought to be involved in metal binding, metal detoxification, or metal resistance. Among the genes with lowered expression in strain MSR33, we noted the *pelG* gene (Rmet_4161), which is part of the *pelABCDEFG* operon required to produce an extracellular polysaccharide that has been implicated in biofilm development [66]. Our sequence data confirmed that the IS*1088* element transposed into the *pelF* gene (Table 2), thereby disrupting expression of the *pelG* gene. This could explain the complete lack of biofilm formation in strain MSR33 reported to us by P. Alviz in a personal communication.

In conclusion, the genome of MSR33 underwent eight insertions, three deletions, and nine SNPs. At least seven of the insertions were due to the action of mobile genetic elements, with their presence fully confirmed by sequence data (implicating IS*1088* in six cases), whereas one small insertion and all three deletions in strain MSR33 may have been the result of DNA recombination or transposition events. The *C. metallidurans* genome is known to be ridden with a very high number of mobile genetic elements, with 57 IS elements, 19 other transposable elements, and 16 genomic islands for its type strain CH34 [12,32,34]. In concordance with this genomic fluidity, *C. metallidurans* displays a highly versatile metabolism and an inherent ability to inhabit a variety of harsh environments [9,62–65]. This adaptability has not come about overnight but is the wonderful result of microbial evolution over long periods of time. In a time in which large chunks of DNA were retrieved from the environment (e.g., by plasmid transfer or gene exchanges), adaptation was brought about by DNA mutations and natural selection and molecular inventions took place, steadily moulding the genome into its present large (6.9 Mb) and highly malleable form, providing the bacterium with a vast array of possibilities for rapid genetic responses (hence its well-chosen epithet as "Master Survivalist") [12]. However, tinkering with this hugely evolved and dynamic genome holds intrinsic dangers. Although the plasmid pTP6 was maintained stably in strain CH34 (i.e., MSR33) for over 70 generations under nonselective conditions [31], it has now become clear from our study at hand that the receiving host's genome underwent multiple changes in the form of 11 INDELs and 9 SNPs, affecting the physiology and heavy metal resistance of the host.

It would be wrong to point the finger at the extra plasmid as the "usual suspect" for these genetic changes, but rather we hold the actual process of conjugation responsible. Conjugative interaction appears to be a strong stimulus for transposition [67–69], and hence it is easy to envisage that, as a result of conjugation procedures, some elements of the extensive mobilome of *C. metallidurans* (with nearly 100 mobile elements) were triggered into action and "moved around", causing genetic changes that led to clearly perceptible but also less visible (and less understood) effects alike. The take-home message here is that the genetic engineering of bacteria with large complex and dynamic genomes should be carried out with much caution and that a strong preference should be given to the new generation of small broad-host-range cloning vectors and CRISPR-based technologies nowadays available [70–74].

Supplementary Materials: The following are available online at http://www.mdpi.com/2073-4425/9/11/551/s1. Figure S1: Genetic map of plasmid pTP6; Table S1: Primers designed for this report; Table S2: Transcriptional changes observed in MSR33 versus CH34 under basal conditions; Table S3: Homology analysis of *mer* gene products present in plasmid pTP6; Table S4: Effect of various mixtures of Hg^{2+} and Cd^{2+} on *C. metallidurans* strains MSR33 and CH34 growth; Table S5: Plasmid copy number (PCN) for *C. metallidurans* strains MSR33 and CH34, determined by quantitative PCR; Table S6: *mer* gene occurrences on the replicons of strains CH34 and MSR33; Table S7: Expression changes of *C. metallidurans* MSR33 with respect to *C. metallidurans* CH34, both grown under nonselective conditions.

Author Contributions: F.A.M., P.J.J., R.V.H., P.M., and L.A.R. conceived and designed the experiments; F.A.M., A.J., and A.P. performed the experiments; F.A.M., R.V.H., and P.M. analyzed the data; L.A.R. and P.J.J. contributed reagents, materials, and analysis tools; F.A.M., P.J.J., R.V.H., P.M., and L.A.R. wrote the paper.

Funding: SCK•CEN EE0630012-09 (to P.J.J., R.V.H., A.J., A.P., and P.M.), CONICYT/FONDECYT 11130117 (L.A.R.), and CONICYT/BC-PhD 72170403 (F.M.)

Acknowledgments: Authors acknowledge research funding by SCK•CEN EE0630012-09 (to P.J.J., R.V.H., A.J., A.P., and P.M.), CONICYT/FONDECYT 11130117 (L.A.R.), and CONICYT/BC-PhD 72170403 (F.M.). Genome sequencing was provided by MicrobesNG (http://www.microbesng.uk), supported by BBSRC grant number BB/L024209/1. We thank D. Vallenet and Z. Rouy of Génoscope (Centre National de Séquençage, Evry, France) for implementing additional features in MaGe and their essential advice in genome annotation.

Conflicts of Interest: The authors declare no conflicts of interest. The founding sponsors had no role in the design of the study; in the collection, analyses, or interpretation of data; in the writing of the manuscript; or in the decision to publish the results.

References

1. Bakermans, C. *Microbial Evolution under Extreme Conditions*; De Gruyter: Altoona, PA, USA, 2015; Volume 2.
2. Knoll, A.H. Paleobiological perspectives on early microbial evolution. *CSH Perspect. Biol.* **2015**, *7*, a018093. [CrossRef] [PubMed]
3. Springael, D.; Top, E.M. Horizontal gene transfer and microbial adaptation to xenobiotics: New types of mobile genetic elements and lessons from ecological studies. *Trends Microbiol.* **2004**, *12*, 53–58. [CrossRef] [PubMed]
4. Cao, L.; Liu, H.; Zhang, H.; Huang, K.; Gu, T.; Ni, H.; Hong, Q.; Li, S. Characterization of a newly isolated highly effective 3,5,6-trichloro-2-pyridinol degrading strain *Cupriavidus pauculus* P2. *Curr. Microbiol.* **2012**, *65*, 231–236. [CrossRef] [PubMed]
5. Ilori, M.O.; Picardal, F.W.; Aramayo, R.; Adebusoye, S.A.; Obayori, O.S.; Benedik, M.J. Catabolic plasmid specifying polychlorinated biphenyl degradation in *Cupriavidus* sp. strain SK-4: Mobilization and expression in a pseudomonad. *J. Basic Microb.* **2015**, *55*, 338–345. [CrossRef] [PubMed]
6. Mergeay, M.; Van Houdt, R. Adaptation to xenobiotics and toxic compounds by *Cupriavidus* and *Ralstonia* with special reference to *Cupriavidus metallidurans* CH34 and mobile genetic elements. In *Biodegradative Bacteria: How Bacteria Degrade, Survive, Adapt, and Evolve*; Nojiri, H., Tsuda, M., Fukuda, M., Kamagata, Y., Eds.; Springer: Tokyo, Japan, 2014; pp. 105–127.
7. Nies, D.H. Microbial heavy-metal resistance. *Appl. Microbiol. Biotechnol.* **1999**, *51*, 730–750. [CrossRef] [PubMed]
8. Mergeay, M. Bacteria adapted to industrial biotopes: The metal resistant *Ralstonia*. In *Bacterial Stress Responses*; Storz, G., Hengge-Aronis, R., Eds.; ASM Press: Washington, DC, USA, 2000; pp. 403–414.

9. Sobecky, P.A.; Coombs, J.M. Horizontal gene transfer in metal and radionuclide contaminated soils. In *Horizontal Gene Transfer: Genomes in Flux*; Gogarten, M.B., Gogarten, J.P., Olendzenski, L.C., Eds.; Humana Press: Totowa, NJ, USA, 2009; pp. 455–472.
10. Pohlmann, A.; Fricke, W.F.; Reinecke, F.; Kusian, B.; Liesegang, H.; Cramm, R.; Eitinger, T.; Ewering, C.; Potter, M.; Schwartz, E.; et al. Genome sequence of the bioplastic-producing "knallgas" bacterium *Ralstonia eutropha* H16. *Nat. Biotechnol.* **2006**, *24*, 1257–1262. [CrossRef] [PubMed]
11. Amadou, C.; Pascal, G.; Mangenot, S.; Glew, M.; Bontemps, C.; Capela, D.; Carrere, S.; Cruveiller, S.; Dossat, C.; Lajus, A.; et al. Genome sequence of the beta-rhizobium *Cupriavidus taiwanensis* and comparative genomics of rhizobia. *Genome Res.* **2008**, *18*, 1472–1483. [CrossRef] [PubMed]
12. Janssen, P.J.; Van Houdt, R.; Moors, H.; Monsieurs, P.; Morin, N.; Michaux, A.; Benotmane, M.A.; Leys, N.; Vallaeys, T.; Lapidus, A.; et al. The complete genome sequence of *Cupriavidus metallidurans* strain CH34, a master survivalist in harsh and anthropogenic environments. *PLoS ONE* **2010**, *5*, e10433. [CrossRef] [PubMed]
13. Lykidis, A.; Perez-Pantoja, D.; Ledger, T.; Mavromatis, K.; Anderson, I.J.; Ivanova, N.N.; Hooper, S.D.; Lapidus, A.; Lucas, S.; Gonzalez, B.; et al. The complete multipartite genome sequence of *Cupriavidus necator* JMP134, a versatile pollutant degrader. *PLoS ONE* **2010**, *5*, e9729. [CrossRef] [PubMed]
14. Poehlein, A.; Kusian, B.; Friedrich, B.; Daniel, R.; Bowien, B. Complete genome sequence of the type strain *Cupriavidus necator* N-1. *J. Bacteriol.* **2011**, *193*, 5017. [CrossRef] [PubMed]
15. Cserhati, M.; Kriszt, B.; Szoboszlay, S.; Toth, A.; Szabo, I.; Tancsics, A.; Nagy, I.; Horvath, B.; Nagy, I.; Kukolya, J. De novo genome project of *Cupriavidus basilensis* OR16. *J. Bacteriol.* **2012**, *194*, 2109–2110. [CrossRef] [PubMed]
16. Hong, K.W.; Thinagaran, D.A.L.; Gan, H.M.; Yin, W.F.; Chan, K.G. Whole-genome squence of *Cupriavidus* sp. strain BIS7, a heavy-metal-resistant bacterium. *J. Bacteriol.* **2012**, *194*, 6324. [CrossRef] [PubMed]
17. Li, L.G.; Cai, L.; Zhang, T. Genome of *Cupriavidus* sp. HMR-1, a heavy metal-resistant bacterium. *Genome Announc.* **2013**, *1*, e00202-12. [CrossRef] [PubMed]
18. Ray, J.; Waters, R.J.; Skerker, J.M.; Kuehl, J.V.; Price, M.N.; Huang, J.; Chakraborty, R.; Arkin, A.P.; Deutschbauer, A. Complete genome sequence of *Cupriavidus basilensis* 4G11, isolated from the oak ridge field research center site. *Genome Announc.* **2015**, *3*, e00322-15. [CrossRef] [PubMed]
19. Wang, X.Y.; Chen, M.L.; Xiao, J.F.; Hao, L.R.; Crowley, D.E.; Zhang, Z.W.; Yu, J.; Huang, N.; Huo, M.X.; Wu, J.Y. Genome sequence analysis of the naphthenic acid degrading and metal resistant bacterium *Cupriavidus gilardii* CR3. *PLoS ONE* **2015**, *10*, e0132881. [CrossRef] [PubMed]
20. Fang, L.C.; Chen, Y.F.; Zhou, Y.L.; Wang, D.S.; Sun, L.N.; Tang, X.Y.; Hua, R.M. Complete genome sequence of a novel chlorpyrifos degrading bacterium, *Cupriavidus nantongensis* X1. *J. Biotechnol.* **2016**, *227*, 1–2. [CrossRef] [PubMed]
21. Mergeay, M.; Monchy, S.; Vallaeys, T.; Auquier, V.; Benotmane, A.; Bertin, P.; Taghavi, S.; Dunn, J.; van der Lelie, D.; Wattiez, R. *Ralstonia metallidurans*, a bacterium specifically adapted to toxic metals: Towards a catalogue of metal-responsive genes. *FEMS Microbiol. Rev.* **2003**, *27*, 385–410. [CrossRef]
22. Monchy, S.; Benotmane, M.A.; Janssen, P.; Vallaeys, T.; Taghavi, S.; van der Lelie, D.; Mergeay, M. Plasmids pMOL28 and pMOL30 of *Cupriavidus metallidurans* are specialized in the maximal viable response to heavy metals. *J. Bacteriol.* **2007**, *189*, 7417–7425. [CrossRef] [PubMed]
23. Avoscan, L.; Untereiner, G.; Degrouard, J.; Carriere, M.; Gouget, B. Uranium and selenium resistance in *Cupriavidus metallidurans* CH34. *Toxicol. Lett.* **2007**, *172*, S157. [CrossRef]
24. Monsieurs, P.; Moors, H.; Van Houdt, R.; Janssen, P.J.; Janssen, A.; Coninx, I.; Mergeay, M.; Leys, N. Heavy metal resistance in *Cupriavidus metallidurans* CH34 is governed by an intricate transcriptional network. *Biometals* **2011**, *24*, 1133–1151. [CrossRef] [PubMed]
25. Ben Salem, I.; Sghaier, H.; Monsieurs, P.; Moors, H.; Van Houdt, R.; Fattouch, S.; Saidi, M.; Landolsi, A.; Leys, N. Strontium-induced genomic responses of *Cupriavidus metallidurans* and strontium bioprecipitation as strontium carbonate. *Ann. Microbiol.* **2013**, *63*, 833–844. [CrossRef]
26. Van Houdt, R.; Mergeay, M. Genomic context of metal response genes in *Cupriavidus metallidurans* with a focus on strain CH34. In *Metal Response in Cupriavidus Metallidurans: Volume I: From Habitats to Genes and Proteins*; Mergeay, M., Van Houdt, R., Eds.; Springer International Publishing: Cham, Switzerland, 2015; pp. 21–44.

27. Vandenbussche, G.; Mergeay, M.; Van Houdt, R. Metal response in *Cupriavidus metallidurans*: Insights into the structure-function relationship of proteins. In *Metal Response in Cupriavidus Metallidurans: Volume II: Insights into the Structure-Function Relationship of Proteins*; Springer International Publishing: Cham, Switzerland, 2015; pp. 1–70.
28. Nies, D.H. The biological chemistry of the transition metal "transportome" of *Cupriavidus metallidurans*. *Metallomics* **2016**, *8*, 481–507. [CrossRef] [PubMed]
29. Millacura, F.A.; Cardenas, F.; Mendez, V.; Seeger, M.; Rojas, L.A. Degradation of benzene by the heavy-metal resistant bacterium *Cupriavidus metallidurans* CH34 reveals its catabolic potential for aromatic compounds. *bioRxiv* **2017**. [CrossRef]
30. Smalla, K.; Haines, A.S.; Jones, K.; Krogerrecklenfort, E.; Heuer, H.; Schloter, M.; Thomas, C.M. Increased abundance of IncP-1 beta plasmids and mercury resistance genes in mercury-polluted river sediments: First discovery of IncP-1 beta plasmids with a complex *mer* transposon as the sole accessory element. *Appl. Environ. Microb.* **2006**, *72*, 7253–7259. [CrossRef] [PubMed]
31. Rojas, L.A.; Yanez, C.; Gonzalez, M.; Lobos, S.; Smalla, K.; Seeger, M. Characterization of the metabolically modified heavy metal-resistant *Cupriavidus metallidurans* strain MSR33 generated for mercury bioremediation. *PLoS ONE* **2011**, *6*, e17555. [CrossRef] [PubMed]
32. Van Houdt, R.; Monchy, S.; Leys, N.; Mergeay, M. New mobile genetic elements in *Cupriavidus metallidurans* CH34, their possible roles and occurrence in other bacteria. *Antonie Leeuwenhoek* **2009**, *96*, 205–226. [CrossRef] [PubMed]
33. Mijnendonckx, K.; Provoost, A.; Monsieurs, P.; Leys, N.; Mergeay, M.; Mahillon, J.; Van Houdt, R. Insertion sequence elements in *Cupriavidus metallidurans* CH34: Distribution and role in adaptation. *Plasmid* **2011**, *65*, 193–203. [CrossRef] [PubMed]
34. Van Houdt, R.; Monsieurs, P.; Mijnendonckx, K.; Provoost, A.; Janssen, A.; Mergeay, M.; Leys, N. Variation in genomic islands contribute to genome plasticity in *Cupriavidus metallidurans*. *BMC Genom.* **2012**, *13*, 111. [CrossRef] [PubMed]
35. Mergeay, M.; Nies, D.; Schlegel, H.G.; Gerits, J.; Charles, P.; Van Gijsegem, F. *Alcaligenes eutrophus* CH34 is a facultative chemolithotroph with plasmid-bound resistance to heavy metals. *J. Bacteriol.* **1985**, *162*, 328–334. [PubMed]
36. Maniatis, T.; Fritsch, E.F.; Sambrook, J. *Molecular Cloning: A Laboratory Manual*; Cold Spring Harbor Laboratory: Cold Spring Harbor, NY, USA, 1982.
37. Kovach, M.E.; Elzer, P.H.; Hill, D.S.; Robertson, G.T.; Farris, M.A.; Roop, R.M.; Peterson, K.M. Four new derivatives of the broad-host-range cloning vector pBBR1MCS, carrying different antibiotic-resistance cassettes. *Gene* **1995**, *166*, 175–176. [CrossRef]
38. Taghavi, S.; Vanderlelie, D.; Mergeay, M. Electroporation of *Alcaligenes eutrophus* with (mega) plasmids and genomic DNA fragments. *Appl. Environ. Microb.* **1994**, *60*, 3585–3591.
39. Lee, C.; Kim, J.; Shin, S.G.; Hwang, S. Absolute and relative qPCR quantification of plasmid copy number in *Escherichia coli*. *J. Biotechnol.* **2006**, *123*, 273–280. [CrossRef] [PubMed]
40. Bolger, A.M.; Lohse, M.; Usadel, B. Trimmomatic: A flexible trimmer for Illumina sequence data. *Bioinformatics* **2014**, *30*, 2114–2120. [CrossRef] [PubMed]
41. Li, H.; Handsaker, B.; Wysoker, A.; Fennell, T.; Ruan, J.; Homer, N.; Marth, G.; Abecasis, G.; Durbin, R.; 1000 Genome Project Data Processing Subgroup. The sequence alignment/map format and SAMtools. *Bioinformatics* **2009**, *25*, 2078–2079. [CrossRef] [PubMed]
42. Quinlan, A.R. BEDTools: The Swiss-army tool for genome feature analysis. *Curr. Protoc. Bioinform.* **2014**, *47*, 11–34. [CrossRef] [PubMed]
43. Li, H. Aligning sequence reads, clone sequences and assembly contigs with BWA-MEM. *arXiv*, 2013; arXiv:1303.3997.
44. Norberg, P.; Bergstrom, M.; Jethava, V.; Dubhashi, D.; Hermansson, M. The IncP-1 plasmid backbone adapts to different host bacterial species and evolves through homologous recombination. *Nat. Commun.* **2011**, *2*, 268. [CrossRef] [PubMed]
45. Siguier, P.; Gourbeyre, E.; Chandler, M. Bacterial insertion sequences: Their genomic impact and diversity. *FEMS Microbiol. Rev.* **2014**, *38*, 865–891. [CrossRef] [PubMed]
46. Grass, G.; Grosse, C.; Nies, D.H. Regulation of the *cnr* cobalt and nickel resistance determinant from *Ralstonia* sp. strain CH34. *J. Bacteriol.* **2000**, *182*, 1390–1398. [CrossRef] [PubMed]

47. Tibazarwa, C.; Wuertz, S.; Mergeay, M.; Wyns, L.; van Der Lelie, D. Regulation of the *cnr* cobalt and nickel resistance determinant of *Ralstonia eutropha* (*Alcaligenes eutrophus*) CH34. *J. Bacteriol.* **2000**, *182*, 1399–1409. [CrossRef] [PubMed]
48. Kim, E.H.; Nies, D.H.; McEvoy, M.M.; Rensing, C. Switch or funnel: How RND-type transport systems control periplasmic metal homeostasis. *J. Bacteriol.* **2011**, *193*, 2381–2387. [CrossRef] [PubMed]
49. Collard, J.M.; Provoost, A.; Taghavi, S.; Mergeay, M. A new type of *Alcaligenes eutrophus* CH34 zinc resistance generated by mutations affecting regulation of the *cnr* cobalt-nickel resistance system. *J. Bacteriol.* **1993**, *175*, 779–784. [CrossRef] [PubMed]
50. Vandecraen, J.; Monsieurs, P.; Mergeay, M.; Leys, N.; Aertsen, A.; Van Houdt, R. Zinc-induced transposition of insertion sequence elements contributes to increased adaptability of *Cupriavidus metallidurans*. *Front. Microbiol.* **2016**, *7*, 359. [CrossRef] [PubMed]
51. Zheng, H.; Chruszcz, M.; Lasota, P.; Lebioda, L.; Minor, W. Data mining of metal ion environments present in protein structures. *J. Inorg. Biochem.* **2008**, *102*, 1765–1776. [CrossRef] [PubMed]
52. Thilakaraj, R.; Raghunathan, K.; Anishetty, S.; Pennathur, G. In silico identification of putative metal binding motifs. *Bioinformatics* **2007**, *23*, 267–271. [CrossRef] [PubMed]
53. Kirsten, A.; Herzberg, M.; Voigt, A.; Seravalli, J.; Grass, G.; Scherer, J.; Nies, D.H. Contributions of five secondary metal uptake systems to metal homeostasis of *Cupriavidus metallidurans* CH34. *J. Bacteriol.* **2011**, *193*, 4652–4663. [CrossRef] [PubMed]
54. Anton, A.; Grosse, C.; Reissmann, J.; Pribyl, T.; Nies, D.H. Czcd is a heavy metal ion transporter involved in regulation of heavy metal resistance in *Ralstonia* sp. strain CH34. *J. Bacteriol.* **1999**, *181*, 6876–6881. [PubMed]
55. Munkelt, D.; Grass, G.; Nies, D.H. The chromosomally encoded cation diffusion facilitator proteins DmeF and FieF from *Wautersia metallidurans* CH34 are transporters of broad metal specificity. *J. Bacteriol.* **2004**, *186*, 8036–8043. [CrossRef] [PubMed]
56. Barber-Zucker, S.; Shaanan, B.; Zarivach, R. Transition metal binding selectivity in proteins and its correlation with the phylogenomic classification of the cation diffusion facilitator protein family. *Sci. Rep.* **2017**, *7*, 16381. [CrossRef] [PubMed]
57. Scherer, J.; Nies, D.H. CzcP is a novel efflux system contributing to transition metal resistance in *Cupriavidus metallidurans* CH34. *Mol. Microbiol.* **2009**, *73*, 601–621. [CrossRef] [PubMed]
58. Delmar, J.A.; Su, C.-C.; Yu, E.W. Bacterial multi-drug efflux transporters. *Annu. Rev. Biophys.* **2014**, *43*, 93–117. [CrossRef] [PubMed]
59. De Angelis, F.; Lee, J.K.; O' Connell, J.D.; Miercke, L.J.W.; Verschueren, K.H.; Srinivasan, V.; Bauvois, C.; Govaerts, C.; Robbins, R.A.; Ruysschaert, J.M.; et al. Metal-induced conformational changes in ZneB suggest an active role of membrane fusion proteins in efflux resistance systems. *Proc. Natl. Acad. Sci. USA* **2010**, *107*, 11038–11043. [CrossRef] [PubMed]
60. Long, F.; Su, C.C.; Lei, H.T.; Bolla, J.R.; Do, S.V.; Yu, E.W. Structure and mechanism of the tripartite CusCBA heavy-metal efflux complex. *Philos. Trans. R. Soc. Lond. B Biol. Sci.* **2012**, *367*, 1047–1058. [CrossRef] [PubMed]
61. Pak, J.E.; Ekende, E.N.; Kifle, E.G.; O'Connell, J.D.; De Angelis, F.; Tessema, M.B.; Derfoufi, K.M.; Robles-Colmenares, Y.; Robbins, R.A.; Goormaghtigh, E.; et al. Structures of intermediate transport states of ZneA, a Zn(II)/proton antiporter. *Proc. Natl. Acad. Sci. USA* **2013**, *110*, 18484–18489. [CrossRef] [PubMed]
62. Sota, M.; Tsuda, M.; Yano, H.; Suzuki, H.; Forney, L.J.; Top, E.M. Region-specific insertion of transposons in combination with selection for high plasmid transferability and stability accounts for the structural similarity of IncP-1 plasmids. *J. Bacteriol.* **2007**, *189*, 3091–3098. [CrossRef] [PubMed]
63. Leys, N.; Baatout, S.; Rosier, C.; Dams, A.; s' Heeren, C.; Wattiez, R.; Mergeay, M. The response of *Cupriavidus metallidurans* CH34 to spaceflight in the international space station. *Antonie Leeuwenhoek* **2009**, *96*, 227–245. [CrossRef] [PubMed]
64. Mijnendonckx, K.; Provoost, A.; Ott, C.M.; Venkateswaran, K.; Mahillon, J.; Leys, N.; Van Houdt, R. Characterization of the survival ability of *Cupriavidus metallidurans* and *Ralstonia pickettii* from space-related environments. *Microb. Ecol.* **2013**, *65*, 347–360. [CrossRef] [PubMed]
65. Byloos, B.; Coninx, I.; Van Hoey, O.; Cockell, C.; Nicholson, N.; Ilyin, V.; Van Houdt, R.; Boon, N.; Leys, N. The impact of space flight on survival and interaction of *Cupriavidus metallidurans* CH34 with basalt, a volcanic moon analog rock. *Front. Microbiol.* **2017**, *8*, 671. [CrossRef] [PubMed]

66. Vasseur, P.; Vallet-Gely, I.; Soscia, C.; Genin, S.; Filloux, A. The *pel* genes of the *Pseudomonas aeruginosa* PAK strain are involved at early and late stages of biofilm formation. *Microbiology* **2005**, *151*, 985–997. [CrossRef] [PubMed]
67. Godoy, V.G.; Fox, M.S. Transposon stability and a role for conjugational transfer in adaptive mutability. *Proc. Natl. Acad. Sci. USA* **2000**, *97*, 7393–7398. [CrossRef] [PubMed]
68. Christie-Oleza, J.A.; Lanfranconi, M.P.; Nogales, B.; Lalucat, J.; Bosch, R. Conjugative interaction induces transposition of ISPst9 in *Pseudomonas stutzeri* AN10. *J. Bacteriol.* **2009**, *191*, 1239–1247. [CrossRef] [PubMed]
69. Baharoglu, Z.; Bikard, D.; Mazel, D. Conjugative DNA transfer induces the bacterial SOS response and promotes antibiotic resistance development through integron activation. *PLoS Genet.* **2010**, *6*, e1001165. [CrossRef] [PubMed]
70. Jain, A.; Srivastava, P. Broad host range plasmids. *FEMS Microbiol. Lett.* **2013**, *348*, 87–96. [CrossRef] [PubMed]
71. Obranic, S.; Babic, F.; Maravic-Vlahovicek, G. Improvement of pBBR1MCS plasmids, a very useful series of broad-host-range cloning vectors. *Plasmid* **2013**, *70*, 263–267. [CrossRef] [PubMed]
72. Tian, P.; Wang, J.; Shen, X.; Rey, J.F.; Yuan, Q.; Yan, Y. Fundamental CRISPR-cas9 tools and current applications in microbial systems. *Synth. Syst. Biotechnol.* **2017**, *2*, 219–225. [CrossRef] [PubMed]
73. Cook, T.B.; Rand, J.M.; Nurani, W.; Courtney, D.K.; Liu, S.A.; Pfleger, B.F. Genetic tools for reliable gene expression and recombineering in *Pseudomonas putida*. *J. Ind. Microbiol. Biotechnol.* **2018**, *45*, 517–527. [CrossRef] [PubMed]
74. Xiong, B.; Li, Z.; Liu, L.; Zhao, D.; Zhang, X.; Bi, C. Genome editing of *Ralstonia eutropha* using an electroporation-based CRISPR-cas9 technique. *Biotechnol. Biofuels* **2018**, *11*, 172. [CrossRef] [PubMed]

© 2018 by the authors. Licensee MDPI, Basel, Switzerland. This article is an open access article distributed under the terms and conditions of the Creative Commons Attribution (CC BY) license (http://creativecommons.org/licenses/by/4.0/).

Article

Genomic and Transcriptomic Changes That Mediate Increased Platinum Resistance in *Cupriavidus metallidurans*

Md Muntasir Ali [1,2], Ann Provoost [1], Laurens Maertens [1,3], Natalie Leys [1], Pieter Monsieurs [1], Daniel Charlier [2] and Rob Van Houdt [1,*]

1 Microbiology Unit, Belgian Nuclear Research Centre (SCK•CEN), 2400 Mol, Belgium; md.muntasir.ali@sckcen.be (M.M.A.); ann.provoost@sckcen.be (A.P.); laurens.maertens@sckcen.be (L.M.); natalie.leys@sckcen.be (N.L.); pieter.monsieurs@sckcen.be (P.M.)
2 Research Group of Microbiology, Department of Bioengineering Sciences, Vrije Universiteit Brussel, 1050 Brussel, Belgium; dcharlie@vub.be
3 Research Unit in Biology of Microorganisms (URBM), Faculty of Sciences, UNamur, 5000 Namur, Belgium
* Correspondence: rob.van.houdt@sckcen.be; Tel.: +32-14-33-2728

Received: 7 December 2018; Accepted: 15 January 2019; Published: 18 January 2019

Abstract: The extensive anthropogenic use of platinum, a rare element found in low natural abundance in the Earth's continental crust and one of the critical raw materials in the EU innovation partnership framework, has resulted in increased concentrations in surface environments. To minimize its spread and increase its recovery from the environment, biological recovery via different microbial systems is explored. In contrast, studies focusing on the effects of prolonged exposure to Pt are limited. In this study, we used the metal-resistant *Cupriavidus metallidurans* NA4 strain to explore the adaptation of environmental bacteria to platinum exposure. We used a combined Nanopore–Illumina sequencing approach to fully resolve all six replicons of the *C. metallidurans* NA4 genome, and compared them with the *C. metallidurans* CH34 genome, revealing an important role in metal resistance for its chromid rather than its megaplasmids. In addition, we identified the genomic and transcriptomic changes in a laboratory-evolved strain, displaying resistance to 160 µM Pt^{4+}. The latter carried 20 mutations, including a large 69.9 kb deletion in its plasmid pNA4_D (89.6 kb in size), and 226 differentially-expressed genes compared to its parental strain. Many membrane-related processes were affected, including up-regulation of cytochrome c and a lytic transglycosylase, down-regulation of flagellar and pili-related genes, and loss of the pNA4_D conjugative machinery, pointing towards a significant role in the adaptation to platinum.

Keywords: platinum resistance; RNA-Seq; multireplicon; Nanopore; adaptive laboratory evolution

---

1. Introduction

Platinum (Pt) is a rare element that is found in low natural abundance (0.4 parts per billion) in the Earth's continental crust [1,2]. It is extensively used in industry, vehicle exhaust catalysts (VECs), and anticancer drugs [3], with cisplatin being one of the potent anti-cancer drugs in use [4]. Anthropogenic uses and emissions of platinum have resulted in increased concentrations (0.5–1.4 ton $year^{-1}$) in surface environments, which could negatively impact natural habitats, especially because of the solubility of some forms of platinum [3,5]. It can enter waters, sediments, and soils and eventually reach the food chain [3]. Therefore, effective measures must be taken to minimize its spread and increase its recovery from the environment.

Platinum is also one of the critical raw materials in the EU innovation partnership framework, which were selected because of their high economic importance and high supply risk [6]. This makes platinum an important candidate for biological recovery from waste streams and other environmental niches. As bacterial communities have been naturally associated with platinum-group mineral grains [7], efforts have been made to explore the usability of microorganisms for the recovery of platinum group metals from the environment [8]. For instance, biological recovery of platinum has been shown for halophilic microbial communities, indicating that Pt from waste streams can be transformed into Pt-rich biomass, which in turn can be used as input for the refinery of precious metals [9]. In addition, Pt biosorption has also been studied for axenic bacterial cultures, including sulfate-reducing bacteria such as *Desulfovibrio desulfuricans*, *Desulfovibrio fructosivorans*, and *Desulfovibrio vulgaris* [10], as well as *Shewanella oneidensis*, *Cupriavidus metallidurans*, *Geobacter metallireducens*, *Pseudomonas stutzeri*, and *Bacillus toyonensis* [11,12]. The sulfate-reducing *Desulfovibrio* spp. and metal-ion reducing *Shewanella algae* have the capability to reduce Pt to zero state and form Pt nanoparticles in their periplasmic space [13,14]. It has been hypothesized that *Desulfovibrio* spp. can use Pt, as well as palladium, as terminal electron acceptors in their energy production pathway via cytochrome *c3* [15–17]. This promotes nanoparticle formation on the cell surface, preventing re-entry and acting as catalysts for further metal reduction [17]. *Cupriavidus* sp. also showed similar nanoparticle formation in the presence of palladium, according to an equivalent strategy [18].

Studying the interaction between platinum and bacteria showed that platinum inhibits cell division and enhances filamentous growth of *Escherichia coli* [19], *Caulobacter crescentus* and *Hyphomicrobium* sp. [20]. It inhibits DNA synthesis and DNA repair functions were shown to be essential for growth in the presence of platinum, as *E. coli* mutants deficient in DNA repair functions are unable to grow in the presence of platinum [21,22]. Similar to other DNA synthesis inhibitors, Pt also induces prophages from lysogenic *E. coli* strains [23]. The mutagenic ability of different platinum compounds has also been demonstrated in *Salmonella enterica* subsp. *enterica* serovar Typhimurium strains [24].

It is clear that most studies have analyzed the biological immobilization of Pt and its possible applications. However, only a limited number of studies focused on the effect of prolonged exposure to Pt, as would be the case for environmental bacteria in Pt-contaminated waters, soils, and sediments. For instance, Maboeta et al. showed that enzymatic activities and viable biomass were impacted in a platinum tailing disposal facility associated with mining activities [25].

In this study, we used *C. metallidurans* NA4 as a model to explore the adaptation of environmental bacteria to platinum exposure. It has been extensively studied for its resistance to a variety of metal (oxyan)ions [26–28]. We used a combinatorial sequencing approach to fully resolve the *C. metallidurans* NA4 genome consisting of six replicons [29], compared its genome with that of *C. metallidurans* CH34, and identified the genomic and transcriptomic changes in a laboratory-evolved strain, displaying increased resistance to platinum.

2. Materials and Methods

2.1. Strains, Media, and Culture Conditions

C. metallidurans NA4 was routinely cultured at 30 °C in Lysogeny broth (LB) or Tris-buffered mineral medium (MM284) supplemented with 0.2% (*w/v*) gluconate [30]. For culturing on solid medium, 1.5% agar (Thermo Scientific, Oxoid, Hampshire, UK) was added; liquid cultures were grown in the dark on a rotary shaker at 150 rpm. Metal salts used included $PtCl_4$, Na_2PdCl_4, $ZnSO_4 \cdot 7H_2O$, $NiCl_2 \cdot 6H_2O$, $CuSO_4 \cdot 5H_2O$ and $AgNO_3$. (Sigma-Aldrich, Overijse, Belgium).

2.2. Determination of the Minimal Inhibitory Concentration and Generation of Pt-Resistant Mutants

The minimal inhibitory concentration (MIC) of Pt^{4+}, Pd^{2+}, Zn^{2+}, Ni^{2+}, Cu^{2+}, and Ag^+ was determined using the broth dilution method in a 96-well plate containing a concentration gradient of the corresponding metals [31]. To select for *C. metallidurans* mutants displaying increased platinum resistance, a serial passage experiment was performed by continuous exposure to subinhibitory concentrations of Pt^{4+} using the gradient MIC method [32].

2.3. Plasmid Isolation and Restriction Digestion

The extraction of megaplasmids was based on the method proposed by Andrup et al. [33]. Extracted plasmid DNA was separated by horizontal gel electrophoresis (23 cm-long 0.5% Certified Megabase agarose gel (Bio-Rad, Temse, Belgium) in 1X Tris-Borate-EDTA buffer, 100 V, 20 h) in a precooled (4 °C) electrophoresis chamber. After GelRed staining (30 min + overnight destaining at 4 °C in ultrapure water), DNA was visualized and images captured under UV light transillumination (Fusion Fx, Vilber Lourmat, Collégien, France). To confirm the presence and size of the smaller plasmid in NA4Pt (Pt^{4+} resistant mutant of NA4), plasmid DNA was isolated with the Wizard® Plus SV Miniprep DNA Purification System (Promega, Leiden, The Netherlands). The isolated DNA was used for restriction digestion with *PagI* (Fisher Scientific, Merelbeke, Belgium). The products were separated on a 0.6% agarose (Molecular Biology Grade, Eurogentec, Belgium) gel to visualize the individual fragments together with the GeneRuler 1 kb plus ladder (Fisher Scientific, Merelbeke, Belgium).

2.4. Motility, Scanning Electron Microscopy (SEM), and Flow Cytometry

For testing motility, *C. metallidurans* NA4 and NA4Pt were grown in LB media until the OD_{600} reached 0.6. Five µL of the culture was then stab inoculated onto a LB plate containing 0.3% agar. The radius of the growth pattern was measured after 24 h.

For SEM, cells were grown in MM284 in normal growth conditions, centrifuged (5000 rpm for 8 min), washed in Milli-Q water, and fixed with 3% glutaraldehyde solution at 4 °C (3 h). Cells were sputter coated (22 nm) with gold and examined under SEM at an accelerating voltage of 10 kV.

For flow cytometry, bacterial cell suspensions (OD_{600} = 0.6) were diluted 1000 times in 0.2 µm filtered Tris-buffered mineral medium (MM284), Next, SYBR green (Sigma Aldrich) dye was added and incubated at 37 °C for 20 min. Stained bacterial suspensions were analyzed on the Accuri C6 flow cytometer (BD, Erembodegem, Belgium).

2.5. Genome Sequencing

Total DNA from *C. metallidurans* NA4 and NA4Pt was isolated using the QIAamp DNA mini kit (Qiagen, Venlo, the Netherlands). The parental strain NA4 [29] was resequenced using a combination of Illumina and Nanopore sequencing. Illumina sequencing was performed on the Illumina HiSeq 2500 platform using 2×75 bp paired-end sequencing (Baseclear, Leiden, Netherlands). Nanopore sequencing was performed in-house using the MinION device with an R9 flow cell and the Rapid Sequencing kit. Strain NA4Pt was sequenced using the Illumina Miseq platform ($40\times$ coverage; MicrobesNG, Birmingham, UK).

2.6. Genome Assembly

Genome assembly was performed using the pre-assembled contigs based on the 454 sequencing data as "trusted contigs" combined with the illumina and nanopore sequencing data as input for the SPAdes algorithm (version 3.11.1, default parameter settings) [34,35]. Subsequent genome polishing was performed by consecutive runs of an in-house Perl script, where the original reads were realigned against the resulting assembly using the Burrows-Wheeler Aligner (BWA). Based on this output, the genome assembly was updated accordingly until no further single-nucleotide polymorphisms (SNPs) and indels were detected.

The *C. metallidurans* NA4Pt genome was compared to the parental strain at two levels. For the small SNPs and indels, the output of two algorithms was combined: BreSeq 0.32.0 and the Genome Analysis Toolkit (GATK) [36,37]. BreSeq was run using the default parameter settings. Before running GATK, we first converted the raw BWA output to a sorted and index Binary Alignment Map (BAM) file using the view, sort and index command of the SAMtools package version 0.1.18 [38].

SNP prediction was performed on this BAM file by following the pipeline described in Van der Auwera et al. [37], with default parameters, and using ploidy = 1 when running the HaplotyperCaller command. For larger structural variations (insertions and deletion), an in-house developed Python script, specifically focused on the identification of structural variations caused by mobile genetic elements, was used. This program exploits the paired-end information and insert size distributions to predict these variations.

2.7. Transcriptomic Analysis Using RNA-Seq

Gene expression in NA4Pt was compared with the parental strain NA4 under non-selective growth conditions. Three independent *C. metallidurans* NA4 and NA4Pt cultures were allowed to grow until an OD_{600} of 0.6 was reached. Each culture was subdivided in 2 mL portions and cells were harvested by centrifugation for 2 min at $10,000\times g$. Bacterial pellets were flash frozen by immersion into liquid nitrogen and kept frozen at $-80\ °C$ at all times. Total RNA was extracted using the Promega SV Total RNA Isolation System kit (Promega, Leiden, The Netherlands). RNA sequencing (directional mRNA library, RiboZero rRNA depletion and 2×125 bp paired-end sequencing) was performed by Eurofins genomics (Ebersberg, Germany).

2.8. RNA-Seq Data Analysis

Obtained RNA-Seq reads were aligned using BWA software and the default parameters [39]. Raw counts per gene were calculated based on the latest genome annotation of *C. metallidurans* NA4, as available on the MaGe platform. Reads were allowed to map 50 bp upstream of the start codon or 50 bp downstream of the stop codon. Reads that were mapped to ribosomal or transfer RNA were removed from the raw count data to prevent bias in detecting differential expression. Differential expression was calculated using the edgeR package (version 3.2.4) [40] in BioConductor (release 3.0, R version 3.1.2), resulting in a fold change value and a corresponding *p* value corrected for multiple testing for each gene. Genes were found to be differentially expressed if they show an absolute log2 fold change higher than 0.80 and a false discovery rate (FDR) value lower than 0.05.

2.9. Functional Analysis

Homologous genes and synteny groups were computed via the MaGe platform [41]. Homologous genes were based on the bidirectional best hit criterion and a blastP alignment threshold (at least 35% amino-acid identity on 80% of the length of the smallest protein). Synteny, orthologous gene sets that have the same local organization are based on the bidirectional best hit criterion or a blastP alignment threshold (at least 30% amino-acid identity on 80% of the length of the smallest protein), and co-localization (with the maximum number of consecutive genes not involved in a synteny group being five).

Distribution of insertion sequence (IS) elements was determined by identification and annotation of IS elements with ISsaga [42] and manual curation.

MOB typing of plasmids was performed using the mob_typer script, part of the MOB-suite [43]. All analyses were performed using standard parameters. The most recent reference database was downloaded on the 19th of November, 2018.

Plasmids were aligned to pNA4_D by the AliTV perl interface in the AliTV package [44] using standard parameters. Output json files were visualized with the AliTV web service (http://alitvteam.github.io/AliTV/d3/AliTV.html). Subsystem categories were assigned via the online implementation of RASTtk (Rapid Annotation using Subsystem Technology) (http://rast.nmpdr.org/) [45]. Circos plots from genomic data were constructed with Circa 1.2.1 (http://omgenomics.com/circa/).

3. Results and Discussion

3.1. Genome Analysis of C. metallidurans NA4

The *C. metallidurans* NA4 genome was previously sequenced via the 454 GS-FLX sequencing platform [29] and assembled into 109 contigs. Since all *Cupriavidus* genomes have a multipartite organization composed of at least a chromosome, a chromid and one or more megaplasmids [27], we tested if combining sequencing platforms could produce completely closed replicon sequences. *C. metallidurans* NA4, in particular, contains six replicons [29,46]. Complete assembly of multipartite genomes is necessary to fully understand the genomic structure and invaluable to correctly assess genomic events. An additional Illumina and Nanopore sequencing was performed. Integration of the pre-assembled contigs (based on the 454 pyrosquencing data) and the Illumina and Nanopore sequencing data, using the SPAdes platform, resulted in complete assembly and closure of the six *C. metallidurans* NA4 replicons (Table 1). The long reads from Nanopore sequencing resolved the complex repeat regions, while Illumina sequencing provided more accurate sequencing results.

Table 1. Overall characterization of the replicons constituting the *C. metallidurans* NA4 genome.

Replicon	Length (bp)	% GC [1]	# CDSs [2]
Chromosome	3,838,195	63.84	3818
Chromid	2,776,395	63.59	2774
Plasmid pNA4_A	294,575	59.45	416
Plasmid pNA4_B	227,796	60.57	274
Plasmid pNA4_C	155,041	58.90	209
Plasmid pNA4_D	89,606	61.56	120

[1] Guanine-cytosine content; [2] Coding sequences.

The size, number of coding sequences (CDSs) and GC content of the *C. metallidurans* NA4 chromosome and chromid were very similar to that of type strain *C. metallidurans* CH34 [26] (Figure 1). To compare the different replicons of CH34 and NA4, homologous genes were computed and visualized by Circa (Figure 2). Both the chromosome (CHR1) and chromid (CHR2) were well conserved. However, this comparison revealed an extensive number of homologous genes between plasmid pMOL30 from CH34 and the chromid of NA4 (Figure 2). To zoom in on this, large (>15 genes) synteny groups between pMOL30 and the chromid of NA4 were visualized separately (Figure 3), as well as homologous genes shared between pMOL28 and pMOL30, and pNA4_A and pNA4_B (Figure 4).

The latter indicated that the main large metal resistance clusters on pMOL30 are syntenic with gene clusters on NA4's chromid instead of NA4's plasmids. Previously, we showed that pMOL28 and pMOL30 contain large genomic islands that harbor all plasmid-borne genes involved in the response to heavy metals [27,47,48]. Plasmid pMOL30 carries two genomic islands, CMGI-30a and -30b, that convey resistance to cadmium, zinc, cobalt, lead, and mercury, and copper and silver, respectively. A 25-gene cluster within CMGI-30a, containing the *czc* cluster (related to cadmium, zinc and cobalt resistance) and genes involved in membrane-related functions, and highly conserved (>99% nucleotide similarity) in all *C. metallidurans* strains [48], was found on NA4's chromid. Interestingly, this cluster is always flanked by a tyrosine-based site-specific recombinase (TBSSR) associated with a conserved protein of unknown function making up a bipartite module (BIM) [48,49]. Our observation adds evidence to the mobility of this cluster. The cluster on NA4's chromid related to the *pbr* (lead resistance) and *mer* (mercury resistance) cluster (Tn4380) of pMOL30 was flanked at both sides by (remnants of) TBSSRs (only at one side for pMOL30). Three clusters on NA4's chromid were homologous to pMOL30's *cop* cluster (copper resistance) (Figure 3, Supplementary Figure S1). The latter was part of genomic island CMGI-30b of pMOL30 that contains a 33-gene copper-related cluster almost completely induced by Cu^{2+} (coding for the efflux P_I-type ATPase CopF, the heavy metal efflux (HME) resistance nodulation cell-division (RND) system SilCBA, the periplasmic detoxification system CopABCDI, and accessory and membrane-related functions) as well as the *nre/ncc* cluster. One NA4 cluster comprised almost the complete CMGI30-b island except *copV* and *copT*. A second cluster contained the copper-related cluster as well as some gene fragments and was flanked by TBSSRs. The third cluster comprised the *cop* cluster except *copV* and *copT* (Supplementary Figure S1). In addition, a cluster carrying resistance to chromate and cobalt/nickel, similar to that on genomic island CMGI-28a of pMOL28, was also found on NA4's chromid (next to plasmid pNA4_B). The cluster was delimited by IS*Rme1* and a TBBSR, and was located between the *pbr* and *mer* cluster, and the second *cop* cluster. It did not contain the additional five genes that are carried by the pMOL28 and pNA4_B cluster [27,50,51].

Although most metal resistance determinants are conserved between *C. metallidurans* NA4 and CH34, which results in similar growth profiles in the presence of metals [52], they are harbored by different replicons. We showed that, unlike the characteristic megaplasmid pMOL30 in *C. metallidurans* CH34 that is specialized in heavy metal resistance, pNA4_A does not fulfill this role in *C. metallidurans* NA4. Plasmid pNA4_B does harbor the pMOL28-like metal resistance determinants and, in addition, a second *nccYXHCBAN* locus coding for an RND-driven efflux system homologous to that of *C. metallidurans* 31A and KT02, which mediates resistance to 40 mM Ni^{2+}.

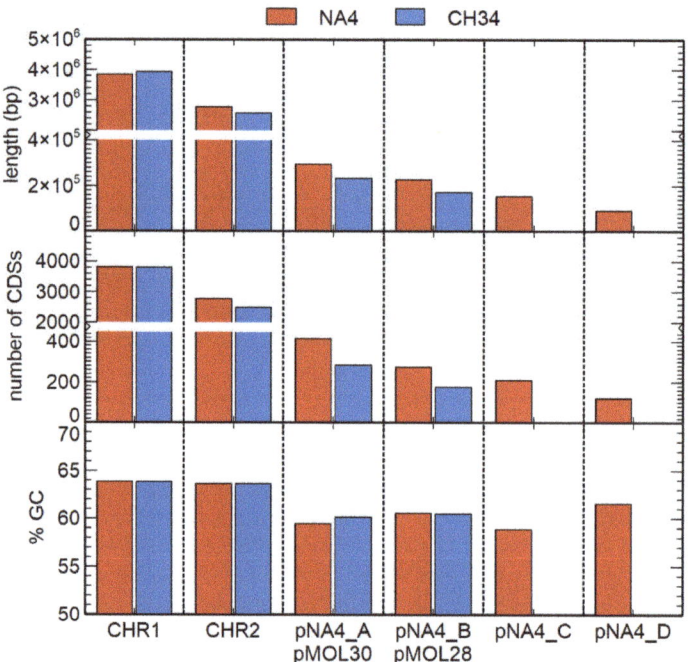

Figure 1. Comparison of the size (bp), number of coding sequences (CDSs), and GC content (% GC) of the *C. metallidurans* NA4 (red) and CH34 (blue) genome replicons (CHR1: chromosome; CHR2: chromid; p: plasmids).

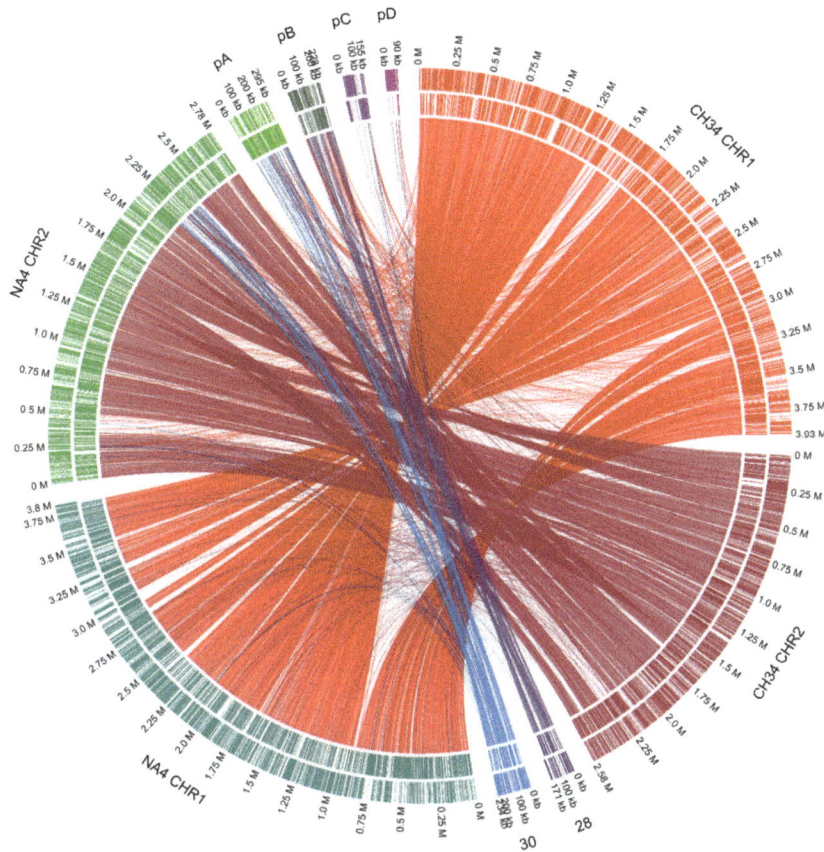

Figure 2. Circa plot of the *C. metallidurans* CH34 and NA4 genomes (CHR1: chromosome; CHR2: chromid) and plasmids (28: pMOL28; 30: pMOL30; pA: pNA4_A; pB: pNA4_B; pC: pNA4_C; pD: pNA4_D). Connections correspond to homologous genes based on the bidirectional best hit criterion and a blastP alignment threshold (at least 35% amino-acid identity on 80% of the length of the smallest protein).

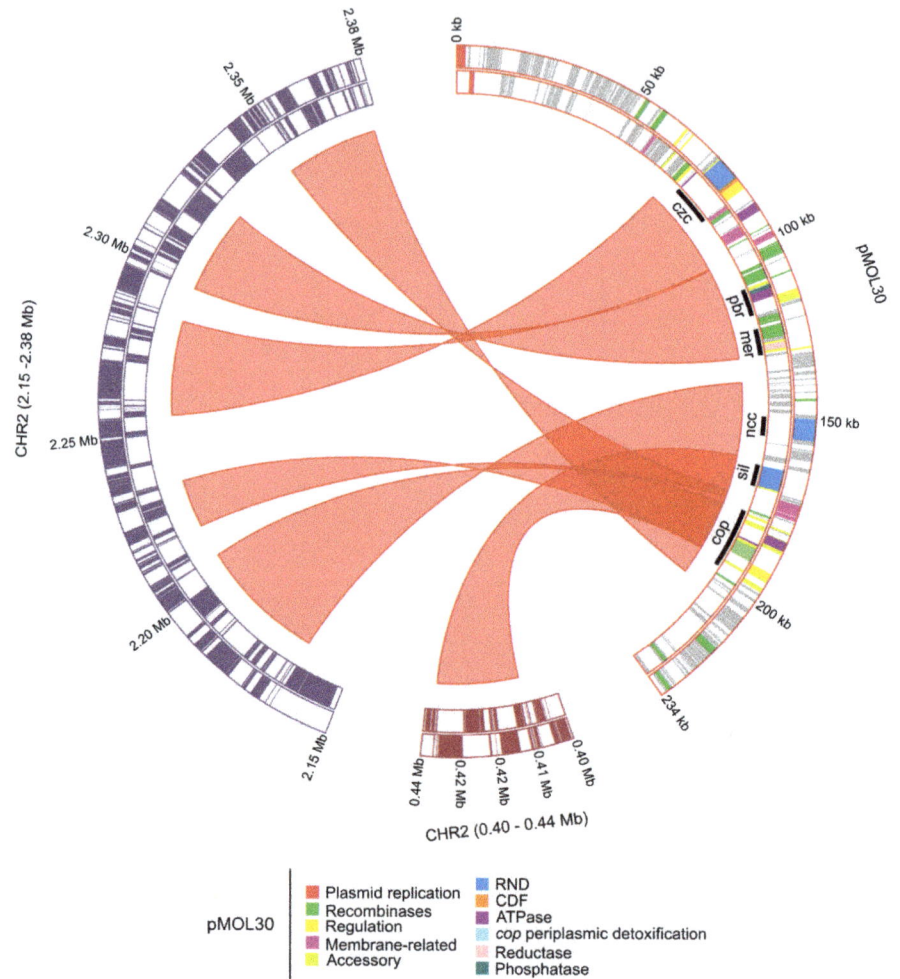

Figure 3. Circa plot of pMOL30 from *C. metallidurans* CH34 and sections of the *C. metallidurans* NA4 chromid (CHR2). Ribbons correspond to synteny groups, orthologous gene sets that have the same local organization, based on the bidirectional best hit criterion or a blastP alignment threshold (at least 30% amino-acid identity on 80% of the length of the smallest protein) and co-localization (with the maximum number of consecutive genes not involved in a synteny group being five). The pMOL30 genes are colored according to their function.

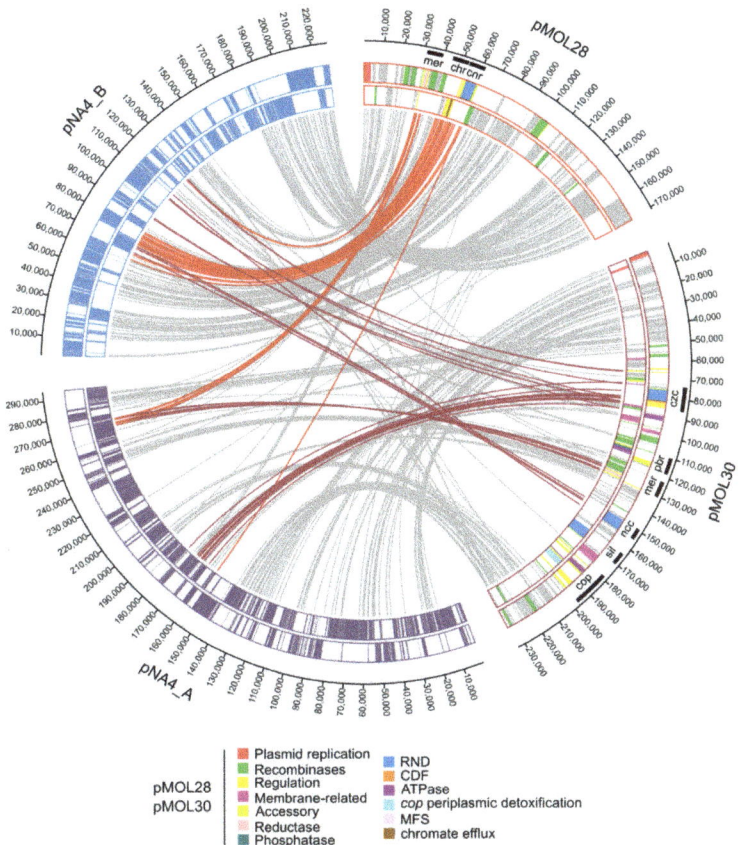

Figure 4. Circa plot of pMOL30 and pMOL28 from *C. metallidurans* CH34 and pNA4_A and pNA4_B from *C. metallidurans* NA4. Connections correspond to homologous genes based on the bidirectional best hit criterion and a blastP alignment threshold (at least 35% amino-acid identity on 80% of the length of the smallest protein). Colored connections correspond to genes involved in metal resistance. The pMOL28 and pMOL30 genes are colored according to their function.

3.1.1. Characterization of (Mega)Plasmids

As our sequencing efforts resulted in the complete closure of the four plasmid replicons, we were able to characterize them in more detail. Numerous proteins are involved in the horizontal transmission of plasmids via conjugation and establishment in the recipient cell. Conjugative plasmids contain an origin of transfer (*oriT*), a DNA relaxase, a Type IV coupling protein (T4CP), and a membrane-associated mating pair formation (MPF) complex, which is a form of Type IV secretion system (T4SS). Transmissible (mobilizable) plasmids require an *oriT* and a relaxase that can be provided in trans [53–57]. Different classification or typing schemes for plasmids have been developed, but the principal ones are replicon and MOB typing, relying on plasmid replication and mobility, respectively [55,56,58]. The phylogenetic relationship among relaxases has been thoroughly studied and resulted in the classification of conjugative systems into six MOB families: MOB_F, MOB_H, MOB_Q, MOB_C, MOB_P, and MOB_V [43,56]. The archetype plasmids defining the families are R388 (MOB_F), R27 (MOB_H), RP4 (MOB_P), RSF1010 (MOB_Q), pMV158 (MOB_V), and CloDF13 (MOB_C). We used MOB-suite to type and characterize pNA4_A, pNA4_B, pNA4_C, and pNA4_D, as well as pMOL28 and pMOL30 (Table 2).

Table 2. Plasmid characterization.

Replicon	Rep Type	Relaxase Type [1]	MPF Type [2]	Predicted Mobility	MASH Nearest Neighbor [3]
pMOL28	591	MOB_H	MPF_T	Conjugative	-
pMOL30	332	(MOB_P)	MPF_F	Conjugative	-
pNA4_A	332	-	MPF_F	Non-mobilizable	pMOL30
pNA4_B	591	MOB_H	MPF_T	Conjugative	pMOL28
pNA4_C	-	-	-	Non-mobilizable	pHS87a (*Pseudomonas aeruginosa*)
pNA4_D	1864	MOB_F	MPF_F	Conjugative	pACP3.3 (*Acidovorax* sp. P3)

[1] Archetype relaxase: plasmid R388 (MOB_F), plasmid R27 (MOB_H), and plasmid RP4 (MOB_P). [2] Archetype mating pair formation (MPF) complex: plasmid F (MPF_F) and plasmid Ti (MPF_T). [3] MASH Nearest Neighbor are not included for pMOL28 and pMOL30, as the closest database match was a self-hit.

Our analysis indicated that pNA4_A and pMOL30, and pNA4_B and pMOL28 were similar based on replicon, relaxase and mating pair formation (MPF) family. Plasmid pNA4_A was classified as being non-mobilizable, whereas pMOL30 as conjugative. Although low-frequency transfer of pMOL30 has been observed, this transfer could be mediated via other conjugative systems. For instance, plasmid RP4 can enhance transfer frequency to 10^{-3} by cointegrate formation via transposition of Tn*4380* [47]. Plasmid pNA4_A carried 416 CDSs, most of them code for unknown proteins (71.6%). Identifiable functions were, next to plasmid replication, maintenance and conjugation, related to metal resistance (partial *czc* cluster and mercury transposon; see Figure 4) and alkaline phosphatase. Plasmid pNA4_B carried 274 CDSs, most of them also code for unknown proteins (59.5%). Identifiable functions were, next to plasmid replication, maintenance and conjugation, related to metal resistance (cluster carrying resistance to chromate and cobalt/nickel, similar to that on genomic island CMGI-28a of pMOL28; see Figure 4).

Plasmid pNA4_C was classified as non-mobilizable and carried 209 CDSs, most of them also code for unknown proteins (74.6%). No accessory plasmid functions could be identified. Plasmid pNA4_D was classified as conjugative and was very similar to plasmids from *Acidovorax carolinensis* P3 (plasmid pACP3.3) and P4 (plasmid pACP4.4), *Acidovorax* sp. JS42 (plasmid pAOVO01), *Alicycliphilus denitrificans* K601 (plasmid pALIDE201), and *Pandoraea pnomenusa* MCB032 (unnamed plasmid) (Supplementary Figure S2). Characterized proteins encoded by these plasmids were mainly related to conjugational transfer and replication.

3.1.2. Insertion Sequence Elements Distribution

Insertion sequences (IS) are simple mobile genetic elements that play an important role in genome plasticity and activity of these IS elements are often correlated with the adaptive potential to promote genetic variability under different environmental challenges [59,60]. An initial assessment of the number and identity of IS elements in NA4 was previously performed based on the draft genome assembly [52]. However, this could lead to an underestimation of the number of IS elements because possible identical IS elements will only be represented as one contig [61]. Therefore, we reanalyzed and determined the correct number of IS elements in NA4 (Table 3). In total, 21 intact IS elements were identified. *C. metallidurans* NA4 carried much less IS elements in comparison with type strain *C. metallidurans* CH34, which carried 57 intact IS elements.

Table 3. Distribution of insertion sequence elements in *C. metallidurans* NA4.

Element	Family	Size (%) [1]	CHR1	CHR2	pNA4_A	pNA4_B	pNA4_C	pNA4_D
IS1071	Tn3	3204 (99.9%)						1
		2991 (93.4%)			2			
ISRme4	IS21	2469 (100%)	4	3				
ISRme9	IS21	2674 (94.8%)					1	
ISRme10	IS30	1063 (100%)		1				
ISRme3	IS3	1288 (100%)	1	2		2		
ISPst3	IS21	2605 (97.8%)	1	2				
ISPa45	IS4	1637 (100%)	1					

[1] Size (bp) of the element and % nucleotide sequence similarity to the insertion sequence (IS) element as defined in ISFinder [62].

3.2. Analysis of C. metallidurans NA4Pt

3.2.1. Determination of Minimal Inhibitory Concentration and Generation of a Pt^{4+} Resistant Mutant

We used *C. metallidurans* NA4, which is able to survive in oligotrophic conditions for many months [46], to scrutinize adaptation to Pt^{4+}. The minimal inhibitory concentration (MIC) of Pt^{4+} for *C. metallidurans* NA4 in Tris-buffered mineral medium was 70 μM (Table 1), which was similar to that of *C. metallidurans* CH34. This already indicated that NA4 has a high level of resistance to Pt^{4+} when compared to other strains such as *Klebsiella pneumoniae* (20 μM), *Acinetobacter baumannii* (30 μM), and *Enterococcus faecium* (60 μM) [63]. Furthermore, the MIC determinations in that study were performed in rich broth medium, which could affect the metal bioavailability and lead to an overestimation of the MIC [63]. Next, *C. metallidurans* NA4 was exposed to a subinhibitory concentration of 62.5 μM Pt^{4+} during 30 days (eight serial passages). After passage on non-selective medium, a mutant (designated NA4Pt) that displayed a higher resistance to Pt^{4+} (MIC of 160 μM) was obtained. No differences were observed between NA4 and NA4Pt when grown in non-selective conditions (Figure 5). In addition, no differences in MIC of Pd^{2+}, Ag^+, Zn^{2+}, Ni^{2+}, and Cu^{2+} were observed (Table 4). From this data it is evident that the NA4Pt mutant is specifically resistant to Pt^{4+} and does not have a higher resistance to Pd^{2+}, in contrast to what was described for *Desulfovibrio* sp. preference during biosorption [14].

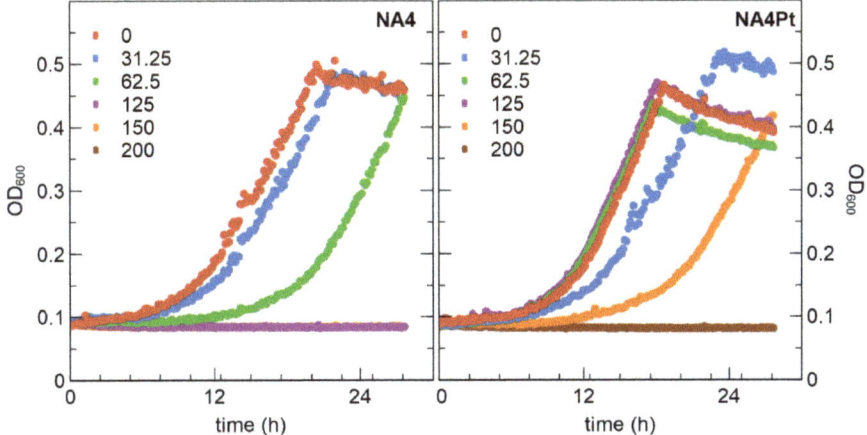

Figure 5. Growth of *C. metallidurans* NA4 and NA4Pt in the presence of different $PtCl_4$ concentrations (in μM).

Table 4. Minimal inhibitory concentration of selected metals for *C. metallidurans* NA4 and NA4Pt.

	Pt^{4+} (µM)	Pd^{2+} (µM)	Ag^+ (µM)	Zn^{2+} (mM)	Ni^{2+} (mM)	Cu^{2+} (mM)
NA4	70	12.5	1	12	40	6
NA4Pt	160	12.5	1	12	40	6

3.2.2. Sequence Analysis of *C. metallidurans* NA4Pt

The laboratory-evolved mutant NA4Pt was sequenced to identify genomic changes such as insertions, deletions and SNPs (Supplementary Table S1). These genomic changes were not observed in other adaptive laboratory evolution experiments with NA4 (unpublished results). In the chromosome, an insertion (+56 bp) in the upstream region of a metal-dependent hydrolase and a point mutation in pseudouridine synthase (*rluB*) (resulting in R64H substitution) were observed. In the chromid, several point mutations (mostly synonymous mutations) and two deletions (a single bp and a 119-bp region), both located in the *copB* gene coding for an outer membrane protein involved in copper resistance, were found. The latter did not affect copper resistance of NA4Pt (Table 4). The biggest change, a large deletion of 69.9 kbp, was observed in plasmid pNA4_D. The 19.7 kbp remaining fragment (positions 52,551 to 72,345) included genes encoding a DNA primase, a C-5 cytosine-specific DNA methylase, a single-stranded DNA-binding (ssb) protein, and the replication initiator protein RepA (Supplementary Table S2), which is responsible for plasmid replication in bacteria [64,65]. This suggested that the remaining part could be maintained as a smaller plasmid instead of being integrated in one of the other replicons. The absence of the native pNA4_D and presence of the smaller plasmid were confirmed by plasmid DNA extraction and *Pag*I digestion (Figure 6).

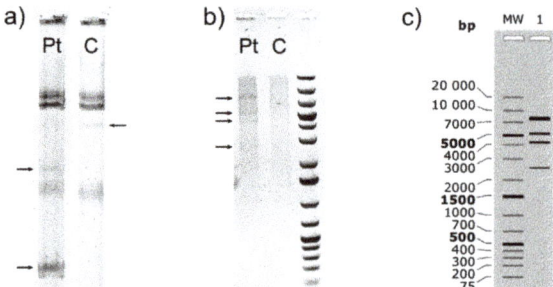

Figure 6. Agarose gel electrophoresis of *C. metallidurans* NA4 (C) and NA4Pt (Pt) plasmid DNA. (**a**) Megaplasmids of NA4 vs. NA4Pt; (**b**) *Pag*I digest of pNA4_D in NA4 vs. NA4Pt (only small plasmids were extracted; therefore, no discrete bands are visible for NA4); and (**c**) theoretical *Pag*I digest of NA4Pt pNA4_D.

The large deletion, which could probably be mediated by the presence of a 221 bp direct repeat (Supplementary Figure S3), resulted in the loss of 89 CDSs. (Mega)plasmids are nonessential and dispensable for cell viability in most environments [66], and large deletions have also been observed in other laboratory-evolved strains. For instance, a large deletion occurred in the megaplasmid pAtC58 from *Agrobacterium tumefaciens* laboratory-evolved strains, resulting in increased virulence gene expression and reduced fitness cost [67]. Prolonged cultivation of the Gram-positive actinobacterium *Rhodococcus opacus* 1CP containing megaplasmid p1CP (740 kb) under non-selective conditions led to the isolation of mutants 1CP.01 and 1CP.02, harboring the shortened plasmid variants p1CP.01 (500 kb) and p1CP.02 (400 kb) [68]. *Methylobacterium extorquens* AM1 lost 10% of its megaplasmid in an evolution experiment, which were beneficial in the applied selective environment, but disadvantageous in alternative environments [69]. Gene loss is a very common evolutionary process in bacteria and provides an increased fitness under one or several growth conditions [70]. The large deletion in pNA4_D resulted in loss of the conjugative machinery (T4SS), which is known to impose a burden [71,72]. For instance, the growth rate of an experimentally evolved *E. coli* increased by IS26-mediated loss of the T4SS on its plasmid pKP33 [73]. Several mechanisms are put forward to explain the reduced fitness costs, such as ribosome occupancy, reduction of energy demands for DNA replication, transcription and translation, and negative interactions between chromosomal pathways and (mega)plasmid-encoded proteins [66,67,74]. In addition, selective processes favoring adaptation to specific stressors/environments can be a driving force behind gene loss [66,70,75,76]. Therefore, although no growth differences were observed between NA4 and NA4Pt, this loss could have an effect in challenging environments with increased selection pressure (e.g., high platinum concentration).

3.2.3. Transcriptome Analysis

Resistance is the result of natural selection for resistance-conferring mutations (i.e., random mutations that allow growth under selection, outcompeting the parent, and subsequent isolation of the adapted mutant) [77]. Therefore, and similar to other studies [78–80], the global shift in transcriptome, resulting from the altered genotype of the evolved strain (NA4Pt) as compared with the parental strain (NA4), was examined by RNA-Seq in non-selective conditions (average total number of reads was 4,324,281 ± 463,666). Up- and down-regulated genes were selected based on log2 fold change (<−0.8 and >0.8) and significance ($p < 0.05$), which resulted in 111 up- and 115 down-regulated genes (Figure 7 and Supplementary Tables S2 and S3).

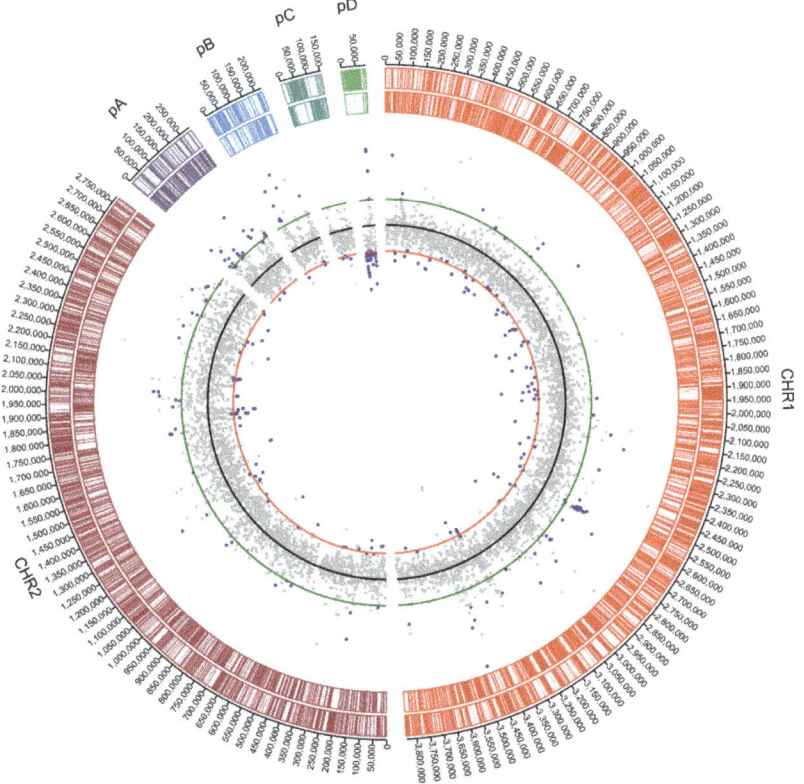

Figure 7. Scatter plot of RNA-Seq-derived gene expression of *C. metallidurans* NA4Pt compared to its parental strain in non-selective conditions. Dots (blue $p < 0.05$) represent Log2 ratios with red, black, and green lines corresponding to −0.8, 8, and 0.8, respectively.

The functional relevance of differentially-expressed genes was explored by using functional categories from the eggNOG classification system [81]. Genes of different categories were found to be differently expressed (Figure 8). However, none of the categories were significantly over-represented (Fisher's exact test).

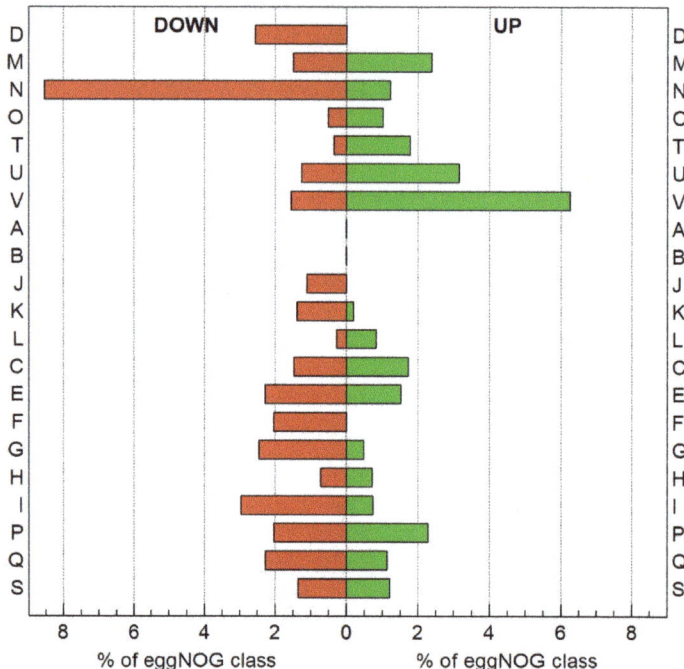

Figure 8. EggNOG classification of differentially expressed genes in *C. metallidurans* NA4Pt, based on RNA-Seq-derived gene expression of *C. metallidurans* NA4Pt compared to its parental strain in non-selective conditions. Percentages were calculated by normalizing the number of up- and down-regulated genes in each category by the total number of genes in the NA4 genome grouped in the corresponding category. (D: Cell cycle control, cell division, and chromosome partitioning; M: Cell wall/membrane/envelope biogenesis; N: Cell motility; O: Posttranslational modification, protein turnover, and chaperones; T: Signal transduction mechanisms; U: Intracellular trafficking, secretion, and vesicular transport; V: Defense mechanisms; A: RNA processing and modification; B: Chromatin structure and dynamics; J: Translation, ribosomal structure, and biogenesis; K: Transcription; L: Replication, recombination, and repair; C: Energy production and conversion; E: Amino acid transport and metabolism; F: Nucleotide transport and metabolism; G: Carbohydrate transport and metabolism; H: Coenzyme transport and metabolism; I: Lipid transport and metabolism; P: Inorganic ion transport and metabolism; Q: Secondary metabolites biosynthesis, transport, and catabolism; S: Poorly characterized).

Overall, genes involved in defense mechanisms, intracellular trafficking, signal transduction and membrane-related genes were more up-regulated than down-regulated. On the other hand, genes involved in carbohydrate transport, nucleotide transport, cell motility and cell-cycle-related genes were more down-regulated than up-regulated. Up-regulated defense mechanism genes included mainly RND-driven efflux systems, which are abundant in the NA4 genome [52]. Next to systems putatively involved in the efflux of chemicals (acridines), the expression of two HME-RND genes was increased. The *cusC* gene is part of the CusCBA efflux pump responsible for copper and silver resistance [82]. In *C. metallidurans*, *cusDCBAF* genes are also up-regulated by silver ions [28], and induction of the

CusC protein synthesis was also observed in the presence of 1 µM silver or 0.85 mM copper ions [83]. The *cnrC* gene is part of *cnrCBAT* efflux system that mediates nickel and cobalt resistance [84]. However, the resistance of NA4Pt to Cu^{2+}, Zn^{2+}, Ag^+ and Ni^{2+} was unaffected compared to that of the parental strain NA4 (Table 4). No other metal resistance genes were up- or down-regulated.

Up-regulated genes of the signal transducing pathways belong to different two-component-system-related proteins (Supplementary Table S2). Namely, *ompR*-family regulators, members of the largest response regulator family involved in many signal transduction processes [85], and *cheY*, which is involved in chemotaxis and modulates motility by regulating flagellar motor switch proteins [86]. In contrast to *cheY*, cell-motility-related genes were down-regulated (*flgC*, *flgD*, *flgG*, *pilW*, and *pilX*). FlgC and FlgG proteins form the rod part of the flagellar basal body [87] and FlgG polymerizes to form the distal rod on top of the proximal rod, acting as a hook cap [88]. PilX and PilW are involved in biogenesis of Type IV fimbriae, which are surface filaments mediating attachment to host epithelial cells and flagella-independent twitching motility [89]. These differences did not affect motility as no significant differences were observed between NA4Pt and NA4 in motility assays (Supplementary Figure S4).

The class of membrane-related up-regulated genes contains a highly up-regulated (36.3-fold) lytic transglycosylase (LT), which usually plays an important role in shaping the periplasmic space and is tightly regulated, as over activity can have deleterious effects [90]. LT is responsible for creating space within the peptidoglycan layer for cell division and the insertion of cell-envelope spanning structures such as flagella and secretion systems [90]. In *C. metallidurans*, different LTs are induced in the presence of Cu^{2+}, Cd^{2+}, Pb^{2+} and Zn^{2+} [91]. Therefore, it may be hypothesized that the LTs might play a role in coping with toxic metal ion concentrations. Next to LTs, membrane-bound cytochrome c, which enables electron transfer as well as the catalysis of various redox reactions [92], is also up-regulated. In *D. vulgaris*, this periplasmic protein is directly involved in H_2-mediated metal reduction [16]. *Desulfovibrio* sp. cytochrome c3 mediates electron transfer to palladium and platinum complexes, thereby reducing them to zero-valent nanoparticles [10,17]. The latter supports a role of this overexpressed gene in the Pt-resistant phenotype. Furthermore, nm-scale colloidal platinum was found in *C. metallidurans* primarily along the cell envelope, where energy generation/electron transport occurs [12].

To scrutinize the impact of differentially-expressed membrane-related genes (LT, flagellar- and pili-related) on morphology, scanning electron microscopy (SEM) was performed and showed that NA4Pt cells appeared to be more elongated (Figure 9). This correlates with the induction of filaments during platinum metal stress [19]. Nevertheless, the presence of a subpopulation of elongated cells could not be confirmed by flow cytometry (Supplementary Figure S5).

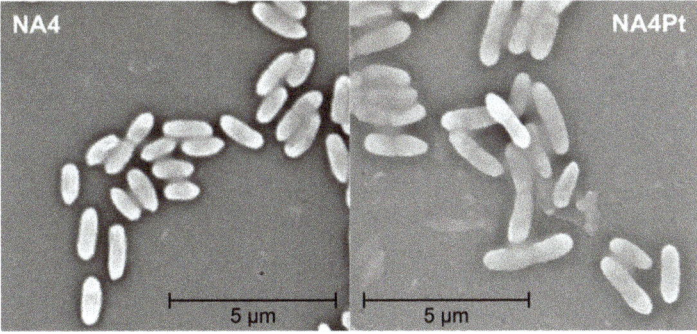

Figure 9. Scanning electron microscopy (SEM) image of *C. metallidurans* NA4 and NA4Pt.

Up-regulated intracellular trafficking genes were mainly related to a Type II secretion system (T2SS), which secretes proteins to the extracellular environment [93]. In *C. metallidurans*, the T2SS secretes alkaline phosphatase in the extracellular environment [94].

Finally, none of the CDSs affected by the genomic changes in NA4Pt (either directly or via alteration in their promoter region) were differentially expressed (Supplementary Tables S1–S3). The latter points towards a significant role of the large deletion in pNA4_D on the overall expression pattern, as well as adaptation to platinum.

4. Conclusions

We resolved the multireplicon genome of *C. metallidurans* NA4 by combining Illumina and Nanopore sequencing and; thereby, paved the way for genetic, genomic and evolution-related studies. Further analysis of the replicons showed a distinctive pattern in the location of metal resistance determinants, with the chromid playing a pivotal role in contrast to type strain CH34, for which the megaplasmids pMOL28 and pMOL30 are the main actors. An NA4 derivative (NA4Pt) that showed increased resistance to platinum was generated via an adaptive laboratory evolution experiment. On the basis of our observations, one might speculate that the increased resistance to Pt^{4+} in *C. metallidurans* NA4Pt is not mediated by a dedicated mechanism, but by pleiotropic alterations in membrane-related processes, such as pili, peptidoglycan turnover and electron transfer, probably elicited by the large deletion in its 98-kb plasmid.

Supplementary Materials: The following are available online at http://www.mdpi.com/2073-4425/10/1/63/s1, Figure S1: *C. metallidurans* chromid clusters homologous to the *cop* cluster on pMOL30; Figure S2: Whole plasmid alignments of pNA4_D from *C. metallidurans* NA4 and pACP3.3 and pACP4.4 from *A. carolinensis* P3 and P4, pAOVO01 from *Acidovorax* sp. JS42, pALIDE201 from *A. denitrificans* K601, and an unnamed plasmid from *P. pnomenusa* MCB032; Figure S3: Alignment of 221 bp direct repeat flanking large deletion in pNA4_D; Figure S4: Motility assay in soft (0.3%) agar. Diameter measurement of the haloes formed by *C. metallidurans* NA4 and NA4Pt after 24 h; Figure S5: Flow cytometry histogram of *C. metallidurans* NA4 and NA4Pt based on forward-scattered light (FSC) and side-scattered light (SSC); Table S1: List of genomic changes in the laboratory-evolved *C. metallidurans* NA4Pt; Table S2: RNA-Seq-derived up-regulated genes (log2 fold > 0.8; p < 0.05) in *C. metallidurans* NA4Pt, compared to its parental strain in non-selective conditions; Table S3: RNA-Seq-derived down-regulated genes (log2 fold < −0.8; p < 0.05) in *C. metallidurans* NA4Pt, compared to its parental strain in non-selective conditions.

Author Contributions: Conceptualization, R.V.H.; methodology, M.M.A., A.P., and R.V.H.; software, M.M.A, L.M., P.M., and R.V.H.; validation, M.M.A., P.M., D.C., and R.V.H.; formal analysis, M.M. A., L.M., A.P., P.M., and R.V.H.; investigation, M.M.A., A.P., and R.V.H.; resources, M.M.A. and R.V.H.; writing—original draft preparation, M.M.A. and R.V.H.; writing—review and editing, M.M.A., P.M., D.C., and R.V.H.; visualization, L.M. and R.V.H.; supervision, N.L., D.C., and R.V.H.; project administration, N.L. and R.V.H.; funding acquisition, N.L. and R.V.H.

Funding: This research received no external funding.

Conflicts of Interest: The authors declare no conflicts of interest.

References

1. Rauch, S.; Morrison, G.M. Environmental relevance of the platinum-group elements. *Elements* **2008**, *4*, 259–263. [CrossRef]
2. Wedepohl, K.H. The Composition of the Continental-Crust. *Geochim. Cosmochim. Acta* **1995**, *59*, 1217–1232. [CrossRef]
3. Ravindra, K.; Bencs, L.; Van Grieken, R. Platinum group elements in the environment and their health risk. *Sci. Total Environ.* **2004**. [CrossRef]
4. Florea, A.M.; Busselberg, D. Cisplatin as an anti-tumor drug: Cellular mechanisms of activity, drug resistance and induced side effects. *Cancers* **2011**, *3*, 1351–1371. [CrossRef] [PubMed]
5. Maes, S.; Props, R.; Fitts, J.P.; Smet, R.D.; Vilchez-Vargas, R.; Vital, M.; Pieper, D.H.; Vanhaecke, F.; Boon, N.; Hennebel, T. Platinum Recovery from Synthetic Extreme Environments by Halophilic Bacteria. *Environ. Sci. Technol.* **2016**, *50*, 2619–2626. [CrossRef] [PubMed]
6. Hennebel, T.; Boon, N.; Maes, S.; Lenz, M. Biotechnologies for critical raw material recovery from primary and secondary sources: R&D priorities and future perspectives. *New Biotechnol.* **2015**, *32*, 121–127. [CrossRef]

7. Reith, F.; Zammit, C.M.; Shar, S.S.; Etschmann, B.; Bottrill, R.; Southam, G.; Ta, C.; Kilburn, M.; Oberthür, T.; Ball, A.S.; et al. Biological role in the transformation of platinum-group mineral grains. *Nat. Geosci.* **2016**, *9*, 294–298. [CrossRef]
8. Zereini, F.; Wiseman, C.L.S. *Platinum Metals in the Environment*; Springer-Verlag: Berlin/Heidelberg, Germany, 2015.
9. Maes, S.; Claus, M.; Verbeken, K.; Wallaert, E.; De Smet, R.; Vanhaecke, F.; Boon, N.; Hennebel, T. Platinum recovery from industrial process streams by halophilic bacteria: Influence of salt species and platinum speciation. *Water Res.* **2016**, *105*, 436–443. [CrossRef]
10. de Vargas, I.; Macaskie, L.E.; Guibal, E. Biosorption of palladium and platinum by sulfate-reducing bacteria. *J. Chem. Technol. Biotechnol.* **2004**, *79*, 49–56. [CrossRef]
11. Maes, S.; Props, R.; Fitts, J.P.; De Smet, R.; Vanhaecke, F.; Boon, N.; Hennebel, T. Biological Recovery of Platinum Complexes from Diluted Aqueous Streams by Axenic Cultures. *PLoS ONE* **2017**. [CrossRef]
12. Campbell, G.; MacLean, L.; Reith, F.; Brewe, D.; Gordon, R.A.; Southam, G. Immobilisation of Platinum by *Cupriavidus metallidurans*. *Minerals* **2018**, *8*, 10. [CrossRef]
13. Konishi, Y.; Ohno, K.; Saitoh, N.; Nomura, T.; Nagamine, S.; Hishida, H.; Takahashi, Y.; Uruga, T. Bioreductive deposition of platinum nanoparticles on the bacterium *Shewanella algae*. *J. Biotechnol.* **2007**, *128*, 648–653. [CrossRef] [PubMed]
14. Yong, P.; Rowson, N.A.; Farr, J.P.G.; Harris, I.R.; Macaskie, L.E. Bioaccumulation of palladium by *Desulfovibrio desulfuricans*. *J. Chem. Technol. Biotechnol.* **2002**, *77*, 593–601. [CrossRef]
15. Rapp-Giles, B.J.; Casalot, L.; English, R.S.; Ringbauer, J.A.; Dolla, A.; Wall, J.D. Cytochrome c3 Mutants of *Desulfovibrio desulfuricans*. *Appl. Environ. Microbiol.* **2000**, *66*, 671–677. [CrossRef] [PubMed]
16. Elias, D.A.; Suflita, J.M.; McInerney, M.J.; Krumholz, L.R. Periplasmic cytochrome c3 of *Desulfovibrio vulgaris* is directly involved in H2-mediated metal but not sulfate reduction. *Appl. Environ. Microbiol.* **2004**, *70*, 413–420. [CrossRef] [PubMed]
17. Capeness, M.J.; Edmundson, M.C.; Horsfall, L.E. Nickel and platinum group metal nanoparticle production by *Desulfovibrio alaskensis* G20. *New Biotechnol.* **2015**, *32*, 727–731. [CrossRef] [PubMed]
18. Gauthier, D.; Sobjerg, L.S.; Jensen, K.M.; Lindhardt, A.T.; Bunge, M.; Finster, K.; Meyer, R.L.; Skrydstrup, T. Environmentally benign recovery and reactivation of palladium from industrial waste by using gram-negative bacteria. *ChemSusChem* **2010**, *3*, 1036–1039. [CrossRef]
19. Rosenberg, B.; Renshaw, E.; Vancamp, L.; Hartwick, J.; Drobnik, J. Platinum-induced filamentous growth in *Escherichia coli*. *J. Bacteriol.* **1967**, *93*, 716–721.
20. Moore, R.L.; Brubaker, R.R. Effect of cis-platinum(II)diamminodichloride on cell division of *Hyphomicrobium* and *Caulobacter*. *J. Bacteriol.* **1976**, *125*, 317–323.
21. Beck, D.J.; Brubaker, R.R. Effect of cis-platinum(II)diamminodichloride on wild type and deoxyribonucleic acid repair deficient mutants of *Escherichia coli*. *J. Bacteriol.* **1973**, *116*, 1247–1252.
22. Drobnik, J.; Urbankova, M.; Krekulova, A. The effect of cis-dichlorodiammineplatinum(II) on *Escherichia coli* B. The role of fil, exr and hcr markers. *Mutat. Res.* **1973**, *17*, 13–20. [CrossRef]
23. Reslova, S. The induction of lysogenic strains of *Escherichia coli* by cis-dichlorodiammineplatinum (II). *Chem. Biol. Interact.* **1971**, *4*, 66–70. [CrossRef]
24. Beck, D.J.; Fisch, J.E. Mutagenicity of platinum coordination complexes in *Salmonella typhimurium*. *Mutat. Res.* **1980**, *77*, 45–54. [CrossRef]
25. Maboeta, M.S.; Claassens, S.; Van Rensburg, L.; Van Rensburg, P.J.J. The effects of platinum mining on the environment from a soil microbial perspective. *Water Air Soil Pollut.* **2006**, *175*, 149–161. [CrossRef]
26. Janssen, P.J.; Van Houdt, R.; Moors, H.; Monsieurs, P.; Morin, N.; Michaux, A.; Benotmane, M.A.; Leys, N.; Vallaeys, T.; Lapidus, A.; et al. The complete genome sequence of *Cupriavidus metallidurans* strain CH34, a master survivalist in harsh and anthropogenic environments. *PLoS ONE* **2010**, *5*, e10433. [CrossRef] [PubMed]
27. Mergeay, M.; Van Houdt, R. *Metal Response in Cupriavidus metallidurans, Volume I: FROM Habitats to Genes and Proteins*; Springer International Publishing: Basel, Switzerland, 2015; p. 89.
28. Monsieurs, P.; Moors, H.; Van Houdt, R.; Janssen, P.J.; Janssen, A.; Coninx, I.; Mergeay, M.; Leys, N. Heavy metal resistance in *Cupriavidus metallidurans* CH34 is governed by an intricate transcriptional network. *Biometals* **2011**, *24*, 1133–1151. [CrossRef]

29. Monsieurs, P.; Mijnendonckx, K.; Provoost, A.; Venkateswaran, K.; Ott, C.M.; Leys, N.; Van Houdt, R. Genome Sequences of *Cupriavidus metallidurans* Strains NA1, NA4, and NE12, Isolated from Space Equipment. *Genome Announc.* **2014**. [CrossRef]
30. Mergeay, M.; Nies, D.; Schlegel, H.G.; Gerits, J.; Charles, P.; Van Gijsegem, F. *Alcaligenes eutrophus* CH34 is a facultative chemolithotroph with plasmid-bound resistance to heavy metals. *J. Bacteriol.* **1985**, *162*, 328–334.
31. Wiegand, I.; Hilpert, K.; Hancock, R.E. Agar and broth dilution methods to determine the minimal inhibitory concentration (MIC) of antimicrobial substances. *Nat. Protoc.* **2008**, *3*, 163–175. [CrossRef]
32. Randall, C.P.; Oyama, L.B.; Bostock, J.M.; Chopra, I.; O'Neill, A.J. The silver cation (Ag+): Antistaphylococcal activity, mode of action and resistance studies. *J. Antimicrob. Chemother.* **2013**, *68*, 131–138. [CrossRef]
33. Andrup, L.; Barfod, K.K.; Jensen, G.B.; Smidt, L. Detection of large plasmids from the *Bacillus cereus* group. *Plasmid* **2008**, *59*, 139–143. [CrossRef] [PubMed]
34. Bankevich, A.; Nurk, S.; Antipov, D.; Gurevich, A.A.; Dvorkin, M.; Kulikov, A.S.; Lesin, V.M.; Nikolenko, S.I.; Pham, S.; Prjibelski, A.D.; et al. SPAdes: A new genome assembly algorithm and its applications to single-cell sequencing. *J. Comput. Biol.* **2012**, *19*, 455–477. [CrossRef] [PubMed]
35. Antipov, D.; Korobeynikov, A.; McLean, J.S.; Pevzner, P.A. hybridSPAdes: An algorithm for hybrid assembly of short and long reads. *Bioinformatics* **2016**, *32*, 1009–1015. [CrossRef] [PubMed]
36. Deatherage, D.E.; Barrick, J.E. Identification of mutations in laboratory-evolved microbes from next-generation sequencing data using *breseq*. *Methods Mol. Biol.* **2014**, *1151*, 165–188. [CrossRef] [PubMed]
37. Van der Auwera, G.A.; Carneiro, M.O.; Hartl, C.; Poplin, R.; Del Angel, G.; Levy-Moonshine, A.; Jordan, T.; Shakir, K.; Roazen, D.; Thibault, J.; et al. From FastQ data to high confidence variant calls: The Genome Analysis Toolkit best practices pipeline. *Curr. Protoc. Bioinform.* **2013**. [CrossRef]
38. Li, H.; Handsaker, B.; Wysoker, A.; Fennell, T.; Ruan, J.; Homer, N.; Marth, G.; Abecasis, G.; Durbin, R.; 1000 Genome Project Data Processing Subgroup. The Sequence Alignment/Map format and SAMtools. *Bioinformatics* **2009**, *25*, 2078–2079. [CrossRef]
39. Li, H.; Durbin, R. Fast and accurate long-read alignment with Burrows-Wheeler transform. *Bioinformatics* **2010**, *26*, 589–595. [CrossRef]
40. Robinson, M.D.; McCarthy, D.J.; Smyth, G.K. edgeR: A Bioconductor package for differential expression analysis of digital gene expression data. *Bioinformatics* **2010**, *26*, 139–140. [CrossRef]
41. Vallenet, D.; Belda, E.; Calteau, A.; Cruveiller, S.; Engelen, S.; Lajus, A.; Le Fevre, F.; Longin, C.; Mornico, D.; Roche, D.; et al. MicroScope—An integrated microbial resource for the curation and comparative analysis of genomic and metabolic data. *Nucleic Acids Res.* **2013**, *41*, D636–D647. [CrossRef]
42. Varani, A.M.; Siguier, P.; Gourbeyre, E.; Charneau, V.; Chandler, M. ISsaga is an ensemble of web-based methods for high throughput identification and semi-automatic annotation of insertion sequences in prokaryotic genomes. *Genome Biol.* **2011**, *12*, R30. [CrossRef]
43. Robertson, J.; Nash, J.H.E. MOB-suite: Software tools for clustering, reconstruction and typing of plasmids from draft assemblies. *Microb. Genom.* **2018**. [CrossRef] [PubMed]
44. Ankenbrand, M.J.; Hohlfeld, S.; Hackl, T.; Forster, F. AliTV-interactive visualization of whole genome comparisons. *PEERJ Comput. Sci.* **2017**. [CrossRef]
45. Brettin, T.; Davis, J.J.; Disz, T.; Edwards, R.A.; Gerdes, S.; Olsen, G.J.; Olson, R.; Overbeek, R.; Parrello, B.; Pusch, G.D.; et al. RASTtk: A modular and extensible implementation of the RAST algorithm for building custom annotation pipelines and annotating batches of genomes. *Sci. Rep.* **2015**, *5*, 8365. [CrossRef] [PubMed]
46. Mijnendonckx, K.; Provoost, A.; Ott, C.M.; Venkateswaran, K.; Mahillon, J.; Leys, N.; Van Houdt, R. Characterization of the survival ability of *Cupriavidus metallidurans* and *Ralstonia pickettii* from space-related environments. *Microb. Ecol.* **2013**, *65*, 347–360. [CrossRef] [PubMed]
47. Mergeay, M.; Monchy, S.; Janssen, P.; Houdt, R.V.; Leys, N. Megaplasmids in *Cupriavidus* Genus and Metal Resistance. In *Microbial Megaplasmids*; Schwartz, E., Ed.; Springer Berlin Heidelberg: Berlin/Heidelberg, Germany, 2009.
48. Van Houdt, R.; Monsieurs, P.; Mijnendonckx, K.; Provoost, A.; Janssen, A.; Mergeay, M.; Leys, N. Variation in genomic islands contribute to genome plasticity in *Cupriavidus metallidurans*. *BMC Genom.* **2012**, *13*, 111. [CrossRef] [PubMed]
49. Van Houdt, R.; Monchy, S.; Leys, N.; Mergeay, M. New mobile genetic elements in *Cupriavidus metallidurans* CH34, their possible roles and occurrence in other bacteria. *Antonie Van Leeuwenhoek* **2009**, *96*, 205–226. [CrossRef] [PubMed]

50. Henne, K.L.; Nakatsu, C.H.; Thompson, D.K.; Konopka, A.E. High-level chromate resistance in *Arthrobacter sp.* strain FB24 requires previously uncharacterized accessory genes. *BMC Microbiol.* **2009**, *9*, 199. [CrossRef]
51. Henne, K.L.; Turse, J.E.; Nicora, C.D.; Lipton, M.S.; Tollaksen, S.L.; Lindberg, C.; Babnigg, G.; Giometti, C.S.; Nakatsu, C.H.; Thompson, D.K.; et al. Global proteomic analysis of the chromate response in *Arthrobacter* sp. strain FB24. *J. Proteome Res.* **2009**, *8*, 1704–1716. [CrossRef]
52. Van Houdt, R.; Provoost, A.; Van Assche, A.; Leys, N.; Lievens, B.; Mijnendonckx, K.; Monsieurs, P. *Cupriavidus metallidurans* Strains with Different Mobilomes and from Distinct Environments Have Comparable Phenomes. *Genes* **2018**, *9*, 507. [CrossRef]
53. Shintani, M.; Sanchez, Z.K.; Kimbara, K. Genomics of microbial plasmids: Classification and identification based on replication and transfer systems and host taxonomy. *Front. Microbiol.* **2015**, *6*, 242. [CrossRef]
54. Ramsay, J.P.; Kwong, S.M.; Murphy, R.J.; Yui Eto, K.; Price, K.J.; Nguyen, Q.T.; O'Brien, F.G.; Grubb, W.B.; Coombs, G.W.; Firth, N. An updated view of plasmid conjugation and mobilization in *Staphylococcus*. *Mob. Genet. Elements* **2016**, *6*, e1208317. [CrossRef] [PubMed]
55. Carattoli, A.; Bertini, A.; Villa, L.; Falbo, V.; Hopkins, K.L.; Threlfall, E.J. Identification of plasmids by PCR-based replicon typing. *J. Microbiol. Methods* **2005**, *63*, 219–228. [CrossRef]
56. Garcillan-Barcia, M.P.; Francia, M.V.; de la Cruz, F. The diversity of conjugative relaxases and its application in plasmid classification. *FEMS Microbiol. Rev.* **2009**, *33*, 657–687. [CrossRef] [PubMed]
57. Ilangovan, A.; Connery, S.; Waksman, G. Structural biology of the Gram-negative bacterial conjugation systems. *Trends Microbiol.* **2015**, *23*, 301–310. [CrossRef]
58. Orlek, A.; Stoesser, N.; Anjum, M.F.; Doumith, M.; Ellington, M.J.; Peto, T.; Crook, D.; Woodford, N.; Walker, A.S.; Phan, H.; et al. Plasmid Classification in an Era of Whole-Genome Sequencing: Application in Studies of Antibiotic Resistance Epidemiology. *Front. Microbiol.* **2017**, *8*, 182. [CrossRef] [PubMed]
59. Vandecraen, J.; Chandler, M.; Aertsen, A.; Van Houdt, R. The impact of insertion sequences on bacterial genome plasticity and adaptability. *Crit. Rev. Microbiol.* **2017**, *43*, 709–730. [CrossRef]
60. Siguier, P.; Gourbeyre, E.; Chandler, M. Bacterial insertion sequences: Their genomic impact and diversity. *FEMS Microbiol. Rev.* **2014**, *38*, 865–891. [CrossRef]
61. Ricker, N.; Qian, H.; Fulthorpe, R.R. The limitations of draft assemblies for understanding prokaryotic adaptation and evolution. *Genomics* **2012**, *100*, 167–175. [CrossRef]
62. Siguier, P.; Perochon, J.; Lestrade, L.; Mahillon, J.; Chandler, M. ISfinder: The reference centre for bacterial insertion sequences. *Nucleic Acids Res.* **2006**, *34*, D32–D36. [CrossRef]
63. Vaidya, M.Y.; McBain, A.J.; Butler, J.A.; Banks, C.E.; Whitehead, K.A. Antimicrobial Efficacy and Synergy of Metal Ions against *Enterococcus faecium*, *Klebsiella pneumoniae* and *Acinetobacter baumannii* in Planktonic and Biofilm Phenotypes. *Sci. Rep.* **2017**, *7*, 5911. [CrossRef]
64. Chattoraj, D.K.; Snyder, K.M.; Abeles, A.L. P1 Plasmid Replication—Multiple Functions of Repa Protein at the Origin. *Proc. Natl. Acad. Sci. USA* **1985**, *82*, 2588–2592. [CrossRef] [PubMed]
65. Díaz-López, T.; Lages-Gonzalo, M.; Serrano-López, A.; Alfonso, C.; Rivas, G.; Díaz-Orejas, R.; Giraldo, R. Structural changes in RepA, a plasmid replication initiator, upon binding to origin DNA. *J. Biol. Chem.* **2003**, *278*, 18606–18616. [CrossRef] [PubMed]
66. diCenzo, G.C.; Finan, T.M. The Divided Bacterial Genome: Structure, Function, and Evolution. *Microbiol. Mol. Biol. Rev.* **2017**, *81*, e00019-17. [CrossRef]
67. Morton, E.R.; Merritt, P.M.; Bever, J.D.; Fuqua, C. Large Deletions in the pAtC58 Megaplasmid of *Agrobacterium tumefaciens* Can Confer Reduced Carriage Cost and Increased Expression of Virulence Genes. *Genome Biol. Evol.* **2013**, *5*, 1353–1364. [CrossRef] [PubMed]
68. Konig, C.; Eulberg, D.; Groning, J.; Lakner, S.; Seibert, V.; Kaschabek, S.R.; Schlomann, M. A linear megaplasmid, p1CP, carrying the genes for chlorocatechol catabolism of *Rhodococcus opacus* 1CP. *Microbiology* **2004**, *150*, 3075–3087. [CrossRef] [PubMed]
69. Lee, M.C.; Marx, C.J. Repeated, selection-driven genome reduction of accessory genes in experimental populations. *PLoS Genet.* **2012**, *8*, e1002651. [CrossRef]
70. Koskiniemi, S.; Sun, S.; Berg, O.G.; Andersson, D.I. Selection-driven gene loss in bacteria. *PLoS Genet.* **2012**, *8*, e1002787. [CrossRef]
71. Fernandez-Lopez, R.; Del Campo, I.; Revilla, C.; Cuevas, A.; de la Cruz, F. Negative feedback and transcriptional overshooting in a regulatory network for horizontal gene transfer. *PLoS Genet.* **2014**, *10*, e1004171. [CrossRef]

72. Zahrl, D.; Wagner, M.; Bischof, K.; Koraimann, G. Expression and assembly of a functional type IV secretion system elicit extracytoplasmic and cytoplasmic stress responses in *Escherichia coli*. *J. Bacteriol.* **2006**, *188*, 6611–6621. [CrossRef]
73. Porse, A.; Schonning, K.; Munck, C.; Sommer, M.O. Survival and Evolution of a Large Multidrug Resistance Plasmid in New Clinical Bacterial Hosts. *Mol. Biol. Evol.* **2016**, *33*, 2860–2873. [CrossRef]
74. Romanchuk, A.; Jones, C.D.; Karkare, K.; Moore, A.; Smith, B.A.; Jones, C.; Dougherty, K.; Baltrus, D.A. Bigger is not always better: Transmission and fitness burden of approximately 1MB *Pseudomonas syringae* megaplasmid pMPPla107. *Plasmid* **2014**, *73*, 16–25. [CrossRef] [PubMed]
75. Moran, N.A. Microbial minimalism: Genome reduction in bacterial pathogens. *Cell* **2002**, *108*, 583–586. [CrossRef]
76. Dufresne, A.; Garczarek, L.; Partensky, F. Accelerated evolution associated with genome reduction in a free-living prokaryote. *Genome Biol.* **2005**, *6*, R14. [CrossRef] [PubMed]
77. Lenski, R.E. What is adaptation by natural selection? Perspectives of an experimental microbiologist. *PLoS Genet.* **2017**, *13*, e1006668. [CrossRef] [PubMed]
78. LaCroix, R.A.; Sandberg, T.E.; O'Brien, E.J.; Utrilla, J.; Ebrahim, A.; Guzman, G.I.; Szubin, R.; Palsson, B.O.; Feist, A.M. Use of Adaptive Laboratory Evolution To Discover Key Mutations Enabling Rapid Growth of *Escherichia coli* K-12 MG1655 on Glucose Minimal Medium. *Appl. Environ. Microbiol.* **2015**, *81*, 17–30. [CrossRef] [PubMed]
79. McCloskey, D.; Xu, S.; Sandberg, T.E.; Brunk, E.; Hefner, Y.; Szubin, R.; Feist, A.M.; Palsson, B.O. Growth Adaptation of *gnd* and *sdhCB Escherichia coli* Deletion Strains Diverges From a Similar Initial Perturbation of the Transcriptome. *Front. Microbiol.* **2018**, *9*, 1793. [CrossRef] [PubMed]
80. Sandberg, T.E.; Pedersen, M.; LaCroix, R.A.; Ebrahim, A.; Bonde, M.; Herrgard, M.J.; Palsson, B.O.; Sommer, M.; Feist, A.M. Evolution of *Escherichia coli* to 42 degrees C and subsequent genetic engineering reveals adaptive mechanisms and novel mutations. *Mol. Biol. Evol.* **2014**, *31*, 2647–2662. [CrossRef]
81. Powell, S.; Szklarczyk, D.; Trachana, K.; Roth, A.; Kuhn, M.; Muller, J.; Arnold, R.; Rattei, T.; Letunic, I.; Doerks, T.; et al. eggNOG v3.0: Orthologous groups covering 1133 organisms at 41 different taxonomic ranges. *Nucleic Acids Res.* **2012**, *40*, D284–D289. [CrossRef]
82. Franke, S.; Grass, G.; Rensing, C.; Nies, D.H. Molecular analysis of the copper-transporting efflux system CusCFBA of *Escherichia coli*. *J. Bacteriol.* **2003**, *185*, 3804–3812. [CrossRef]
83. Mergeay, M.; Monchy, S.; Vallaeys, T.; Auquier, V.; Benotmane, A.; Bertin, P.; Taghavi, S.; Dunn, J.; van der Lelie, D.; Wattiez, R. *Ralstonia metallidurans*, a bacterium specifically adapted to toxic metals: Towards a catalogue of metal-responsive genes. *FEMS Microbiol. Rev.* **2003**, *27*, 385–410. [CrossRef]
84. Liesegang, H.; Lemke, K.; Siddiqui, R.A.; Schlegel, H.G. Characterization of the inducible nickel and cobalt resistance determinant cnr from pMOL28 of *Alcaligenes eutrophus* CH34. *J. Bacteriol.* **1993**, *175*, 767–778. [CrossRef] [PubMed]
85. Nguyen, M.P.; Yoon, J.M.; Cho, M.H.; Lee, S.W. Prokaryotic 2-component systems and the OmpR/PhoB superfamily. *Can. J. Microbiol.* **2015**, *61*, 799–810. [CrossRef] [PubMed]
86. Szurmant, H.; Ordal, G.W. Diversity in chemotaxis mechanisms among the bacteria and archaea. *Microbiol. Mol. Biol. Rev.* **2004**, *68*, 301–319. [CrossRef] [PubMed]
87. Fujii, T.; Kato, T.; Hiraoka, K.D.; Miyata, T.; Minamino, T.; Chevance, F.F.; Hughes, K.T.; Namba, K. Identical folds used for distinct mechanical functions of the bacterial flagellar rod and hook. *Nat. Commun.* **2017**, *8*, 14276. [CrossRef] [PubMed]
88. Cohen, E.J.; Hughes, K.T. Rod-to-hook transition for extracellular flagellum assembly is catalyzed by the L-ring-dependent rod scaffold removal. *J. Bacteriol.* **2014**, *196*, 2387–2395. [CrossRef]
89. Alm, R.A.; Hallinan, J.P.; Watson, A.A.; Mattick, J.S. Fimbrial biogenesis genes of *Pseudomonas aeruginosa*: PILW and PILX increase the similarity of type 4 fimbriae to the GSP protein-secretion systems and pilY1 encodes a gonococcal PilC homologue. *Mol. Microbiol.* **1996**, *22*, 161–173.
90. Scheurwater, E.; Reid, C.W.; Clarke, A.J. Lytic transglycosylases: Bacterial space-making autolysins. *Int. J. Biochem. Cell Biol.* **2008**, *40*, 586–591. [CrossRef]
91. Monchy, S.; Benotmane, M.A.; Janssen, P.; Vallaeys, T.; Taghavi, S.; van der Lelie, D.; Mergeay, M. Plasmids pMOL28 and pMOL30 of *Cupriavidus metallidurans* are specialized in the maximal viable response to heavy metals. *J. Bacteriol.* **2007**, *189*, 7417–7425. [CrossRef]

92. Bertini, I.; Cavallaro, G.; Rosato, A. Cytochrome c: Occurrence and functions. *Chem. Rev.* **2006**, *106*, 90–115. [CrossRef]
93. Cianciotto, N.P.; White, R.C. Expanding Role of Type II Secretion in Bacterial Pathogenesis and Beyond. *Infect. Immun.* **2017**. [CrossRef]
94. Xu, H.; Denny, T.P. Native and Foreign Proteins Secreted by the *Cupriavidus metallidurans* Type II System and an Alternative Mechanism. *J. Microbiol. Biotechnol.* **2017**, *27*, 791–807. [CrossRef] [PubMed]

 © 2019 by the authors. Licensee MDPI, Basel, Switzerland. This article is an open access article distributed under the terms and conditions of the Creative Commons Attribution (CC BY) license (http://creativecommons.org/licenses/by/4.0/).

Article

Using a Chemical Genetic Screen to Enhance Our Understanding of the Antibacterial Properties of Silver

Natalie Gugala [1,†], Joe Lemire [1,†], Kate Chatfield-Reed [1,†], Ying Yan [2], Gordon Chua [1] and Raymond J. Turner [1,*]

1. Department of Biological Sciences, University of Calgary, 2500 University Dr. NW, Calgary, AB T2N 1N4, Canada; ngugala@ucalgary.ca (N.G.); jalemire@ucalgary.ca (J.L.); kchatfieldreed@gmail.com (K.C.-R.); gchua@ucalgary.ca (G.C.)
2. Department of Mathematics and Statistics, University of Calgary, 2500 University Dr. NW, Calgary, AB T2N 1N4, Canada; ying.yan@ucalgary.ca
* Correspondence: turnerr@ucalgary.ca; Tel.: +1-403-220-7484
† These authors contributed equally to this study.

Received: 7 June 2018; Accepted: 3 July 2018; Published: 6 July 2018

Abstract: It is essential to understand the mechanisms by which a toxicant is capable of poisoning the bacterial cell. The mechanism of action of many biocides and toxins, including numerous ubiquitous compounds, is not fully understood. For example, despite the widespread clinical and commercial use of silver (Ag), the mechanisms describing how this metal poisons bacterial cells remains incomplete. To advance our understanding surrounding the antimicrobial action of Ag, we performed a chemical genetic screen of a mutant library of *Escherichia coli*—the Keio collection, in order to identify Ag sensitive or resistant deletion strains. Indeed, our findings corroborate many previously established mechanisms that describe the antibacterial effects of Ag, such as the disruption of iron-sulfur clusters containing proteins and certain cellular redox enzymes. However, the data presented here demonstrates that the activity of Ag within the bacterial cell is more extensive, encompassing genes involved in cell wall maintenance, quinone metabolism and sulfur assimilation. Altogether, this study provides further insight into the antimicrobial mechanism of Ag and the physiological adaption of *E. coli* to this metal.

Keywords: silver; silver toxicity; silver resistance; Keio collection; *Escherichia coli*; antimicrobials

1. Introduction

For centuries, metal compounds have been deployed as effective antimicrobial agents [1]. The use of silver (Ag) for antimicrobial purposes is a practice that dates back thousands of years [2] and is still implemented for medical purposes in an effort to curtail the rise of antimicrobial resistant pathogens [3–6], a threat that has once again surfaced as a clinical challenge [7–10].

Applications of Ag-based antimicrobials include: wound dressings [11] and other textiles [12], antiseptic formulations [13], nanoparticles [14], coatings [15], nanocomposites [16], polymers [17], and part of antibiotic combination therapies [18]. Many of these approaches have proven to be effective in controlling and eradicating pathogenic microorganisms.

Presently, research in this field focuses on finding new formulations and utilities for Ag-based antimicrobials. Despite this, the identity of the cellular targets that are involved in Ag antimicrobial activities are known to a far lesser degree [19]. This current knowledge gap hinders the potential utility of Ag-based antimicrobials, and in turn the expansion of this metal as a therapeutic agent.

Previous studies examining the mechanisms of Ag resistance and toxicity have not provided a complete understanding of the global cellular effects of Ag exposure on the bacterial cell. Further, several studies fail to build upon preceding work and the literature is replete with contradicting reports, in part due to non-standardized conditions of study. Furthermore, it has been demonstrated that the speciation/oxidation state of Ag has substantial influence on toxicity, a factor that is dependent on the source of Ag ions [4], growth conditions, and is further complicated by the organism (species and strain) of interest [6].

Proposed mechanisms of metal toxicity include the production and propagation of reactive oxygen species through Fenton chemistry and antioxidant depletion, the disruption of iron-sulfur clusters, thiol coordination and the exchange of a catalytic/structural metal that leads to protein dysfunction, interference with nutrient uptake, and genotoxicity [19]. Microorganisms are able to withstand metal toxicity through several mechanisms such as reduced uptake, efflux, extracellular and intracellular sequestration, repair, metabolic by-pass and chemical modification [20]. Whether these mechanisms are solely responsible for cell death or resistance has yet to be determined. Still, what is understood is that metals demonstrate broad-spectrum activity and decreased target specificity [19] when compared to conventional antimicrobials.

In this work, we hypothesized that Ag exerts its effects on multiple targets both directly and indirectly, and thus various cellular systems may be altered by Ag exposure. To test this, we performed a genotypic screening workflow of a mutant library composed of 3985 strains, each containing a different inactivated non-essential gene in *Escherichia coli*. Using a comparable genome-wide workflow [21] and by use of transcriptomic profiling [22,23], similar approaches have been implemented in order to study the mechanisms of action caused by Ag. Despite this, genes conferring resistance to Ag when absent have been studied and compared to a far lesser degree than those that result in sensitivity when absent. Further, many previous approaches aimed at studying Ag toxicity and resistance have primarily examined the effects of Ag shock or rapid pulses of exposure, followed by the evaluation of gene expression. Hence, as a means of complementing existing work, we have identified a number of genes that are implicated in prolonged Ag resistance and/or toxicity, and mapped their metabolic function to their respective cellular system.

2. Materials and Methods

2.1. Escherichia coli Strains and Storage

The Keio collection [24]—a mutant library of 3985 single-gene *E. coli* BW25113 mutants (*lacI*^{q}, *rrnB*$_{T14}$, Δ*lacZ*$_{WJ19}$, *hsdR*514, Δ*ara*BAD$_{AH33}$, Δ*rha*BAD$_{LD78}$)—was obtained from the National BioResource Project *E. coli*, (National Institute of Genetics, Shizuoka, Japan). All strains were initially stored at $-80\ °$C in vials containing Lysogeny Broth (LB) media (VWR International, Mississauga, ON, Canada) with 30% glycerol (VWR International). For chemical genetic screening, the Keio collection was transferred, and subsequently arrayed into 96-well microtiter plates containing LB medium with 30% glycerol. Construction of the arrayed Keio collection and pre-culturing of the *E. coli* strains was carried out on LB agar (1.0%). For chemical genetic screening, M9 minimal media (6.8 g/L Na$_2$HPO$_4$, 3 g/L KH$_2$PO$_4$, 1 g/L NH$_4$Cl, 0.5 g/L NaCl, 4 mg/L glucose, 0.5 mg/L MgSO$_4$ and 0.1 mg/L CaCl$_2$) containing Noble agar (1.0%) with and without silver nitrate (AgNO$_3$) was used for treatment and control testing, respectively (all obtained from VWR International).

Although the Keio collection strains are engineered with a kanamycin resistance cassette in place of the gene of interest [24], for our experiments here, we did not include this antibiotic in the growth media. Synergistic antimicrobial effects were found when bacterial cells were grown in the presence of Ag and kanamycin (data not included).

2.2. Stock Ag Solution

Silver nitrate (AgNO$_3$) was obtained from Sigma-Aldrich (St. Louis, MO, USA). Stock solutions of Ag were made at equivalent molarities of Ag in distilled and deionized (dd)H$_2$O and stored in glass vials for no longer than two weeks.

2.3. Determination of the Minimal Inhibitory Concentration and Controls

The minimal inhibitory concentration was determined using a known Ag sensitive strain (*cusB*) and negative control strains (*lacA* and *lacY*). CusB is a part of the CusCFBA copper/silver efflux system [25]; therefore, it was anticipated that the absence of this gene would confer toxicity, denoted as a Ag sensitive hit. Further, LacA and LacY, are not expected to be involved in Ag resistance or toxicity. The aforementioned strains, along with the parent strain (wild type (WT)) were grown at 37 °C on M9 minimal media and Noble agar (1%) in the presence or absence of Ag at varying concentrations. The Ag concentration found to visibly decrease colony size in the *cusB* mutant and demonstrate no changes in colony size in the *lacA* and *lacY* mutants was selected. Furthermore, the latter mutants and the WT were grown in the presence of 100 µM ionic nitrate to ensure growth was not impeded by the accompanying counter ion. The full chemical genetic screen was challenged at time zero of inoculation in the presence of 100 µM AgNO$_3$.

2.4. Screening

M9 minimal media Noble agar (1%) plates were prepared two days prior to use. Colony arrays in 96-format were produced and processed using a BM3 robot (S&P Robotics Inc., Toronto, ON, Canada). The strains were spotted using a 96-pin replicator, allowing for uniform application. Cells were transferred from the arrayed microtiter plates using the replicator onto LB agar plates. These plates were then grown overnight at 37 °C. Once grown, the colonies were spotted using the replicator onto two sets—with and without 100 µM AgNO$_3$—of M9 minimal media Noble agar plates, and subsequently grown overnight at 37 °C. Images of both sets of plates were acquired using the spImager (S&P Robotics Inc.) and colony size, which is a measure of fitness, was determined using integrated image processing software. For each 96-colony array, four technical trials per strain were combined onto a single plate in 384-colony array format and three biological trials were performed. Therefore, each strain was tested a total of 12 times.

2.5. Normalization

Experimental factors such as incubation time and temperature, local nutrient availability, colony location, gradients in the growth medium and neighboring mutant fitness were all considered as independent variables that could contribute to systematic variation, and subsequently affect colony size. As a result, the colonies were normalized and scored using Synthetic Genetic Array Tools 1.0 (SGATools) [26]. Firstly, all of the plates were normalized to establish identical median colony size working on the assumption that most colonies exhibited WT fitness. Next, to ensure the colonies were directly comparable, colonies were rescaled, a factor that is primarily important for colonies close to the edge of the plate. Further, spatial smoothing accounted for partialities in each plate owing to inconsistencies, such as the thickness of the agar. Very large colonies, likely an indication of contamination among other factors, and those that were different from the corresponding technical replicates were removed. Lastly, colonies that were larger than anticipated and located next to colonies that were found to be smaller than anticipated were marked as potential false-positive hits.

Following this normalization, the colonies were scored. Here, paired evaluation was completed by comparing the colony size (in the presence of Ag) to a matched control (in the absence of Ag). Fitness values were established, and the subsequent scores represented deviation from the fitness of the WT strain. Once normalized and scored, colonies displaying a reduction in size were indicative of a Ag sensitive hit and those displaying an increase in colony size qualified as a Ag resistant hit. Finally, the *p*-value was calculated as a two-tailed *t*-test and significance was determined using the Benjamini-Hochberg procedure, as a means of lowering the false discovery rate, which was selected to be 0.1.

2.6. Data Mining and Analyses

Subsequent analyses were conducted using Pathway Tools Omics Dashboard, which surveys against the EcoCyc database [27]. This allowed for clustering of the hits into systems, subsystems, component subsystems, and lastly, into individual objects. It is important to note that genes can be found in multiple systems, since many are involved in a number of cellular processes.

Further, in order to identify biological processes most prominent under Ag challenge, enrichment analyses were conducted for the Ag resistant and Ag sensitive hits. To analyze the gene list, the Database for Annotation, Visualization and Integrated Discovery (DAVID) bioinformatics resource was utilized [28,29]. Lastly, as a means of exposing the direct (physical) and indirect (functional) protein-protein connectivity between the gene hits, the Search Tool for the Retrieval of Interacting Genes/Proteins (STRING) database was used [30]. Interactive node maps based on experimental, co-expression and gene fusion studies were generated based on genes defined in our chemical genetics screen.

3. Results and Discussion

3.1. Genome-Wide Screen of Ag-Resistant and Ag-Sensitive Hits

The chemical genetic screen completed in this work provided a method for genome-wide probing of non-essential genes involved in Ag-sensitivity or -resistance in *E. coli*. A total of 3810 non-essential genes were screened for growth in the presence of 100 µM AgNO$_3$ (Supplementary Materials, Table S1). 3073 mutants displayed little change in colony size in the presence of Ag with a normalized fitness score between ±0.1 (Figure 1). The statistical colony size cutoff that indicated a significant difference in fitness was selected to be ±0.15, or two standard deviations from the mean. This resulted in 225 gene hits, which represents approximately 5% of the open reading frames in the *E. coli* genome. The remaining gene hits were not regarded as significant hits in this work based solely on the cut offs selected. In general, the normalization was performed on the assumption that Ag does not specifically interact with the deleted gene but rather impedes growth due to environmental stress. In short, those displaying hits between the cut off values were assumed to have non-specific or neutral interactions with Ag.

It is important to note that when reflecting on the data generated from our chemical genetic screen, it is the absence of the gene that imparts the Ag-resistant or -sensitive phenotype. Upon Ag exposure, an increase in colony size (>0.15) is suggestive of an Ag-resistant hit, and therefore the presence of this gene is proposed to confer toxicity. On the contrary, a decrease in colony size (<−0.15) was designated to be an Ag-sensitive hit, and therefore the presence of this gene is proposed to confer resistance. In total, the deletion of 106 and 119 genes resulted in Ag-resistant and -sensitive hits, respectively (Tables 1 and 2). These gene hits were mapped to their corresponding cellular systems using EcoCyc (Figure 2 and Table A1). In short, genes were found in multiple cellular systems, validating our hypothesis that Ag cytotoxicity and the corresponding physiological responses of *E. coli* involve a number of cellular mechanisms.

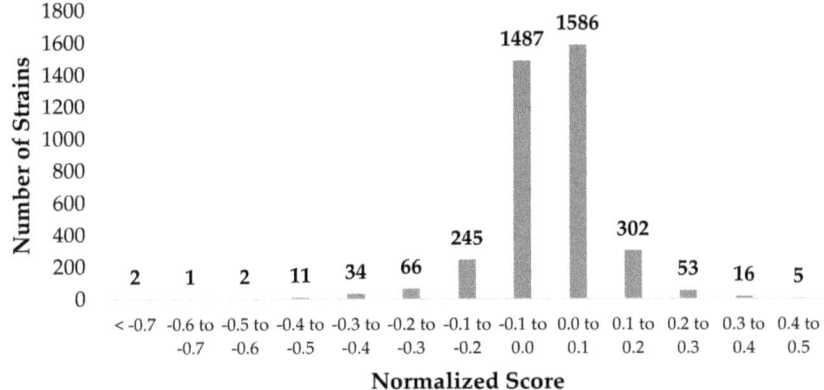

Figure 1. Synthetic Array Tools (version 1.0) was used to normalize and score the silver (Ag)-resistant and -sensitive gene hits as a means of representing the growth differences in *Escherichia coli* K12 BW25113 in the presence of 100 µM silver nitrate (AgNO$_3$). Only those with a score greater or less than ±0.15, respectively, were selected for further analysis. Hits between ±0.15 were regarded as having neutral or non-specific interactions with Ag. The *p*-value was a two-tailed *t*-test and significance was determined using the Benjamini-Hochberg procedure; false discovery rate was selected to be 0.1. Each individual score represents the mean of 12 trials.

Table 1. Ag-resistant hits organized according to system and subsystem mined using the Omics Dashboard (Pathway Tools), which surveys against the EcoCyc Database; genes represent resistant hits, each with a score >0.15 and a false discovery rate of 0.1 [1,2].

System	Subsystem	Gene [3]
Central Dogma	Transcription	alaS crp dicC gadE gcvR lysR putA yciT yhjB yiif yjiR
	Translation	alaS ettA
	DNA Metabolism	cffC dam recT
	RNA Metabolism	rluF alaS gluQ trmL crp dicC gadE gcvR lysR ogrK putA yciT yhjB yiif yjiR yjtD
	Protein Metabolism	argE envZ lipB sdhE ldcA pepB prc rhsB rzpD
Cell Exterior	Transport	malE nhaB exbB btuB dppF glcA ompG lptB mngA yejF
	Cell wall biogenesis/organization	ldcA
	Lipopolysaccharide Metabolism	wcaI
	Pilus	yraK
	Flagellum	fliL fliR
	Outer membrane	bbtuB csgF nlpE ompA ompG rhsB
	Plasma membrane	agaD cyoC cysQ damX dppF envZ ettA exbB fliL fliR glcA lptB malE mngA nhaB ppx prc putA yaiP yccF yejF ygdD yifK yojI yqfA
	Periplasm	malE nlpE prc

Table 1. *Cont.*

System	Subsystem	Gene [3]
Biosynthesis	Amino acid biosynthesis	argE cysk serC proC serA serC metL trpB trpD
	Nucleotide biosynthesis	dcd pyrF
	Amine biosynthesis	gss
	Carbohydrate biosynthesis	mdh
	Secondary metabolite biosynthesis	fldB
	Cofactor biosynthesis	bioC bioF nudB lipB nadA nadB nadC gss thiS serC
	Other	aroC metL argE alaS
Degradation	Amino acid degradation	astA cysK gadA putA
	Carbohydrate degradation	galM yigL glcE
	Secondary metabolite degradation	idcA
	Polymer degradation	idcA
Other pathways	Inorganic nutrient metabolism	cysC cysD cysH cysI
	Detoxification	gadA sodA
	Activation/inactivation/interconversion	cysC cysD
	Other	ahpF bglB cysQ dam gluQ pepB ppx prc purU rluF trmL yfaU yjhG
Energy	TCA cycle	mdh
	Fermentation	mdh
	Aerobic respiration	cyoC putA
	Other	bioC bioF mdh
Cellular process	Cell cycle/Division	dam damX dicC
	Cell death	ldcA
	Genetic transfer	ompA ygcO
	Biofilm formation	csgF
	Adhesion	yraK
	Locomotion	fliL malE rzpD
	Viral response	ompA rzpD
	Bacterial response	rzpD
	Host interaction	ompA rzpD
Response to stimulus	Heat	sodA
	DNA damage	dam malE ompA recT yaiP yciT
	pH	sodA
	Oxidant detoxification	sodA
	Other	ahpF btuB crp cysC cysD cysH cysI dcd dppF envZ exbB fliL nhaB prc putA recT rzpD ybaM yejF yigL yojI

[1] Each individual score represents the mean of 12 trials—three biological and four technical; [2] Two-tailed *t*-test and significance was determined using the Benjamini-Hochberg procedure; [3] Gene hits can be mapped to more than one system and subsystem.

Table 2. Ag-sensitive hits organized according to system and subsystem mined using the Omics Dashboard (Pathway Tools), which surveys against the EcoCyc Database; genes represent resistant hits, each with a score <−0.15 and a false discovery rate of 0.1 [1,2].

System	Subsystem	Gene [3]
Central Dogma	Transcription	arcB exuR fis galR glnL higB hupB rapA rfaH sspA rhoL ybeY yfiR
	Translation	higB prfC rhaH rplI tufB ybeY
	DNA Metabolism	fis hsdS hofM ruvA mutL
	RNA Metabolism	arcB exuR fis galR glnL higB hupB rapA rfah rhoL rsmE rraB sspA ybeY yfiR ygfZ
	Protein Metabolism	arcB glnL higB hybD iadA mobA pflA prfC pqqL rfaH rplI tufB ybeY ygeY yicR
Cell Exterior	Transport	chbB clcA cusB cysA cysP dtpB fepA feoB tdcC tolC trkH tyrP yiaN
	Cell wall biogenesis/organization	amiB rfe
	Lipopolysaccharide metabolism	kdsD rfaD rfe waaG
	Pilus	yfcQ
	Flagellum	flgH
	Outer membrane	fepA flgH lpp tolC yraP
	Plasma membrane	arcB atpB atpE atpF bcsF clcA clcB cstA cysA dtpB feoB glnL glvB hokD hycB ppdB rfe sanA tdcC tolC trkH tufB tyrP ydcV ydjZ ygeY ygiZ yhaH yhjD yiaB yiaN yibN yjiG yqiJ
	Periplasm	amiB cusB cysP hmp lpp sanA tolC yfdX yjfY yraP ytfJ
	Cell wall components	rfe
Biosynthesis	Amino acid biosynthesis	hisA ilvG lysC
	Nucleotide biosynthesis	add
	Fatty acid and lipid biosynthesis	fabF wag clsB
	Carbohydrate biosynthesis	yggF rfaD kdsD
	Cofactor biosynthesis	mobA ubiE gshB
	Other	aroL lysC
Degradation	Amino acid degradation	ilvG pflB
	Nucleotide degradation	add
	Amine degradation	caiC
	Carbohydrate degradation	yidA ulaG
	Secondary metabolite degradation	lsrF
	Aromatic degradation	hcaD mhpC
Other pathways	Other	amiB higB hmp hsdS iadA mutL nfsB nudF pflA qorR rsmE ruvA
Energy	Glycolysis	yggF
	Pentose phosphate pathway	rpiA
	Fermentation	hycB pflB
	ATP synthesis	atpB atpE atpF

Table 2. Cont.

System	Subsystem	Gene [3]
Cellular processes	Cell cycle and division	amiB minC
	Cell death	hokD
	Genetic transfer	ydcV
	Biofilm formation	yfjR
	Adhesion	yfcQ
	Locomotion	flgH
	Viral Response	fis
Response to Stimulus	Starvation	cstA sanA sspA
	Heat	Nudf ybeY yobF
	DNA damage	add feoB hisA mutL pflA ruvA ybiX yiaB yqiJ
	Osmotic stress	flgH
	pH	clcA
	Detoxification	cusB
	Other	arcB cstA dtpB fis glnL hcaD hmp hsdS mhpC sanA sspA tolC tufB yfdS yggX

[1] Each individual score represents the mean of 12 trials—three biological and four technical; [2] Two-tailed t-test and significance was determined using the Benjamini-Hochberg procedure; [3] Gene hits can be mapped to more than one system and subsystem.

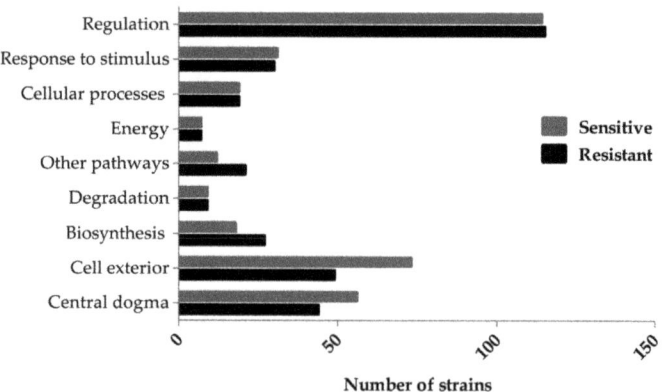

Figure 2. Ag-resistant and -sensitive gene hits mapped to component cellular processes. The cutoff fitness score implemented was −0.15 and 0.15 (two standard deviations from the mean) and the gene hits with a score less or greater than, respectively, were chosen for further analyses. The hits were mined using the Omics Dashboard (Pathway Tools), which surveys against the EcoCyc Database. Several gene hits are mapped to more than one subsystem. The p-value was calculated as a two-tailed t-test and significance was determined using the Benjamini-Hochberg procedure; false discovery rate was selected to be 0.1. Each individual score represents the mean of 12 trials.

Comparable numbers of Ag-resistant and -sensitive hits were mapped in the systems 'Response to stimulus'—starvation, heat, cold, DNA damage, pH, detoxification, osmotic stress, and other, 'Cellular processes'—cell cycle and division, cell death, genetic transfer, biofilm formation, quorum sensing, adhesion, locomotion, viral response, response to bacterium, host interactions with host, other pathogenesis proteins, and 'Degradation'—amino acids, nucleotide, amine, carbohydrate/carboxylate, secondary metabolite, alcohol, polymer and aromatic, the cell exterior, and regulation. A greater number of Ag-resistant than -sensitive hits were mapped to the processes 'Biosynthesis'—amino acids, nucleotides, fatty acid/lipid amines, carbohydrate/carboxylates, cofactors, secondary metabolites, and other pathways and 'Other pathways'—detoxification, inorganic nutrient metabolism, macromolecule modification, activation/inactivation/interconversion and other enzymes.

In total, 49 and 73 resistant and sensitive hits, respectively, were found to be a part of the 'Cell exterior'—transport, cell wall biogenesis and organization, lipopolysaccharide metabolism, pilus, flagellar, outer and inner membrane, periplasm, and cell wall components. Compared to the latter cellular processes, non-essential genes comprising 'Energy' processes—including glycolysis, the pentose phosphate pathway, the tricarboxylic acid (TCA) cycle, fermentation, and aerobic and anaerobic respiration were found to be involved in Ag toxicity or resistance the least, by more than seven-fold when compared to genes mapped to the 'Cell exterior'.

Based on the fold enrichment, metal binding proteins were affected to the same degree in both Ag-resistant and -sensitive groups, displaying an enrichment score <5 (Figure 3). However, when examining proteins involved with specific metals in more detail, such as zinc and magnesium, fold enrichment values were >5, but only for the Ag-sensitive hits (Figure 3). Cellular and anaerobic respiration were represented by the Ag-sensitive hits only, while processes involved in amino acid biosynthesis were heavily enriched for by the Ag-resistant hits.

A number of hits were found to be involved with the cell membrane using EcoCyc's system of classification, but this was not detected in the fold enrichment analysis. Here, cell membrane proteins were affected three-fold less than the most highly represented clusters, which were amino acid biosynthesis and phosphoproteins for the Ag-resistant and -sensitive hits, respectively (Figure 3).

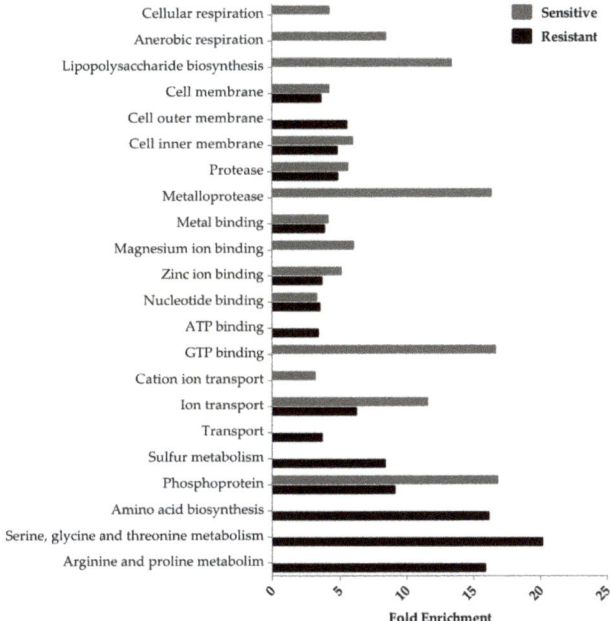

Figure 3. Functional enrichment among the Ag-resistant and -sensitive gene hits. The DAVID gene functional classification (version 6.8) database, a false discovery rate of 0.1 and a score cutoff of −0.15 and 0.15 (two standard deviations from the mean) were used to measure the magnitude of enrichment against the genome of *E. coli*. Processes with a p-value < 0.05, fold enrichment value ≥ 3 and gene hits >3 are included only. Each individual score represents the mean of 12 trials.

3.2. Ag-Resistant Gene Hits

3.2.1. Regulators of Gene Expression

When examining processes of the 'Central dogma'—systems involved in replication, and transcription to translation—in more detail, each subsystem had a mean score between 0.211 and 0.294 (Figure 4a). Despite this consistency, transcription and RNA metabolism contained the greatest number of Ag-resistant hits, 12 and 16, respectively. The protein EttA—energy-dependent translational throttle protein [31], can be found within the subsystems translation and protein metabolism. EttA is sensitive to the energy state of the cell. This protein represses translational elongation in response to high ADP/ATP, stimulating dipeptide bond synthesis in the presence of ATP (cell high energy state) and vice versa. As a result, EttA may inhibit translation in Ag-treated cells due to the occurrence of high ADP/ATP ratios. The absence of EttA might allow for increased translation of proteins, such as RecA [32] or CusB [25], which may result in Ag resistance. Furthermore, six proteins involved in proteolysis were found to confer resistance when absent, such as Prc. This enzyme is a periplasmic protease that processes and degrades specific proteins, has been found to provide resistance against a number of small hydrophilic antibiotics and causes the leakage of periplasmic proteins when absent [33]. Antibiotic resistant mechanisms have been compared to those of metal ions, drawing on similarities such as substrate modification or sequestration. The leakage of the periplasmic proteins in *prc* mutants may result in Ag sequestration, thereby causing metal resistance.

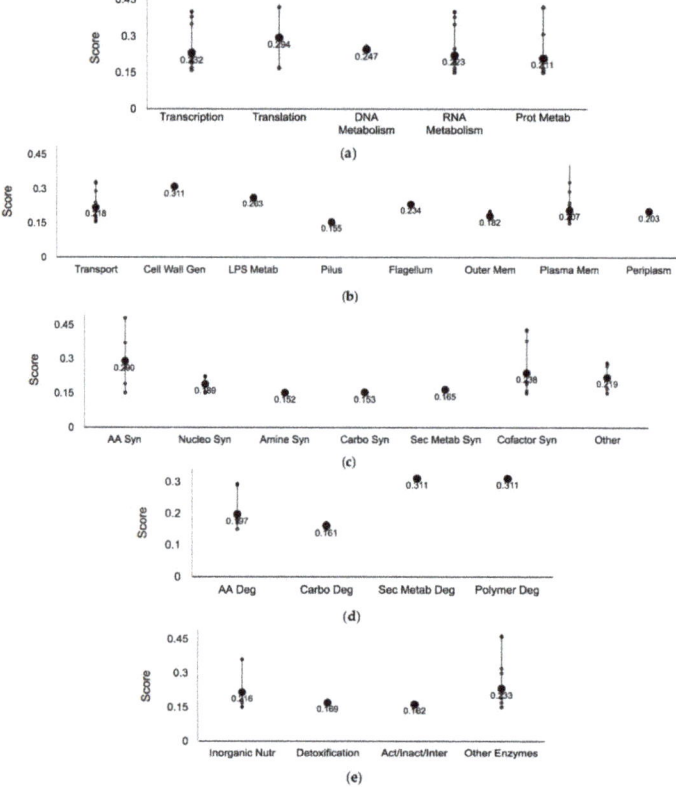

Figure 4. *Cont.*

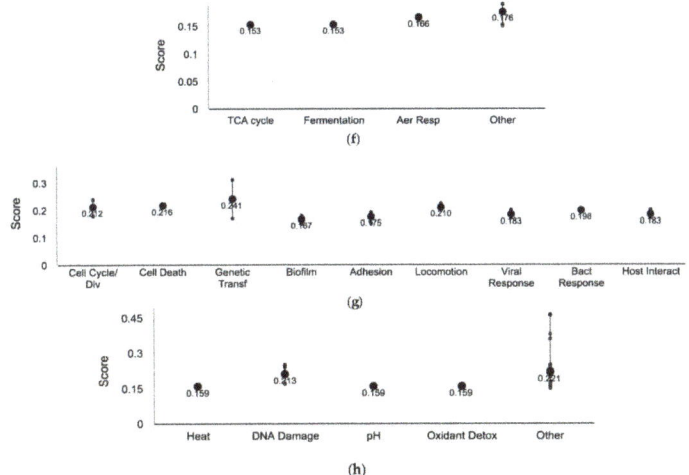

Figure 4. Ag-resistant gene hits plotted against respective cellular processes. Y-axis representative of the normalized score, smaller circles represent the individual hits and the larger circles represent the mean of each subsystem. The *p*-value was calculated as a two-tailed *t*-test and significance was determined using the Benjamini-Hochberg procedure; false discovery rate was selected to be 0.1. Each individual score represents the mean of 12 trials. (**a**) Central Dogma; (**b**) Cell exterior; (**c**) Biosynthesis; (**d**) Degradation; (**e**) Other pathways; (**f**) Energy; (**g**) Cellular processes; and (**h**) Response to stimulus. Plots constructed using Pathway Tools, Omics Dashboard.

3.2.2. Cell Membrane Proteins

It has been demonstrated that Ag may exert toxicity and potentially impede growth by acting on the cell membrane [34,35]. In this study, 49 coding genes that resulted in Ag resistance when absent were determined to be a part of the 'Cell exterior', which includes proteins of the cell membrane, periplasm and extracellular structures (Figures 2 and 4b). Of these, 25 genes coded for plasma membrane proteins, and while Ag was observed to enter bacterial cells [23], the exact mechanism of import has yet to be determined. Loss of the porin genes *ompC* and *ompF* was observed to confer resistance to Ag [36]. While these two genes were not detected within our cut offs, we did recover two additional porin genes (*ompA* and *ompG*) as conferring Ag resistance when deleted. Relative to this, it has been demonstrated that a mechanism of entry for zinc into the cell is co-transport with low molecular weight metabolites via transport proteins found within the membrane [37]. Further, ExbB, a Ag-resistant hit with a score of 0.241, is part of the energy transducing Ton system that transports iron-siderophore complexes and vitamin B12 across the outer membrane [38]. Collectively, these findings provide insight into possible mechanisms of Ag import, such as entry through porins, co-transport with metabolites or the replacement of Ag with other ions predetermined for import. The enrichment analysis offered further evidence for this hypothesis, as a number of ion transport proteins and proteins pertaining to the cell membrane were involved in Ag resistance when absent (Figure 3). Furthermore, MngA, a permease that simultaneously phosphorylates 2-*O*-α-mannosyl-D-glycerate in a process called group translocation, contains two putative phosphorylation sites His^{87} and Cys^{192} [39]. Thiols are regarded as soft bases, and according to the hard-soft acid base theory, which is key to the reactivity and coordination of metals [40], cysteine, and to a lesser degree methionine and imidazole chemically interact with Ag(I) with high affinity. Therefore, proteins with key structural or catalytic thiols/imidazols are possible Ag interacting sites.

3.2.3. Biosynthetic Enzymes

Eight hits were found to be involved in the biosynthesis of amino acids and 10 hits were found to be involved in cofactor/prosthetic group/electron carries catabolism (Figure 4c). When examining the functional enrichment analysis, serine, glycine, threonine, arginine and proline biosynthetic processes were highly enriched, on average three-fold more than the remaining cellular processes (Figure 3). The third step in the synthesis of NAD^+ from L-aspartate occurs via the enzyme NadC—quinolinate phosphoribosyltransferase [41] and based on our data the absence of this protein confers resistance in *E. coli*. In fact, the genes coding for the first and second steps of de novo NAD^+ synthesis, NadB—L-aspartate oxidase and NadA—quinolinate synthase, respectively, were also found to be Ag resistant hits. NadA contains a [4Fe-4S] cluster that is required for activity [42]. Soft metals have the capacity to inactivate dehydratases in vitro via iron-sulfur cluster degradation, possibly leading to the bridging of the sulfur atoms [43]. As a result, proteins with iron-sulfur centers are of possible interest when examining the interactions of Ag with cellular biomolecules. Furthermore, it has been demonstrated that H_2O_2 formation is diminished via the addition of precursors involved in the synthesis of NAD^+ [44]. The absence of one gene involved in NAD^+ biosynthesis may result in metabolite accumulation since there is no evidence of negative precursor feedback inhibition. Therefore, there is a possibility that deletion of the *nadA*, *nadB* or *nadC* may confer resistance if H_2O_2 is generated in the presence of Ag.

Using the STRING database, several points of interaction were revealed. Among the Ag-resistant hits, the latter genes involved in de novo NAD^+ production were connected to proteins a part of amino acid biosynthesis, including *trpB*, *aroC*, and *metL* (Supplementary Materials, Figure S2).

3.2.4. Catabolic Enzymes

Genes encoding enzymes functioning in the catabolism of metabolites, such as amino acids, fatty acids, carbohydrates and polymers, were underrepresented compared to anabolism (Figures 2 and 4d). In fact, in the functional enrichment analysis, degradation processes were not represented within the cutoffs selected (Figure 3). The gene *idcA*—L,D-carboxypeptidase, a component of secondary metabolite and polymer degradation, had an elevated score of 0.311. IdcA is essential for murein turnover [45]. Murein processing is an important energy-conserving activity that transports cell wall components from the exterior of the cell to the cytoplasm [46]. Evidence has demonstrated that during logarithmic growth, the *idcA* mutant strain displays a decrease in the overall cross-linkage of murein, causing a reduction in turnover and the abundance of murein being transported into the cell. In turn, this may result in the transport of fewer Ag ions, which may have bound to the cell wall, into the cell thereby prompting increased resistance in the *idcA* mutant strain. Metal nanoparticles have been proposed to target the outer membrane regions of bacteria due to strong electrostatic interactions and co-coordination of the metal with the lipopolysaccharide or similar cell wall structures [47]. The particles are proposed to release ionic Ag, likely triggering toxicity through membrane damage and facilitating the entry of excess Ag ions.

3.2.5. Sulfur Metabolism Proteins

Within the subsystem inorganic nutrient metabolism, a part of 'Other pathways', which also includes processes such as macromolecule modification and activation/inactivation/interconversion (Figure 4e), one pathway was found to be affected by Ag exposure—sulfur metabolism. CysH—phosphor-adenylsulfate reductase is involved in assimilatory sulfate reduction by catalyzing the reduction of 3'-phospho-adenylylsulfate to sulfite and adenosine 3',5-biphospahte (PAP). This protein contains highly conserved cysteine residues that become oxidized to form a disulfide bond [48]—possible targets based on the affinity of Ag for sulfur. Moreover, the *cysC*, *cysD* and *cysI* genes, also involved in the pathway sulfate reduction I (assimilatory) via phosphorylation, adenylation and reduction, respectively, were also Ag-resistant hits. These sulfate assimilatory proteins are linked to the Ag-resistant hit CysQ, which is involved in the recycling of PAP and has been experimentally determined to be the main target of lithium toxicity [49] (Supplementary Materials, Figure S2). The protein CysI, contains a siroheme and one [4Fe-4S] cluster per polypeptide chain [50]. Comparably, it has been demonstrated that the exposure of Ag nanoparticles upregulates the expression of several genes involved in iron and sulfate homeostasis [22], including those aforementioned. A decrease in the activity of this pathway reduces the amount of hydrogen sulfide required for processes such as L-cysteine biosynthesis, and since Ag interacts with sulfur compounds well, such as hydrogen sulfide—the final product of sulfate reduction I—fewer Ag targets may be available when genes of this pathway are deleted. CysH had the highest score of 0.360 out of all four sulfur assimilatory genes, and since this protein interacts with thioredoxin, the absence of CysH may free reduced thioredoxin, thus providing elevated resistance in presence of reactive oxygen species that may arise under Ag stress.

3.2.6. Biofilm Formation

In total 19 genes in the 'Cellular processes' system, which includes subsystems such as genetic transfer, quorum sensing, adhesion and locomotion, were found to confer resistance when absent (Figure 2). Three hits were involved in cell cycle and division, and two were found to be involved in biofilm formation (Figure 4g), such as CsgF, which is an outer membrane protein that initiates curli subunit polymerization, and therefore involved in the colonization of surfaces and biofilm formation [49]. In the absence of CsgF, less biofilm is formed, and according to our results, Ag resistance is generated. Biofilms commonly provide resistance in the face of fluctuating or threatening environments [51]; however, studies have shown that bacterial residence within a biofilm does not always provide enhanced resistance against metals [6,52,53], an observation supported by this work. An explanation for this may reside in the ability of biofilms to sequester Ag ions by attracting them to varying components of the extracellular polymeric matrix. While this may provide resistance, it may also concentrate ions within a localized area, thereby causing greater sensitivity. Similarly, Ag nanoparticles have been shown to inhibit *E. coli* biofilm formation by potentially targeting curli fibers [54], therefore the absence of curli fibers may promote Ag resistance. Previous studies, which have found that biofilm formation are a source of Ag resistance [32], were completed under differing culture conditions, therefore direct comparisons are challenging.

3.2.7. DNA Damage and Repair

The effect of Ag exposure on DNA damage and repair in *E. coli* has been inconsistent from several studies involving gene deletion strains. Radzig et al. showed that several deletion strains lacking in the ability to excise DNA bases were sensitive to Ag exposure, but not the ΔrecA strain, which is involved in SOS repair [36]. In contrast, the ΔrecA deletion strain showed Ag sensitivity in a previous study [32]. From our list of Ag resistant hits, six mutants were identified within the DNA damage subsystem (Figure 4h) including the Δdam strain. Dam is methyltransferase that functions in mismatch DNA repair in *E. coli* and may also play a role in controlling oxidative damage. Based on this protein's function, we expect that the deletion strain of the gene would exhibit Ag sensitivity, potentially due to a deficiency in DNA repair of oxidative damage. However, the *dam1*Δ strain exhibits an upregulation of RecA and constitutive SOS activity which may be the nature of the Ag resistance exhibited in this mutant [36]. Moreover, we also identified several other Ag-resistant strains from our screens (*purF, damX, dcd, ruvC* and *ompA*) that are also known to possess RecA-mediated constitutive SOS activity [32].

3.3. Ag-Sensitive Hits

3.3.1. Central Dogma and Cell Exterior Proteins

Within the 'Central dogma', 56 mutants resulted in Ag sensitivity (Figure 5a). For example, *ruvA*, a gene found to be involved in DNA repair, had a normalized score of −0.430 [55]. Direct DNA damage has not been attributed to Ag exposure; however, in the presence of reactive oxygen species potentially triggered by Ag exposure, the propagation of Fenton active iron may cause DNA damage [19,56].

In total of 73 genes were mapped to the system 'Cell exterior' (Figure 5b). The gene *ygiZ*, which codes for a putative inner membrane protein, had a score of −0.751, the lowest value of any protein in this screen. A common resistance mechanism employed by microbes is the export of the challenge from the periplasm or interior of the cell to the extracellular space [20]. The fold enrichment analysis supported this finding—cell membrane proteins and those involved in the ion transport were highly enriched (Figure 3). In total, 13 transport proteins conferred Ag sensitivity when absent, such as *cusB*, which encodes for a component of the copper/silver export system CusCFBA in *E. coli*, and contains several methionine residues important for function [57]. In the absence of this protein, sensitivity is anticipated, since the cell is unable to expel Ag ions. Another Ag sensitive hit was Lpp, considered to be the most abundant protein in *E. coli* [56]. Cells lacking Lpp have been found to be hypersensitive to toxic compounds [56], potentially because there is less protein available to sequester the incoming threat. In addition, the protein TolC was an Ag resistant hit. This protein is required for the function of a number of efflux systems including the AcrAB multidrug efflux system, which is involved in the export of a number of toxic exogenous compounds [58]. In contrast to efflux proteins, we identified the *cysA* and *cysP* genes—thiosulfate and sulfate permeases—to be sensitive hits when absent. CysA and CysP function in the first step of cysteine biosynthesis, which may be important in Ag resistance since this metal may target cysteine residues via thiol side chains [32].

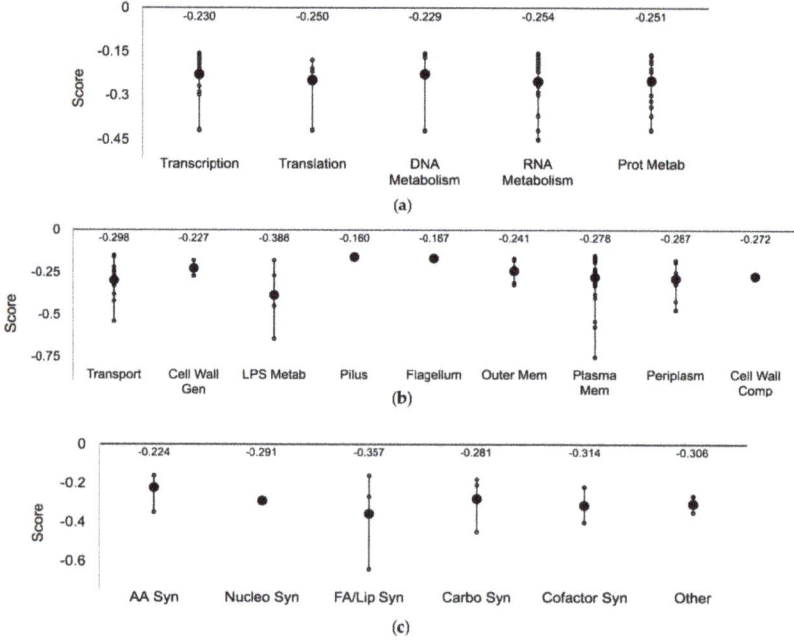

Figure 5. *Cont.*

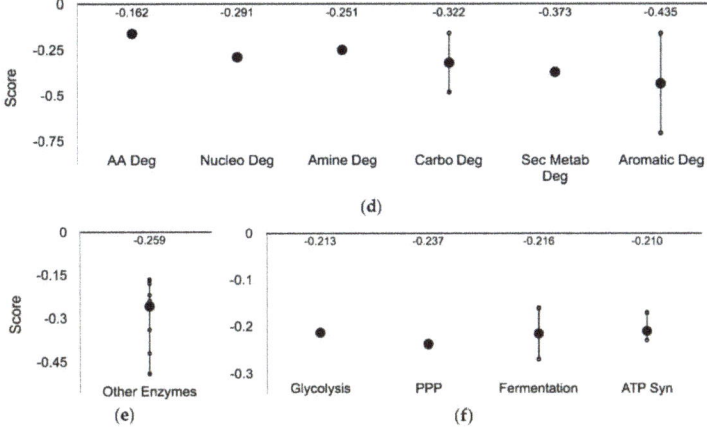

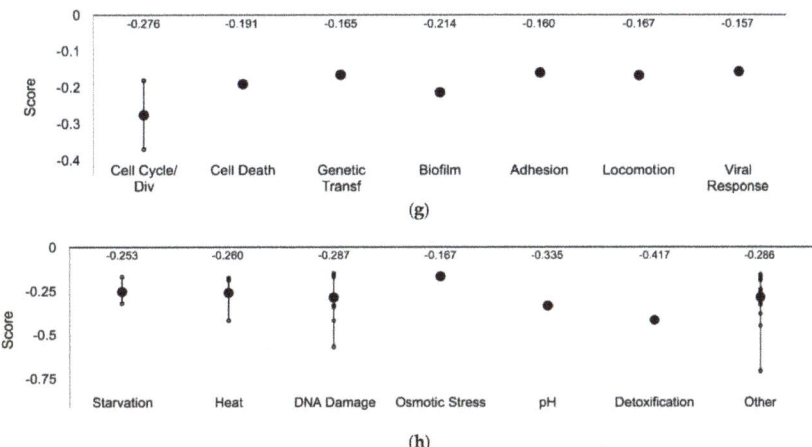

Figure 5. Ag-sensitive gene hits plotted against respective cellular processes. Y-axis representative of the normalized score, smaller circles represent the individual hits and the larger circles represent the mean of each subsystem. The *p*-value was a two-tailed *t*-test and significance was determined using the Benjamini-Hochberg procedure; false discovery rate was selected to be 0.1. Each individual score represents the mean of 12 trials. (**a**) Central Dogma; (**b**) Cell exterior; (**c**) Biosynthesis; (**d**) Degradation; (**e**) Other pathways; (**f**) Energy; (**g**) Cellular processes; and (**h**) Response to stimulus. Plots constructed using Pathway Tools, Omics Dashboard.

3.3.2. Lipopolysaccharide Biosynthetic Genes

In total, 18 Ag-sensitive hits were mapped to 'Biosynthesis processes' (Figure 5c). Processes associated with lipopolysaccharide biosynthesis were highly represented in the enrichment analysis (Figure 3). FabF, a key protein involved in fatty acid biosynthesis, and *clsB*—cardiolipin synthase B were found to be Ag-sensitive hits. If Ag targets the cellular membrane, lipid biosynthesis/regeneration could serve as a mechanism of Ag resistance and consequently, Ag toxicity would be increased if either of these processes were compromised *via* the deletion of these candidate genes.

Processes of biomolecule degradation were affected to a lesser degree than biosynthesis (Figure 2). Only nine hits were mapped to this system (Figure 5d). The mutant *hcaD* had the second lowest score of -0.707 in this screen. This protein is a predicted ferredoxin reductase subunit that is involved in the degradation of aromatic acids as carbon sources.

3.3.3. Three Ag-Sensitive Hits Comprise the ATP Synthase F_o Complex

Seven hits were mapped to 'Energy processes' (Figure 5f). Of these, three are components of the ATP synthase F_o complex—AtpB, AtpE and AtpF. Ag has been suggested to damage the respiratory chain of *E. coli* [59], thereby preventing the efficient pumping of protons across the membrane. Small disruptions to the F_o complex may amplify this consequence and render this biological process hypersensitive. If this mechanism is correct and the cytoplasmic membrane becomes more permeable to protons, than the cell will attempt to compensate for this increase in acidity via several mechanisms, one being the reversal of ATP synthase in order to pump protons outward (if ATP is not limiting) and decrease cytoplasmic proton concentrations [60]. If the ATP synthase complex exhibits decreased activity due to disruptions in any of the subunits, this resistant mechanism may be unable to function properly, resulting in greater Ag sensitivity.

Several nodes of interaction based on the STRING connectivity maps were made evident within this cluster of proteins such as the association of *atpF* and *atpB* to *gshB* and several putative membrane proteins, *tufB*—elongation factor Tu and *ppgL*—a putative zinc peptidase (Supplementary Materials, Figure S3).

3.3.4. Oxidative Stress Response Genes

Out of the 31 proteins mapped to 'Response to Stimulus', 24 were involved in mediating DNA damage and other processes (Figure 5h). The gene coding for glutathione synthetase—*gshB* was found to be an Ag-sensitive hit. Strains overexpressing either GshA or GshB are more resistant to oxidative damage, and this system has been shown to mediate metal resistance [61]. As a result, the deletion of either gene is anticipated to cause Ag sensitivity. Furthermore, the putative Fe^{+2} trafficking protein, YggX was found to have a score of -0.450. This protein is proposed to play a role in preventing the oxidation of iron-sulfur clusters [62], a proposed mechanism of Ag toxicity. The absence of this protective protein may result in sensitivity since it can be found at elevated concentrations in vivo and it is involved in mediating oxidative damage [63]. Further, the protein Hmp, a flavohemoglobin with nitric oxide dioxygenase activity [64] had a score of -0.254. This protein has been shown to protect respiratory cytochromes in *E. coli* [37], which is a possible mechanism of Ag toxicity [59,65].

4. Conclusions

In this work, a chemical genetic screen of a mutant library was performed as a means of drawing insight into the mechanisms of Ag toxicity and resistance in bacteria. In total, 3810 mutant strains containing single deletions of non-essential genes in *E. coli* were screened, and subsequent hits were bioinformatically evaluated in order to highlight processes and pathways that are affected by Ag exposure. This systematic mutant screen involved a low but prolonged concentration of Ag exposure on solid minimal media to avoid indirect secondary and acute responses, while also attempting to directly target relevant genes. Here, resistant hits represented genes involved in enhancing the cytotoxicity of Ag, while in contrast sensitive hits represented genes functioning in tolerance to Ag including physiological responses that mitigate toxicity.

In short, processes involved with the cell exterior and the central dogma were found to be affected by Ag exposure to a greater extent than other processes analyzed. However, when further examining the fold enrichment, the cell membrane and transport were involved in Ag exposure to a lesser degree. In fact, proteins involved in amino acid biosynthesis (Ag sensitivity), phosphoproteins and metalloproteins (Ag resistance) were most densely represented as hits in this work—trends that were supported by the protein-protein interaction networks.

Our work supports many previously proposed mechanisms of Ag toxicity—disruption of iron-sulfur cluster containing proteins and certain cellular redox enzymes, and DNA damage, and Ag resistance—toxin export and sequestration. However, the data presented here also demonstrates that the activity of Ag within the bacterial cell is more extensive than previously suggested, involving genes a part of the cell wall structure, quinone metabolism, ATP synthesis and sulfur reduction.

The use of Ag as an antimicrobial is a practice garnering considerable popularity, as the introduction of Ag-based compounds, such as combination treatments, nanomaterials, and formulations make way. In order to continue the development of this metal as a therapeutic agent, it is imperative that we gather more understanding into the accompanying mechanisms of Ag toxicity and resistance. This study provides a vast number of biomolecular mechanistic hypotheses to the community investigating the mechanisms of action of Ag and other metals.

Supplementary Materials: The following are available online at http://www.mdpi.com/2073-4425/9/7/344/s1, Figure S1: Resistant and sensitive hits involved in regulation; Figure S2: Connectivity map displaying silver resistant hits; Figure S3: Connectivity map displaying silver sensitive hits; Figure S4: Silver resistant metabolic overview; Figure S5: Silver sensitive metabolic overview; Table S1: Normalized and scored gene list.

Author Contributions: Conceptualization, J.A.L., K.C.-R. and R.J.T.; Data curation, N.G., K.C.-R. and Y.Y.; Formal analysis, N.G., K.C.-R. and Y.Y.; Funding acquisition, J.A.L., G.C. and R.J.T.; Investigation, J.A.L. and K.C.-R.; Methodology, J.A.L. and K.C.-R.; Resources, G.C. and R.J.T.; Supervision, G.C.; Visualization, J.A.L. and K.C.-R.; Writing—original draft, N.G.; Writing—review & editing, N.G., J.A.L., K.C.-R., G.C. and R.J.T.

Funding: Natalie Gugala was funded by an Eyes High Doctoral Scholarship from the University of Calgary. Joe A. Lemire. was funded by the Banting Postdoctoral Fellowship Program, Alberta Innovates Health Solutions Postdoctoral Fellowship, and a MITACS Accelerate Fellowship. Raymond J. Turner was funded by a project bridge grant from the Canadian Institutes of Health Research (CIHR bridge: PJT-149009). Raymond J. Turner. and Gordon Chua were also funded by Discovery grants from the Natural Science and Engineering Research Council (NSERC DG: RGPIN/04811-2015) of Canada.

Acknowledgments: The authors of this paper would like to acknowledge Iain George and Connor Westersund for their assistance with the BM3 robot from S&P Robotics Inc. and experimental set-up, respectively.

Conflicts of Interest: The authors declare no conflict of interest.

Appendix

Table A1. Systems and comprising subsystems cited in this study. The Ag resistant and sensitive hits were surveyed against the EcoCyc database permitting the clustering of the hits into systems, subsystems, component subsystems, and lastly into individual objects [1,2].

Systems	Subsystems
Regulation	Signaling, sigma factor regulon, transcription factor, and transcription factor regulons
Response to Stimulus	Starvation, heat, cold, DNA damage, pH, detoxification, osmotic stress, and other
Cellular processes	Cell cycle and division, cell death, genetic transfer, biofilm formation, quorum sensing, adhesion, locomotion, viral response, response to bacterium, host interactions with host, other pathogenesis proteins
Energy	Glycolysis, the pentose phosphate pathway, the TCA cycle, fermentation, and aerobic and anaerobic respiration
Other pathways	Detoxification, inorganic nutrient metabolism, macromolecule modification, activation/inactivation/interconversion, and other enzymes
Degradation	Amino acids, nucleotide, amine, carbohydrate/carboxylate, secondary metabolite, alcohol, polymer and aromatic, the cell exterior, and regulation

Table A1. *Cont.*

Systems	Subsystems
Biosynthesis	Amino acids, nucleotides, fatty acid/lipid amines, carbohydrate/carboxylates, cofactors, secondary metabolites, and other pathways
Cell exterior	Transport, cell wall biogenesis and organization, lipopolysaccharide metabolism, pilus, flagellar, outer and inner membrane, periplasm, and cell wall components
Central Dogma	Transcription, translation, DNA metabolism, RNA metabolism, protein metabolism and protein folding and secretion

[1] Each individual score represents the mean of 12 trials—three biological and four technical; [2] Two-tailed *t*-test and significance was determined using the Benjamini-Hochberg procedure.

References

1. Turner, R.J. Metal-Based Antimicrobial Strategies. *Microb. Biotechnol.* **2017**, *10*, 1062–1065. [CrossRef] [PubMed]
2. Alexander, J.W. History of the Medical Use of Silver. *Surg. Infect.* **2009**, *10*, 289–292. [CrossRef] [PubMed]
3. Melaiye, A.; Youngs, W.J. Silver and Its Application as an Antimicrobial Agent. *Expert Opin. Ther. Pat.* **2005**, *15*, 125–130. [CrossRef]
4. Lemire, J.A.; Kalan, L.; Bradu, A.; Turner, R.J. Silver Oxynitrate, an Unexplored Silver Compound with Antimicrobial and Antibiofilm Activity. *Antimicrob. Agents Chemother.* **2015**, *59*, 4031–4039. [CrossRef] [PubMed]
5. Politano, A.D.; Campbell, K.T.; Rosenberger, L.H.; Sawyer, R.G. Use of Silver in the Prevention and Treatment of Infections: Silver Review. *Surg. Infect.* **2013**, *14*, 8–20. [CrossRef] [PubMed]
6. Gugala, N.; Lemire, J.A.; Turner, R.J. The Efficacy of Different Anti-Microbial Metals at Preventing the Formation of, and Eradicating Bacterial Biofilms of Pathogenic Indicator Strains. *J. Antibiot.* **2017**, *70*, 775–780. [CrossRef] [PubMed]
7. Aminov, R.I. A Brief History of the Antibiotic Era: Lessons Learned and Challenges for the Future. *Front. Microbiol.* **2010**, *1*, 1–7. [CrossRef] [PubMed]
8. Spellberg, B.; Guidos, R.; Gilbert, D.; Bradley, J.; Boucher, H.W.; Scheld, W.M.; Bartlett, J.G.; Edwards, J. The Epidemic of Antibiotic-Resistant Infections: A Call to Action for the Medical Community from the Infectious Diseases Society of America. *Clin. Infect. Dis.* **2008**, *46*, 155–164. [CrossRef] [PubMed]
9. French, G.L. The Continuing Crisis in Antibiotic Resistance. *Int. J. Antimicrob. Agents* **2010**, *36*, S3–S7. [CrossRef]
10. Neu, H.C. The Crisis in Antibiotic Resistance. *Science* **1992**, *257*, 1064–1073. [CrossRef] [PubMed]
11. Rigo, C.; Roman, M.; Munivrana, I.; Vindigni, V.; Azzena, B.; Barbante, C.; Cairns, W.R.L. Characterization and Evaluation of Silver Release from Four Different Dressings Used in Burns Care. *Burns* **2012**, *38*, 1131–1142. [CrossRef] [PubMed]
12. Sataev, M.S.; Koshkarbaeva, S.T.; Tleuova, A.B.; Perni, S.; Aidarova, S.B.; Prokopovich, P. Novel Process for Coating Textile Materials with Silver to Prepare Antimicrobial Fabrics. *Colloids Surf. A Physicochem. Eng. Asp.* **2014**, *442*, 146–151. [CrossRef]
13. George, N.; Faoagali, J.; Muller, M. Silvazine® (Silver Sulfadiazine and Chlorhexidine) Activity against 200 Clinical Isolates. *Burns* **1997**, *23*, 493–495. [CrossRef]
14. Guzman, M.; Dille, J.; Godet, S. Synthesis and Antibacterial Activity of Silver Nanoparticles against Gram-Positive and Gram-Negative Bacteria. *Nanomed. Nanotechnol. Biol. Med.* **2012**, *8*, 37–45. [CrossRef] [PubMed]
15. Paladini, F.; Pollini, M.; Talà, A.; Alifano, P.; Sannino, A. Efficacy of Silver Treated Catheters for Haemodialysis in Preventing Bacterial Adhesion. *J. Mater. Sci. Mater. Med.* **2012**, *23*, 1983–1990. [CrossRef] [PubMed]
16. Kumar, R.; Münstedt, H. Silver Ion Release from Antimicrobial Polyamide/silver Composites. *Biomaterials* **2005**, *26*, 2081–2088. [CrossRef] [PubMed]
17. Gordon, O.; Slenters, T.V.; Brunetto, P.S.; Villaruz, A.E.; Sturdevant, D.E.; Otto, M.; Landmann, R.; Fromm, K.M. Silver Coordination Polymers for Prevention of Implant Infection: Thiol Interaction, Impact on Respiratory Chain Enzymes, and Hydroxyl Radical Induction. *Antimicrob. Agents Chemother.* **2010**, *54*, 4208–4218. [CrossRef] [PubMed]

18. Morones-Ramirez, J.R.; Winkler, J.A.; Spina, C.S.; Collins, J.J. Silver Enhances Antibiotic Activity against Gram-Negative Bacteria. *Sci. Transl. Med.* **2013**, *5*, 1–11. [CrossRef] [PubMed]
19. Lemire, J.A.; Harrison, J.J.; Turner, R.J. Antimicrobial Activity of Metals: Mechanisms, Molecular Targets and Applications. *Nat. Rev. Microbiol.* **2013**, *11*, 371–384. [CrossRef] [PubMed]
20. Harrison, J.J.; Ceri, H.; Turner, R.J. Multimetal Resistance and Tolerance in Microbial Biofilms. *Nat. Rev. Microbiol.* **2007**, *5*, 928–938. [CrossRef] [PubMed]
21. Ivask, A.; Elbadawy, A.; Kaweeteerawat, C.; Boren, D.; Fischer, H.; Ji, Z.; Chang, C.H.; Liu, R.; Tolaymat, T.; Telesca, D.; et al. Toxicity Mechanisms in *Escherichia coli* Vary for Silver Nanoparticles and Differ from Ionic Silver. *ACS Nano* **2014**, *8*, 374–386. [CrossRef] [PubMed]
22. McQuillan, J.S.; Shaw, A.M. Differential Gene Regulation in the Ag Nanoparticle and Ag^+ Induced Silver Stress Response in *Escherichia coli*: A Full Transcriptomic Profile. *Nanotoxicology* **2014**, *8*, 177–184. [CrossRef] [PubMed]
23. Saulou-Bérion, C.; Gonzalez, I.; Enjalbert, B.; Audinot, J.N.; Fourquaux, I.; Jamme, F.; Cocaign-Bousquet, M.; Mercier-Bonin, M.; Girbal, L. *Escherichia coli* under Ionic Silver Stress: An Integrative Approach to Explore Transcriptional, Physiological and Biochemical Responses. *PLoS ONE* **2015**, *10*, 1–25. [CrossRef] [PubMed]
24. Baba, T.; Ara, T.; Hasegawa, M.; Takai, Y.; Okumura, Y.; Baba, M.; Datsenko, K.A.; Tomita, M.; Wanner, B.L.; Mori, H. Construction of *Escherichia coli* K-12 in-Frame, Single-Gene Knockout Mutants: The Keio Collection. *Mol. Syst. Biol.* **2006**, *2*. [CrossRef] [PubMed]
25. Franke, S.; Grass, G.; Nies, D.H. The Product of the *ybdE* Gene of the *Escherichia coli* Chromosome Is Involved in Detoxification of Silver Ions. *Microbiology* **2001**, *147*, 965–972. [CrossRef] [PubMed]
26. Wagih, O.; Usaj, M.; Baryshnikova, A.; VanderSluis, B.; Kuzmin, E.; Costanzo, M.; Myers, C.L.; Andrews, B.J.; Boone, C.M.; Parts, L. SGAtools: One-Stop Analysis and Visualization of Array-Based Genetic Interaction Screens. *Nucleic Acids Res.* **2013**, *41*, 591–596. [CrossRef] [PubMed]
27. Keseler, I.M.; Mackie, A.; Santos-Zavaleta, A.; Billington, R.; Bonavides-Martínez, C.; Caspi, R.; Fulcher, C.; Gama-Castro, S.; Kothari, A.; Krummenacker, M.; et al. The EcoCyc Database: Reflecting New Knowledge about *Escherichia coli* K-12. *Nucleic Acids Res.* **2017**, *45*, D543–D550. [CrossRef] [PubMed]
28. Huang, D.W.; Sherman, B.T.; Lempicki, R.A. Systematic and Integrative Analysis of Large Gene Lists Using DAVID Bioinformatics Resources. *Nat. Protoc.* **2009**, *4*, 44–57. [CrossRef] [PubMed]
29. Huang, D.W.; Sherman, B.T.; Lempicki, R.A. Bioinformatics Enrichment Tools: Paths toward the Comprehensive Functional Analysis of Large Gene Lists. *Nucleic Acids Res.* **2009**, *37*, 1–13. [CrossRef] [PubMed]
30. Szklarczyk, D.; Morris, J.H.; Cook, H.; Kuhn, M.; Wyder, S.; Simonovic, M.; Santos, A.; Doncheva, N.T.; Roth, A.; Bork, P.; et al. The STRING Database in 2017: Quality-Controlled Protein-Protein Association Networks, Made Broadly Accessible. *Nucleic Acids Res.* **2017**, *45*, D362–D368. [CrossRef] [PubMed]
31. Boël, G.; Smith, P.C.; Ning, W.; Englander, M.T.; Chen, B.; Hashem, Y.; Testa, A.J.; Fischer, J.J.; Wieden, H.J.; Frank, J.; et al. The ABC-F Protein EttA Gates Ribosome Entry into the Translation Elongation Cycle. *Nat. Struct. Mol. Biol.* **2014**, *21*, 143–151. [CrossRef] [PubMed]
32. Xiu, Z.; Liu, Y.; Mathieu, J.; Wang, J.; Zhu, D.; Alvarez, P.J. Elucidating the Genetic Basis for *Escherichia coli* Defense against Silver Toxicity Using Mutant Arrays. *Environ. Toxicol. Chem.* **2014**, *33*, 993–997. [CrossRef] [PubMed]
33. Chung, C.H.; Goldberg, A.L. Purification and Characterization of Protease So, a Cytoplasmic Serine Protease in *Escherichia coli*. *J. Bacteriol.* **1983**, *170*, 921–926.
34. Sondi, I.; Salopek-Sondi, B. Silver Nanoparticles as Antimicrobial Agent: A Case Study on *E. coli* as a Model for Gram-Negative Bacteria. *J. Colloid Interface Sci.* **2004**, *275*, 177–182. [CrossRef] [PubMed]
35. Jung, W.K.; Koo, H.C.; Kim, K.W.; Shin, S.; Kim, S.H.; Park, Y.H. Antibacterial Activity and Mechanism of Action of the Silver Ion in *Staphylococcus aureus* and *Escherichia coli*. *Appl. Environ. Microbiol.* **2008**, *74*, 2171–2178. [CrossRef] [PubMed]
36. Radzig, M.A.; Nadtochenko, V.A.; Koksharova, O.A.; Kiwi, J.; Lipasova, V.A.; Khmel, I.A. Antibacterial Effects of Silver Nanoparticles on Gram-Negative Bacteria: Influence on the Growth and Biofilms Formation, Mechanisms of Action. *Colloids Surf. B Biointerfaces* **2013**, *102*, 300–306. [CrossRef] [PubMed]
37. Beard, S.J.; Hashim, R.; Wu, G.; Binet, M.R.B.; Hughes, M.N.; Poole, R.K. Evidence for the Transport of Zinc(II) Ions via the Pit Inorganic Phosphate Transport System in *Escherichia coli*. *FEMS Microbiol. Lett.* **2000**, *184*, 231–235. [CrossRef] [PubMed]

38. Skare, J.T.; Postle, K. Evidence for a TonB-Dependent Energy Transduction Complex in *Escherichia coli*. *Mol. Microbiol.* **1991**, *5*, 2883–2890. [CrossRef] [PubMed]
39. Utsumi, R.; Horie, T.; Katoh, A.; Kaino, Y.; Tanabe, H.; Noda, M. Isolation and Characterization of the Heat-Responsive Genes in *Escherichia coli*. *Biosci. Biotechnol. Biochem.* **1996**, *60*, 309–315. [CrossRef] [PubMed]
40. Parr, R.G.; Pearson, R.G. Absolute Hardness: Companion Parameter to Absolute Electronegativity. *J. Am. Chem. Soc.* **1983**, *105*, 7512–7516. [CrossRef]
41. Bhatia, R.; Calvo, K.C. The Sequencing, Expression, Purification, and Steady-State Kinetic Analysis of Quinolinate Phosphoribosyl Transferase from *Escherichia coli*. *Arch. Biochem. Biophys.* **1996**, *325*, 270–278. [CrossRef] [PubMed]
42. Cicchillo, R.M.; Tu, L.; Stromberg, J.A.; Hoffart, L.M.; Krebs, C.; Booker, S.J. *Escherichia coli* Quinolinate Synthetase Does Indeed Harbor a [4Fe-4S] Cluster. *J. Am. Chem. Soc.* **2015**, *127*, 7310–7311. [CrossRef]
43. Xu, F.F.; Imlay, J.A. Silver(I), Mercury(II), Cadmium(II), and Zinc(II) Target Exposed Enzymic Iron-Sulfur Clusters When They Toxify *Escherichia coli*. *Appl. Environ. Microbiol.* **2012**, *78*, 3614–3621. [CrossRef] [PubMed]
44. Korshunov, S.; Imlay, J.A. Two Sources of Endogenous H_2O_2 in *Escherichia coli*. *Mol. Microbiol.* **2011**, *75*, 1389–1401. [CrossRef] [PubMed]
45. Templin, M.F.; Ursinus, A.; Höltje, J.V. A Defect in Cell Wall Recycling Triggers Autolysis during the Stationary Growth Phase of *Escherichia coli*. *EMBO J.* **1999**, *18*, 4108–4117. [CrossRef] [PubMed]
46. Goodell, E.W.; Schwarz, U. Release of Cell Wall Peptides into Culture Medium by Exponentially Growing *Escherichia coli*. *J. Bacteriol.* **1985**, *162*, 391–397. [PubMed]
47. Raghunath, A.; Perumal, E. Metal Oxide Nanoparticles as Antimicrobial Agents: A Promise for the Future. *Int. J. Antimicrob. Agents* **2017**, *49*, 137–152. [CrossRef] [PubMed]
48. Berendt, U.; Haverkamp, T.; Prior, A.; Schwenn, J.D. Reaction Mechanism of Thioredoxin: 3′-Phospho-adenylylsulfate Reductase Investigated by Site-Directed Mutagenesis. *Eur. J. Biochem.* **1995**, *233*, 347–356. [CrossRef] [PubMed]
49. White, D.C.; Frerman, F.E. Extraction, Characterization, and Cellular Localization of the Lipids of *Staphylococcus aureus*. *J. Bacteriol.* **1967**, *94*, 1854–1867. [PubMed]
50. Siegel, L.M.; Rueger, D.C.; Barber, M.J.; Krueger, R.J.; Orme-Johnson, N.R.; Orme-Johnson, W.H. *Escherichia coli* Sulfite Reductase Hemoprotein Subunit. *J. Biol. Chem.* **1982**, *257*, 6348–6350.
51. Stewart, P.S.; William Costerton, J. Antibiotic Resistance of Bacteria in Biofilms. *Lancet* **2001**, *358*, 135–138. [CrossRef]
52. Harrison, J.J.; Turner, R.J.; Ceri, H. High-Throughput Metal Susceptibility Testing of Microbial Biofilms. *BMC Microbiol.* **2005**, *5*, 1–11. [CrossRef] [PubMed]
53. Harrison, J.J.; Turner, R.J.; Ceri, H. Persister Cells, the Biofilm Matrix and Tolerance to Metal Cations in Biofilm and Planktonic *Pseudomonas aeruginosa*. *Environ. Microbiol.* **2005**, *7*, 981–994. [CrossRef] [PubMed]
54. Shafreen, R.B.; Seema, S.; Ahamed, A.P.; Thajuddin, N.; Ali Alharbi, S. Inhibitory Effect of Biosynthesized Silver Nanoparticles from Extract of Nitzschia Palea Against Curli-Mediated Biofilm of *Escherichia coli*. *Appl. Biochem. Biotechnol.* **2017**, *183*, 1351–1361. [CrossRef] [PubMed]
55. Rice, D.W.; Rafferty, J.B.; Artymiuk, P.J.; Lloyd, R.G. Insights into the Mechanisms of Homologous Recombination from the Structure of RuvA. *Curr. Opin. Struct. Biol.* **1997**, *7*, 798–803. [CrossRef]
56. Linley, E.; Denyer, S.P.; McDonnell, G.; Simons, C.; Maillard, J.Y. Use of Hydrogen Peroxide as a Biocide: New Consideration of Its Mechanisms of Biocidal Action. *J. Antimicrob. Chemother.* **2012**, *67*, 1589–1596. [CrossRef] [PubMed]
57. Su, C.; Yang, F.; Long, F.; Reyon, D.; Routh, M.D.; W, D.; Mokhtari, A.K.; Ornam, J.D. Van; Rabe, K.L.; Hoy, J.A.; et al. Crystal Structure of the Membrane Fusion Protein CusB from *Escherichia coli*. *J. Mol. Biol.* **2009**, *393*, 342–355. [CrossRef] [PubMed]
58. Fralick, A. Evidence That TolC Is Required for Functioning of the Mar/AcrAB Efflux Pump of *Escherichia coli*. *Am. Soc. Microbiol.* **1996**, *178*, 5803–5805. [CrossRef]
59. Holt, K.B.; Bard, A.J. Interaction of Silver(I) Ions with the Respiratory Chain of *Escherichia coli*: An Electrochemical and Scanning Electrochemical Microscopy Study of the Antimicrobial Mechanism of Micromolar Ag. *Biochemistry* **2005**, *44*, 13214–13223. [CrossRef] [PubMed]
60. DiRienzo, J.M.; Nakamura, K.; Inouye, M. The Outer Membrane Porteins of Gram-Negative Bacteria: Biosynthesis, Assembly, and Functions. *Proteins* **1978**, *47*, 481–532. [CrossRef]

61. Helbig, K.; Bleuel, C.; Krauss, G.J.; Nies, D.H. Glutathione and Transition-Metal Homeostasis in *Escherichia coli*. *J. Bacteriol.* **2008**, *190*, 5431–5438. [CrossRef] [PubMed]
62. Pomposiello, P.J.; Koutsolioutsou, A.; Carrasco, D.; Demple, B. SoxRS-Regulated Expression and Genetic Analysis of the *yggX* Gene of *Escherichia coli*. *Society* **2003**, *185*, 6624–6632. [CrossRef]
63. Link, A.J.; Robison, K.; Church, G.M. Comparing the Predicted and Observed Properties of Proteins Encoded in the Genome of *Escherichia coli* K-12. *Electrophoresis* **1997**, *18*, 1259–1313. [CrossRef] [PubMed]
64. Ioannidis, N.; Cooper, C.E.; Poole, R.K. Spectroscopic Studies on an Oxygen-Binding Haemoglobin-like Flavohaemoprotein from *Escherichia Coli*. *Biochem. J.* **1992**, *288*, 649–655. [CrossRef] [PubMed]
65. Percival, S.L.; Hill, K.E.; Williams, D.W.; Hooper, S.J.; Thomas, D.W.; Costerton, J.W. A Review of the Scientific Evidence for Biofilms in Wounds. *Wound Repair Regen.* **2012**, *20*, 647–657. [CrossRef] [PubMed]

© 2018 by the authors. Licensee MDPI, Basel, Switzerland. This article is an open access article distributed under the terms and conditions of the Creative Commons Attribution (CC BY) license (http://creativecommons.org/licenses/by/4.0/).

Article

Using a Chemical Genetic Screen to Enhance Our Understanding of the Antimicrobial Properties of Gallium against *Escherichia coli*

Natalie Gugala [1], Kate Chatfield-Reed [2], Raymond J. Turner [1] and Gordon Chua [1,*]

[1] Department of Biological Sciences, University of Calgary, 2500 University Dr. NW, Calgary, AB T2N 1N4, Canada; ngugala@ucalgary.ca (N.G.); turnerr@ucalgary.ca (R.J.T.)
[2] Seidman Cancer Center, University Hospitals, 11100 Euclid Ave, Cleveland, OH 44106, USA; katherine.chatfield-reed@UHhospitals.org
[*] Correspondence: gchua@ucalgary.ca; Tel.: +1-403-220-7769

Received: 7 November 2018; Accepted: 4 January 2019; Published: 9 January 2019

Abstract: The diagnostic and therapeutic agent gallium offers multiple clinical and commercial uses including the treatment of cancer and the localization of tumors, among others. Further, this metal has been proven to be an effective antimicrobial agent against a number of microbes. Despite the latter, the fundamental mechanisms of gallium action have yet to be fully identified and understood. To further the development of this antimicrobial, it is imperative that we understand the mechanisms by which gallium interacts with cells. As a result, we screened the *Escherichia coli* Keio mutant collection as a means of identifying the genes that are implicated in prolonged gallium toxicity or resistance and mapped their biological processes to their respective cellular system. We discovered that the deletion of genes functioning in response to oxidative stress, DNA or iron–sulfur cluster repair, and nucleotide biosynthesis were sensitive to gallium, while Ga resistance comprised of genes involved in iron/siderophore import, amino acid biosynthesis and cell envelope maintenance. Altogether, our explanations of these findings offer further insight into the mechanisms of gallium toxicity and resistance in *E. coli*.

Keywords: *Escherichia coli*; gallium; antimicrobial agents; metal toxicity; metal resistance; metal-based antimicrobials

---

1. Introduction

The therapeutic capabilities of gallium(III) (Ga) have been and continue to be exploited for a number of clinical applications, which include: the treatment of cancer, autoimmune and infectious diseases, for the localization of tumors, inflammation and infection sites, and the reduction of accelerated bone resorption [1,2]. At the nuclear level, certain characteristics of this abiogenic metal permit essential metal mimicry, owing its similarities to Fe. In particular, the pharmacological characteristics of Ga are likely a result of its Fe(III)-like coordination chemistry and its ability to form stable six-coordinated complexes through ionic bonding [3]. This metal is trivalent and a hard acid in solution, according to the hard-soft acid-base theory [4], binding well with strong Lewis bases. As a result, Ga tends to form bonds with oxygen predominantly forming $Ga(OH)_{4-}$ (gallate) at pH 7.4 [5].

Despite Ga's similarities to the essential metal Fe, these metals share two main differences: (i) Ga cannot be reduced under biologically relevant reduction potentials, whereas Fe can be readily changed to and from a reduced state; and (ii) the concentration of unbound Fe(III) in solution is extremely low, localized primarily as a neutral complex with organic compounds, whereas gallate, which is anionic, can exist at significant concentrations [6].

As an Fe(III) mimetic, Ga(III) can incorporate itself into proteins and enzymes replacing Fe and effectively halting several essential metabolic processes [7–14]. Since the bioavailability of Fe is scarce, organisms, such as bacteria, have produced a variety of biomolecular chelating scavenging systems including siderophores and Fe-chelating proteins. Cells rapidly multiplying are more susceptible to Ga toxicity due to their high Fe demands [0]. As a result, this metal is approved by the US Food and Drug Administration for the treatment of cancer-associated hypercalcemia (Ganite®, Genta, NJ, USA) and has been tested as an antimicrobial agent against a variety of organisms including *Mycobacterium tuberculosis* [15,16], *Pseudomonas aeruginosa* [9,10,17], *Staphylococcus aureus* [18], *Rhodococcus equi* [19], *Acinetobacter baumannii* [20], and *Escherichia coli* [21].

In general, proposed mechanisms of toxicity for metal-based antimicrobials include the production and propagation of reactive oxygen species (ROS), the disruption of Fe-sulfur centers, thiol coordination, the exchange of a catalytic or structural metal, which in turn may lead to protein dysfunction, obstructed nutrient uptake, and genotoxicity [22]. The route by which Ga enters the cells is unknown, although, it is predominantly assumed that this metal crosses the cytoplasmic membrane by exploiting Fe-uptake routes, such as siderophores [23]. Several studies have explored the use of Fe-chelators as "Trojan horses" as a means of improving the delivery and toxicity of this metal in bacterial cells [14]. Still, there is insufficient research demonstrating that complexes of Ga and Fe-chelators/siderophores, such as Ga-citrate, increase the antibacterial abilities of this metal mainly since the import of this metal is not suggested to be the limiting step [23]. Furthermore, Ga exposure has been demonstrated to trigger the production of ROS in vitro [7,8]. Upon the cytoplasmic replacement of Fe with Ga, the available Fe pool is thought to increase, in turn fostering Fenton chemistry [22].

Bacteria have developed mechanisms of resistance as a means of withstanding metal toxicity. Some mechanisms include extracellular and intracellular sequestration, efflux, reduced uptake, repair, metabolic by-pass, and chemical modification [24]. Microbial resistant mechanisms associated with Ga have been studied to a far lesser degree, nonetheless, studies have shown that Ga is not as effective as postulated. For example, Ga resistance in *P. aeruginosa* and *Burkholderia cepacia* has been identified, suggested to be the result of decreased Ga import and the formation of bacterial biofilms [25,26].

Currently, research in this field is directed toward discovering novel utilities for this metal, still, the expansion of Ga as a therapeutic antimicrobial has been delayed compared to other metal-based antimicrobials, such as silver and copper. In short, it is essential that the mechanisms of Ga action in microbes are explored to greater degree in order to further the development of this antimicrobial agent.

In this work, we hypothesized that Ga exerts toxicity on multiple targets. Furthermore, we believe that there are several mechanisms of resistance that are fundamental to an organism's adaptive response under sub-lethal concentrations of Ga. To evaluate this, we performed a genotypic screening workflow of an *E. coli* mutant library composed of 3985 strains. Each strain contains a different inactivated non-essential gene. Genome-wide toxin/stressor-challenge workflows have been used to study silver [27–30], copper [31,32], cadmium [33], cobalt [33], and zinc [34]; however, no such study has been implemented to examine the effects of Ga. Therefore, as a means of complementing existing work, we have identified a number of genes that may be involved in Ga toxicity or resistance and mapped their biological processes to their respective cellular system in *E. coli*.

2. Materials and Methods

All methods are as described previously by Gugala et al. [30] and all chemicals were obtained from VWR International, Mississauga, Canada, unless otherwise stated.

2.1. Escherichia coli Strains

The Keio collection [35] consisting of 3985 single gene *Escherichia coli* BW25113 mutants (*lacI*^{q} *rrnB*$_{T14}$ Δ*lacZ*$_{WJ19}$ *hsdR*514 Δ*ara*BAD$_{AH33}$ Δ*rha*BAD$_{LD78}$), was obtained from the National BioResource Project *E. coli* (National Institute of Genetics, Shizuoka, Japan).

2.2. Determination of the Minimal Inhibitory Concentration and Controls

The sublethal inhibitory concentration, a concentration below the minimal inhibitory concentration that is found to visibly challenge selected mutants under prolonged metal exposure, was determined using Δ*recA*, Δ*lacA* and Δ*lacY* strains from the Keio collection. The protein RecA is involved in a number of processes, including homologues recombination and the induction of the SOS response in reaction to DNA damage [36]. Evidence may suggest that Ga causes the formation of ROS, although the precise mechanism of production is unknown. As a result, the absence of this gene was anticipated to confer the Ga sensitive phenotype, implied by a decrease in colony formation, since it is thought to be involved in mitigating ROS stress. Further, the protein products of *lacA* and *lacY* were not anticipated to be involved in Ga resistance or toxicity, therefore mutant strains of these genes were used as negative controls. Strains Δ*recA*, Δ*lacA*, and Δ*lacY*, and the wild-type (WT) were grown overnight at 37 °C on M9 minimal media plates (6.8 g/L Na_2HPO_4, 3.0 g/L KH_2PO_4, 1.0 g/L NH_4Cl, 0.5 g/L NaCl, 4.0 mg/L glucose, 0.5 mg/L $MgSO_4$ and 0.1 mg/L $CaCl_2$) containing Noble agar (1.0%) in the presence and absence of Ga at varying concentrations. The concentration of Ga that visibly decreased colony formation in the *recA* mutant and produced no growth changes in the negative control strains was selected as the sublethal inhibitory concentration. Furthermore, Δ*recA*, Δ*lacA*, and Δ*lacY* and the WT strain were grown overnight in the presence of ionic nitrate at the equivalent molarity as the sublethal inhibitory concentration to ensure growth was not influenced by the accompanying counter ion. In order to identify Ga-sensitive and -resistant genes in this study, the Keio collection was exposed to 100 μM $Ga(NO_3)_3$ (Ga). Gallium nitrate was obtained from Sigma–Aldrich, St. Louis, MO, USA. Stock solutions of Ga were prepared with deionized H_2O and stored in glass vials for no longer than two weeks.

Similarly, Δ*recA*, Δ*lacA*, and Δ*lacY* and the WT strain were grown on M9 minimal media plates in the presence of varying concentrations of hydroxyurea (HU), obtained from USBiological Salmen, MA, USA, or sulfometuron methyl (SMM) obtained from Chem Service, West Chester, PA, USA, dissolved in ddH_2O and dimethyl sulfoxide, respectively. Select mutants from the Keio collection were exposed to a final concentration of 5.0 mg/mL HU and 5.0 μg/mL SMM in the presence and absence of 100 μM $Ga(NO_3)_3$.

2.3. Screening

M9 minimal media and Noble agar (1.0%) plates, with and without the addition of Ga, were prepared two days prior to use. Here, Ga was added directly to the liquid agar and swirled before solidification. Colony arrays in 96-format were produced and processed using a BM3 robot and spImager (S&P Robotics Inc., Toronto, ON, Canada), respectively. Cells were transferred from the arrayed microtiter plates using a 96-pin replicator onto Luria-Bertani (LB) media agar plates and grown overnight at 37 °C. Colonies were then transferred using the replicator onto two sets of M9 minimal media Noble agar plates, with and without 100 μM $Ga(NO_3)_3$. Plates were then grown overnight at 37 °C. All images were acquired using the spImager and colony size, a measure of Ga sensitivity or resistance, was determined using integrated image processing software. Three biological trials were conducted and each of these trials included four technical replicates originating from the 96-colony array, which were combined and expanded onto a single plate in 384-colony array format; n (trials) ≥ 9. Strains presenting less than nine replicates were excluded (see Section 2.5).

Select mutants were exposed to HU or SMM at sublethal inhibitory concentrations. Identical conditions were maintained to enable direct comparisons between mutants grown in the presence of Ga only, and those grown in the presence of Ga and either HU or SMM. Here HU or SMM were added to the M9 minimal media plates directly before solidification.

2.4. Normalization

In this study, incubation time and temperature, nutrient availability, colony location, agar plate imperfections, batch effects, and neighboring mutant fitness were considered independent variables that could influence colony size and subsequently cause systematic variation. As a result, the colonies were normalized and scored using Synthetic Genetic Array Tools 1.0 (SGATools) [37,38], a tool that associates mutant colony size with fitness, thereby enabling quantitative comparisons. All the plates were normalized to establish average colony size, working on the assumption that the majority of the colonies would exhibit WT fitness since the concentration of Ga used in this study was below the minimal inhibitory concentration.

Mutant colony sizes in the presence (challenge) and absence (control) of Ga were quantified, scored, and compared as deviation from the expected fitness of the WT strain. This assumes a multiplicative model and not an additive effect originating from the challenge. Once scored, mutants displaying a reduction in colony size were indicative of a Ga sensitive hit and those displaying an increase in colony size were recovered as Ga resistant hits. Finally, the *p*-value was calculated as a two-tailed *t*-test and significance was determined using the Benjamini–Hochberg procedure, as a method of lowering the false discovery rate, which was selected to be 10%.

2.5. Data Mining and Analyses

Data mining was performed using Pathway Tools Omics Dashboard, which surveys against the EcoCyc database [39] and Uniport [40]. This allowed for the clustering of the Ga resistant and sensitive data sets into systems, subsystems, and individual objects (Table A1). Here, genes can be found in multiple systems since many are involved in a number of cellular processes.

Enrichment analyses were performed using the DAVID Bioinformatics Resource 6.8 [41,42]. Moreover, as a means of revealing the direct (physical) and indirect (functional) protein interactions amongst the gene hits, the STRING database [43] was utilized. Node maps based on experimental, co-expression, and gene fusion studies were generated using the Ga resistant and sensitive hits found in our screen.

3. Results and Discussion

3.1. Genome-Wide Screen of Ga Resistant and Sensitive Hits

In this work, the chemical genetic screen provided a method for the identification of the non-essential genes that may be involved in Ga resistance or sensitivity. A total of 3985 non-essential genes were screened for growth in the presence of 100 µM Ga(NO$_3$)$_3$ and from here, 3641 hits, in which $n \geq 9$, were used for subsequent statistical analyses (Figure 1 and Supplementary Table S1). The statistical cutoff that suggested a significant difference in fitness when compared to the WT, indicated by a change in colony size, was selected to be two standard deviations from the mean or a normalized score of +0.162 and −0.154. This resulted in 107 gene hits, which represents approximately 2.5% of the open reading frames in the *E. coli* K-12 genome. In general, the normalization was performed with the assumption that hits presenting scores within two standard deviations from the mean had non-specific or neutral interactions with Ga. Therefore, the remaining hits were not regarded as significant based exclusively on the cutoffs selected.

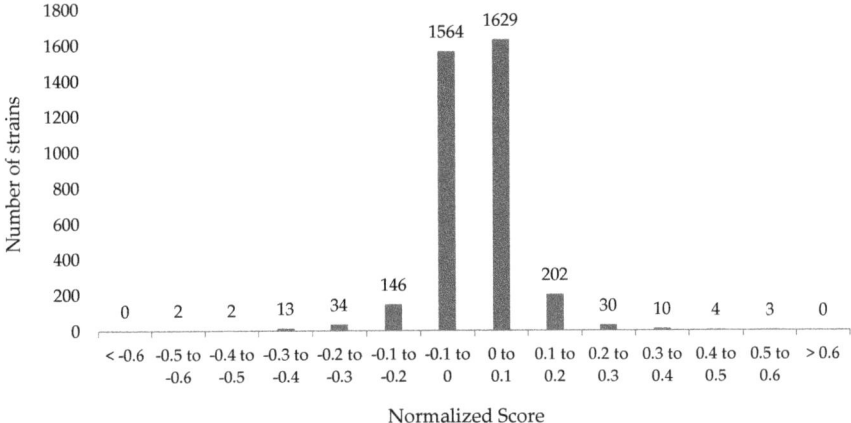

Figure 1. Synthetic Array Tools (version 1.0) was used to normalize and score the Gallium(III) (Ga) resistant and sensitive hits as a means of representing the growth differences in *Escherichia coli* K12 BW25113 in the presence of 100 μM Ga(NO$_3$)$_3$. Each individual score represents the mean of 9–12 trials.

In this work, the absence of the gene was inferred to give rise to the Ga resistant or -sensitive phenotype. A decrease in colony size (normalized score < −0.154) signified a Ga-sensitive hit, which implied that the presence of this gene increased Ga resistance. Here, 58 genes were found to cause Ga sensitivity when absent (Table 1). Likewise, an increase in colony size (normalized score > 0.162) signified a Ga-resistant hit, therefore the presence of this gene may suggest an increase in toxicity. Comparably, 49 genes were found to impart resistance when absent (Table 2), within the cutoffs applied.

Table 1. Ga sensitive hits organized according to system and subsystem mined using the Omics Dashboard (Pathway Tools), which surveys against the EcoCyc Database; genes represent sensitive hits with scores < −0.154.

System	Subsystem	Gene [1]	Score [2,3]
Central dogma	Transcription	evgA	−0.166
		hns	−0.175
		lgoR	−0.401
		nagC	−0.191
		rseA	−0.26
		ulaR	−0.556
	Translation	bipA	−0.204
	DNA metabolism	holC	−0.327
		holD	−0.217
		ruvC	−0.184
		intR	−0.27
		recA	−0.309
		recD	−0.199
	RNA metabolism	rbfA	−0.35
		rim	−0.298
		mnmA	−0.212
		rnt	−0.322
		ygfZ	−0.373
		evgA	−0.166
		hns	−0.175
		lgoR	−0.401
		nagC	−0.191
		rseA	−0.269
		sspA	−0.214
		ulaR	−0.556

Table 1. Cont.

System	Subsystem	Gene [1]	Score [2,3]
	Protein metabolism	lipA	−0.318
		pphA	−0.198
		slyD	−0.273
	Protein folding and secretion	slyD	−0.273
	Transport	zunC	−0.361
		tolC	−0.539
		ugpC	−0.29
	Pilus	ybgO	−0.163
	Flagellum	fliG	−0.235
	Outer membrane	tolC	−0.539
Cell exterior	Plasma membrane	clsA	−0.171
		cysQ	−0.203
		fdnI	−0.251
		fliG	−0.235
		gspA	−0.199
		hokA	−0.181
		nuoK	−0.247
		rseA	−0.269
		ubiG	−0.265
		ugpC	−0.29
		znuC	−0.361
	Periplasm	tolC	−0.539
		yebF	−0.268
Biosynthesis	Amino acid	dmlI	−0.418
		metL	−0.189
		mtn	−0.329
	Nucleoside and nucleotide	purT	−0.216
	Fatty acid/lipid	clsA	−0.171
	Carbohydrate	mdh	−0.287
	Secondary metabolites	mtn	−0.329
		fdx	−0.168
		fdx	−0.168
		gshA	−0.165
	Cofactor	lipA	−0.318
		pabA	−0.224
		pabC	−0.258
		ubiG	−0.265
	Other	metL	−0.189
Degradation	Amino acid	astD	−0.301
	Nucleoside and nucleotide	mtn	−0.329
	Amine	purT	−0.216
	Carbohydrate	garK	−0.173
		dmlA	−0.418
Energy	Glycolysis	gpmA	−0.175
	Tricarboxylic acid cycle	mdh	−0.287
	Fermentation	mdh	−0.287
	Aerobic respiration	nuoK	−0.247
	Anaerobic respiration	fdnI	−0.251
		nuoK	−0.247
	Other	mdh	−0.287
		nuoK	−0.247

Table 1. Cont.

System	Subsystem	Gene [1]	Score [2,3]
Cellular processes	Biofilm	hns	−0.175
	Adhesion	ybgO	−0.163
	Locomotion	fliG	−0.235
		recA	−0.309
	Viral response	intR	−0.27
	Host interaction	intR	−0.27
		slyD	−0.273
	Symbiosis	slyD	−0.273
Response to stimulus	Starvation	sspA	−0.29
		ugpC	−0.214
	Heat	bipA	−0.204
		gloB	−0.297
		slyD	−0.273
	Cold	bipA	−0.204
		rbfA	−0.35
	DNA damage	rbfA	−0.35
		recA	−0.39
		recD	−0.199
		ruvC	−0.184
	Osmotic stress	gshA	−0.165
		ubiG	−0.265
	Other	evgA	−0.166
		fliG	−0.235
		grxD	−0.266
		holC	−0.327
		holD	−0.217
		pphA	−0.198
		rseA	−0.269
		sspA	−0.214
		tolC	−0.539
		ugpC	−0.29
Other pathways	Inorganic nutrient metabolism	fdnI	−0.251
		nuoK	−0.247
	Detoxification	gloB	−0.297
		grxD	−0.266
	Macromolecule modification	mnmA	−0.212
		rnt	−0.322
	Other enzymes	bfr	−0.17
		cysQ	−0.203
		pphA	−0.198
		recD	−0.199
		ruvC	−0.184
		slyD	−0.273

[1] Gene hits can be mapped to more than one system and subsystem. [2] Each individual score represents the mean of 9–12 trials. [3] Two-tailed t-test and significance was determined using the Benjamini–Hochberg procedure; false discovery rate 10%.

Table 2. Ga-resistant hits organized according to system and subsystem mined using the Omics Dashboard (Pathway Tools), which surveys against the EcoCyc Database; genes represent resistant hits with scores >0.162.

System	Subsystem	Gene [1]	Score [2,3]
Central dogma	Transcription	ilvY	0.215
		metR	0.372
		odhR	0.353
	DNA metabolism	hofM	0.62
		xerD	0.168
		cas2	0.177
	RNA metabolism	symE	0.177
		ilvY	0.215
		metR	0.372
		pdhR	0.353
	Protein metabolism	mrcB	0.249
	Protein folding and secretion	yraI	0.18
Cell exterior	Transport	cysU	0.362
		fepG	0.312
		tonB	0.341
		caiT	0.403
		yiaO	0.6
		par	0.266
	Cell wall biogenesis	alr	0.353
		evnC	0.203
		mrcB	0.249
		yraI	0.18
	Lipopolysaccharide metabolism	cspG	0.204
		rfaC	0.201
	Outer membrane	par	0.266
		pqiC	0.345
	Plasma membrane	atpE	0.172
		atpH	0.176
		caiT	0.403
		cycU	0.362
		envU	0.203
		fepG	0.312
		mrcB	0.249
		pqiC	0.345
		tonB	0.341
		torC	0.259
		rfaC	0.201
		yaaU	0.237
		yafU	0.214
		yifK	0.18
	Periplasm	ansB	0.204
		asr	0.247
		envC	0.203
		mrcB	0.249
		pqiC	0.345
		tolB	0.2
		tonB	0.341
		torC	0.259
		yiaO	0.6
		yral	0.18
	Cell wall component	mrcB	0.249
		torC	0.259

Table 2. Cont.

System	Subsystem	Gene [1]	Score [2,3]
Biosynthesis	Amino acid	alr	0.353
		avtA	0.384
		leuA	0.302
		leuC	0.205
		metA	0.241
		proB	0.258
		trpB	0.611
		trpD	0.273
	Fatty acid/lipid	rfaC	0.201
	Carbohydrate	cpsG	0.204
		rfaC	0.201
	Cofactor, prosthetic groups, electron carrier	bioF	0.183
		bioH	0.194
		coaA	0.193
		thiE	0.226
	Cell structure	mrcB	0.249
	Other	aroF	0.236
Degradation	Amino acid	alr	0.353
		ansB	0.204
	Fatty acid/lipid	atoA	0.246
Energy	Glycolysis	pykF	0.169
	Fermentation	pykF	0.169
	Anaerobic respiration	torC	0.259
	Adenosine triphosphate biosynthesis	atpE	0.172
		atpH	0.176
	Other	hydN	0.249
Cellular processes	Cell cycle/division	envC	0.203
		tolB	0.2
		xerD	0.168
	Cell death	envC	0.203
	Adhesion	tonB	0.341
	Viral response	cas2	0.177
		tonB	0.341
	Symbiosis	tonB	0.341
Response to stimulus	Heat	pykF	0.169
	DNA damage	par	0.266
		symE	0.177
		yiaO	0.6
	pH	oxc	0.519
	Other	asr	0.247
		caiT	0.403
		cas2	0.177
		envC	0.203
		mrcB	0.249
		tolB	0.2
		tonB	0.341
		torC	0.259
		xerD	0.168
		yaaU	0.237
Other pathways	Other enzymes	oxc	0.519
		sepG	0.201

[1] Gene hits can be mapped to more than one system and subsystem. [2] Each individual score represents the mean of 9–12 trials. [3] Two-tailed t-test and significance was determined using the Benjamini–Hochberg procedure; false discovery rate 10%.

Using Pathway Tools, which surveys against the EcoCyc database, a number of gene hits were mapped to more than one system and subsystem (Tables 1 and 2). In general, comparable number of hits were mapped to the system "Response to stimulus", "Cellular processes", "Energy", and "Biosynthesis" (Figure 2). Still, "Regulation", "Degradation", and proteins of the "Cell exterior" contained more resistant hits. Whereas "Other pathways" and proteins involved in processes of the "Central dogma" were represented by the Ga-sensitive hits at least two-fold more than the Ga resistant hits (Figure 2). Proteins residing or involved in maintaining cell envelope homeostasis were not enriched in the resistant hits; however, two-fold more hits were mapped to the system "Cell exterior" using EcoCyc's system of classification when compared to the sensitive hits (Figure 2).

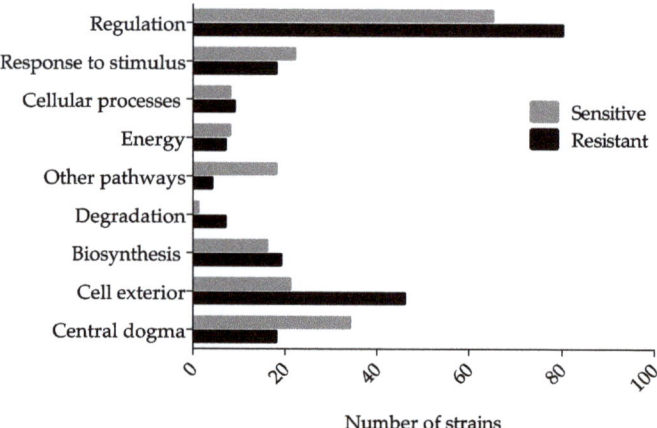

Figure 2. Ga-resistant and -sensitive gene hits mapped to component cellular processes. Several gene hits are mapped to more than one subsystem. The cutoff fitness score selected was two standard deviations from the mean and recovered gene hits with a score outside this range were chosen for further analyses. The hits were mined using the Omics Dashboard (Pathway Tools), which surveys against the EcoCyc database. Each individual score represents the mean of 9–12 trials.

Despite similar numbers of resistant and sensitive hits scored in this screen, a greater number of categories were enriched for by the resistant hits, such as the biosynthesis of the vital coenzyme—biotin—when surveyed using the DAVID gene functional classification (Figure 3).

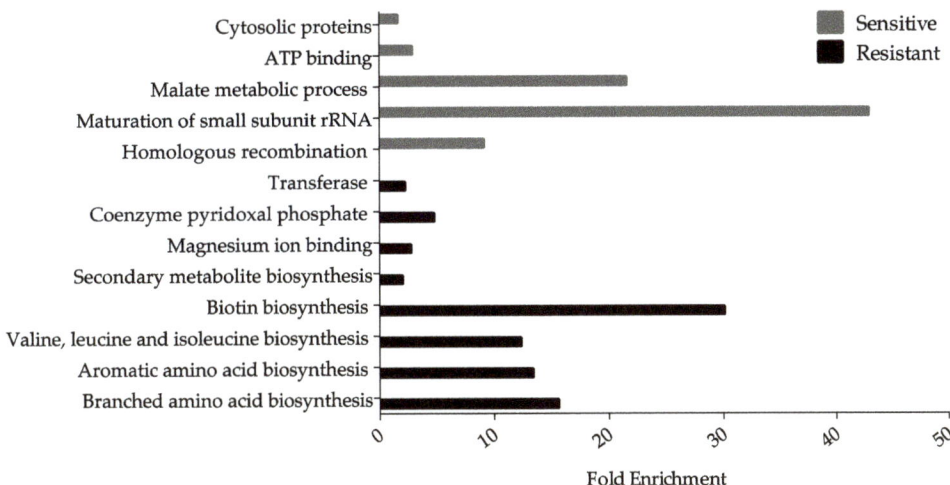

Figure 3. Functional enrichment among the Ga-resistant and -sensitive gene hits. The DAVID gene functional classification (version 6.8) database, a false discovery rate of 10% and a cutoff score two standard deviations from the mean was used to measure the magnitude of enrichment of the selected gene hits against the genome of *E. coli* K-12. Only processes with gene hits ≥3 were included.

In addition, a number of amino acid biosynthetic processes and cytosolic proteins were enriched in the resistant hits, whereas proteins involved in the processing of 20S pre-rRNA and malate metabolic processes were enriched in the sensitive hits (Figure 3). In general, the enrichment profile of the resistant and sensitive hits provides insight into the dissimilarities between the mechanisms of Ga toxicity and resistance since there was no overlap in enrichment (Figure 3). Based on previous reports [44,45] several mutants belonging to the Keio collection, such as those involved in the synthesis of amino acids, did not grow in M9 minimal media, contrary to what we observed in this study. We attribute this observation to the presence of residual resources, such as amino acids, that were carried over from the LB media agar plates onto to the M9 minimal media agar plates. Once these resources are exhausted, dying cells may provide a source of nutrients for surviving cells. Furthermore, previous studies have provided cutoff values as markers of growth, such as one-third the average OD [45]. Mutants displaying growth below the cutoff are regarded as non-growers despite possible survival. As a follow up, we grew a number of mutants overnight, including *leuC*, *metA*, *proA*, *ilvB*, *trpD*, *lacA*, and the WT strain in liquid M9 minimal media from existing culture stocks (see Section 2.1) and transferred 20 µL onto M9 minimal agar plates in the absence and presence of Ga. These strains were then grown overnight. Growth was only observed for the WT strain and the *lacA* mutant. When the same procedure was completed with liquid LB medium and agar plates, colony formation was evident for each mutant tested. As a result, in this study we were able to test mutants that have otherwise been reported to not grow on minimal media due to the lack of essential nutrients, such as amino acids.

3.2. Ga Sensitive Systems

3.2.1. Iron Homeostasis and Transport, and Fe–Sulfur Cluster Proteins

Gallium(III) has been shown to disrupt the function of several enzymes containing Fe–sulfur clusters, likely by competing for Fe-binding sites [7]. *Escherichia coli* contains over 10 Fe-acquisition systems, encoded by over 35 genes [46], providing an abundance of Ga potential targets, such as the sensitive hit *fdx* (ferredoxin). The protein product of *fdx* serves as an electron transfer protein in a wide variety of metabolic reactions, including the assembly of Fe–sulfur clusters [47], consequently, Ga resistance is probable if this metal is damaging Fe–sulfur centers. Ferredoxin may also serve as a binding site since the exchange of Fe may cause Ga sequestration. Furthermore, the sensitive hit *lipA* (lipoyl synthase) codes for an enzyme that uses ferredoxin as a reducing source, and catalytic Fe–sulfur clusters to produce lipoate [48]. This protein's requirement for ferredoxin may provide an explanation for the two-fold score decrease observed in the *lipA* mutant when compared to the *fdx* mutant. Furthermore, our screen recovered the hit *ygfZ*, which codes for a folate-binding protein that is implicated in protein assembly and the repair of Fe–sulfur clusters [49]. The loss of *ygfZ* results in sensitivity to oxidative stress likely due to the generation of ROS, subsequently, this may lead to the inhibition of Fe–sulfur cluster assembly or repair [50]. In addition, the disruption of Fe–sulfur clusters has been found to downregulate the uridine thiolation of particular tRNAs as a means of decreasing sulfur consumption [51]. This process appears to be important in coupling translation with levels of sulfur-containing amino acids. We recovered *trmU*, which encodes a tRNA thiouridylase as a Ga-sensitive hit in this study.

The redox pair Fe(II)/Fe(III) is well suited for a number of redox reactions and electron transfers. Accordingly, bacteria have developed a number of Fe-acquisition systems, such as siderophores and Fe-chelating proteins [52]. Siderophores, such as enterobactin are synthesized internally and exported extracellularly to scavenge Fe(III) from the environment [53]. The ferric-siderophore complex is imported into the cell and then degraded to release Fe(III) [53] and since Ga is an Fe mimetic [54], this metal has been demonstrated to bind certain siderophores [23]. The protein TolC is an outer membrane carrier required for the export of the high-affinity siderophore enterobactin from the periplasm to the external environment [55]. The Ga sensitivity of the Δ*tolC* strain may be due to the periplasmic accumulation of Ga-enterobactin complexes. If TolC is inactivated, then less enterobactin is exported outside the cell in turn providing more Ga targets, and as a result, Ga-enterobactin complexes may accumulate inside the cell. Further, EvgA is part of the EvgAS two-component system involved in the transcriptional regulation of *tolC* [56]. Loss of *evgA* is expected to display a similar defect in enterobactin export as would a *tolC* mutant, thus resulting in Ga sensitivity. Finally, *bfr* (bacterioferritin) was recovered as a sensitive hit in this work. This protein, which binds one heme group per dimer and two Fe atoms per subunit, functions in Fe storage and oxidation [57]. The sensitivity phenotype of the Δ*bfr* strain may be associated with a failure to mitigate Fe-mediated ROS production due to the disruption of Fe homeostasis in the presence of Ga (see Section 3.2.2).

3.2.2. Oxidative Stress

The production of ROS has been shown to be a mechanism of metal toxicity. Exposure to hydrogen peroxide or other agents that catalyze the production of ROS, such as superoxide, causes DNA and protein damage to macromolecules including proteins, lipids, nucleic acids and carbohydrates [58]. This in turn causes the upregulation of genes encoding ROS-scavenging enzymes [58]. An increase in cytoplasmic Fe intensifies ROS toxicity by catalyzing the exchange of electrons from donor to hydrogen peroxide [22]. Consequently, this may require the assistance of cellular antioxidants such as glutathione, and enzymes such as catalase, superoxide dismutase and peroxidase [59]. Gallium(III) is Fenton inactive, and therefore the induction of ROS in the presence of Ga is likely to result in the release of Fe in the cytoplasm. One study observed higher levels of oxidized lipids and proteins in Ga exposed *Pseudomonas fluorescens* [7]. In turn, the oxidative environment stimulated the synthesis of nicotinamide adenine dinucleotide phosphate (NADPH) via the overexpression of NADPH-producing enzymes, invoking a reductive environment.

In this screen, several sensitive Ga hits effective in ROS protection were recovered, including γ-Glutamate-cysteine ligase, or *gshA*. Strains lacking this gene have been shown to be hypersensitive to thiol-specific damage generated through mercury and arsenite exposure [60]. Similarly, strains lacking glyoxalase II (*gloB*), also a sensitive hit in this study, accumulate S-lactoylglutathione and demonstrate depleted glutathione pools [61]. If this antioxidant is depleted, then the potential for ROS-mediated protection is lowered. Furthermore, the gene *grxD*, which codes for a scaffold protein that transfers intact Fe–sulfur clusters to ferredoxin, was also recovered as a Ga-sensitive hit. The presence of this abundant protein is further upregulated during stationary phase [62] and one study demonstrated, using the Keio collection, that a *grxD* mutant is sensitive to Fe depletion [63]. Based on this observation, Ga exposure may prompt toxicity via Fe exhaustion, or the introduction of this toxin may result in ROS production thereby leading to Fe–sulfur damage. Finally, bacterioferritin (*bfr*) was also identified as a sensitive hit in this work. This protein acts to prevent the formation of hydrogen peroxide from the oxidation of Fe(II) atoms [57]. The sensitivity phenotypes of the Δ*gshA*, Δ*gloB*, Δ*grxD*, and Δ*bfr* strains may be associated with Fe-mediated ROS production upon the disruption of Fe-homeostasis in Ga exposed cells.

The sensitive hit *ubiG*, involved in the production of ubiquinol-8, a key electron carrier used in the presence of oxygen or nitrogen, was recovered in this screen. The production of ubiquinol from 4-hydroxybenzoate and trans-octaprenyl diphosphate necessitates the use of six enzymes and UbiG twice [64]. Mutant strains deficient in ubiquinol demonstrate higher levels of ROS in the cytoplasmic membranes, a threat lessened via the addition of exogenous ubiquinol [65]. Furthermore, the Δ*ubiG* strain exhibited reduced fitness when exposed to oxidative stress [65]. Altogether, the presence of this hit may be explained by the exacerbation of the production of ROS due to Ga exposure alongside the compromised oxidative stress response of the Δ*ubiG* strain.

3.2.3. Deoxynucleotide and Cofactor Biosynthesis, and DNA Replication and Repair

Compounds targeting ribonucleotide reductase (RNR), a key enzyme involved in the synthesis of deoxynucleotides from ribonucleotides, have long been regarded as cancer therapeutics [66]. In mammalian cells, Ga targets RNR through at least two mechanisms. These mechanisms include the inhibition of cellular Fe uptake resulting in decreased Fe availability at the M2 subunit of the enzyme [67] and direct inhibition of RNR activity [68], leading to a reduction in the concentration of nucleotides in the cell. This mechanism is not limited to mammalian cells. Gallium(III) has been shown to inhibit RNR and aconitase activity in *M. tuberculosis* [16]. If RNR inhibition is in fact a mechanism of Ga toxicity, then we predict that gene deletions resulting in decreased deoxynucleotide levels may cause hypersensitivity. Consequently, the deletion of the gene *purT*, which is involved in purine nucleotide biosynthesis [69], resulted in Ga sensitivity in this study.

Chromosomal replication is delayed in *E. coli* cells when the deoxynucleotide pool is depleted upon the inhibition of RNR [70]. If this is the case, then a defect in DNA replication may result in hypersensitivity to Ga. Our observation that the loss of the DNA polymerase III subunits HolC and HolD causes Ga sensitivity appears to support this hypothesis. Another potential consequence of RNR inhibition is an increase in stalled replication forks, which are prone to DNA strand breakage [70]. Resumption of stalled replication forks and double strand breaks due to defective RNR function require the activity of recombination repair enzymes such as the RuvABC, RecBCD and RecA [71,72]. Our results support these observations since the deletion of *recA*, *recD* or *ruvC* triggered the Ga sensitive phenotype. It is important to note that genes involved in base and nucleotide excision repair were not retrieved as Ga sensitive hits suggesting that DNA damage associated with Ga exposure may be predominantly in the form double stranded breaks.

A number of sensitive hits were mapped to the subsystem "Biosynthesis of cofactors, prosthetic groups and electron carriers". Processes affected include folate, lipoate, quinol, quinone, ubiquinol and thiamine biosynthesis. The gene products of *pabA* and *pabC*, which encode an aminodeoxychorismate synthase and an aminodeoxychorismate lyase, respectively, are involved in the biosynthesis of p-aminobenzoic acid [73], a precursor of folate. In both prokaryotes and eukaryotes, folate cofactors are necessary for a range of biosynthetic processes including purine and methionine biosynthesis (Figure 4) [74]. Folate biosynthesis has long served as an antibiotic target in prokaryotes since this cofactor is synthesized only in bacteria yet actively imported by eukaryotes using membrane associated processes [75]. Similar to *purT*, the Ga sensitivity of Δ*pabA* and Δ*pabC* strains may be a result of the reduction in deoxynucleotide levels caused by the inactivation of RNR.

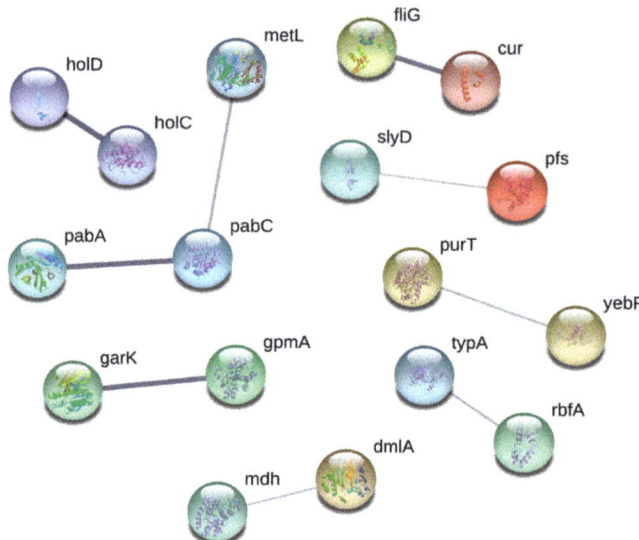

Figure 4. Connectivity map displaying the predicted functional associations between the Ga-sensitive gene hits; disconnected gene hits not shown. The thicknesses of the lines indicate the degree of confidence prediction for the given interaction, based on fusion, curated databases, experimental and co-expression evidence. Figure generated using STRING (version 10.5) and a medium confidence score of 0.4.

To test the potential connection between Ga and RNR activity, we exposed the *holC*, *holD*, *recA*, *recD*, *ruvC* and *purT* mutants to hydroxyurea (HU), which is a known inhibitor of RNR activity [76]. Further, we included a number of mutants involved in DNA synthesis, such as *ruvA* and *recR*, that were not uncovered in our initial screen. In *E. coli*, HU has been shown to increase ribonucleotide pools and decrease total deoxyribonucleotide concentrations, thus negatively affecting the synthesis of DNA [77]. We exposed these mutants to sublethal concentrations of HU and normalized the cellular effect of this agent. Using this reagent, the sensitivity of the *holC*, *ruvC* and *recD* mutants in the presence of HU and Ga was found to increase (Table 3). Furthermore, *ruvA*, which assists in recombinational repair together with *ruvB* [78], was also found to be a sensitive hit in the presence of this inhibitor. The genes *purT* and *holD* were not uncovered as either sensitive or resistant hits based on the cutoffs applied and no changes in the sensitivity or resistance of either *lacA* or *lacY*, negative controls in this work, were statistically identified.

Table 3. Hydroxyurea sensitive and gene hits involved in the synthesis of DNA, normalized to include only the effects of Ga exposure; those with a score two deviations from the mean are included.

Gene	Score without HA	Score with HA [1,2]
ruvA	N/A	−0.257
recA	−0.309	−0.299
ruvC	−0.184	−0.299
holC	−0.327	−0.351
recD	−0.199	−0.561

[1] Each individual score represents the mean of 9–12 trials. [2] Two-tailed *t*-test and significance was determined using the Benjamini–Hochberg; procedure; false discovery rate 10%. HA: Hydroxyurea

3.3. Systems Involved in Ga Resistance

3.3.1. Fe Transport Systems

In *E. coli*, the mechanisms by which Ga is transported into the cell have yet to be identified. In this screen, we identified a number of transport proteins that confer resistance against Ga when absent. Metal resistance mechanisms may involve decreased import or enhanced export of the toxin. Therefore, loss of a gene in which the product mediates import of the toxin into the cell would prevent its accumulation and result in resistance. Both FepG and TonB are proteins that demonstrate close interaction (Figure 5) and fit the latter criterion, both involved in the import of Fe-siderophores. The protein FepG is an inner membrane subunit of the ferric enterobactin ATP-binding cassette transporter complex. When *fepG* is inactivated, *E. coli* cells lose ferric enterobactin uptake abilities [79,80]. The protein product of *tonB* is a component of the Ton system which functions to couple energy from the proton motive force with the active transport of Fe-siderophore complexes and Vitamin B12 across the outer membrane [81]. Since Ga entry into the bacterial cell can occur through siderophore binding and since this metal is an Fe mimetic [23,54], we hypothesize that in the absence of *fepG* and *tonB* Ga import and intracellular accumulation is reduced.

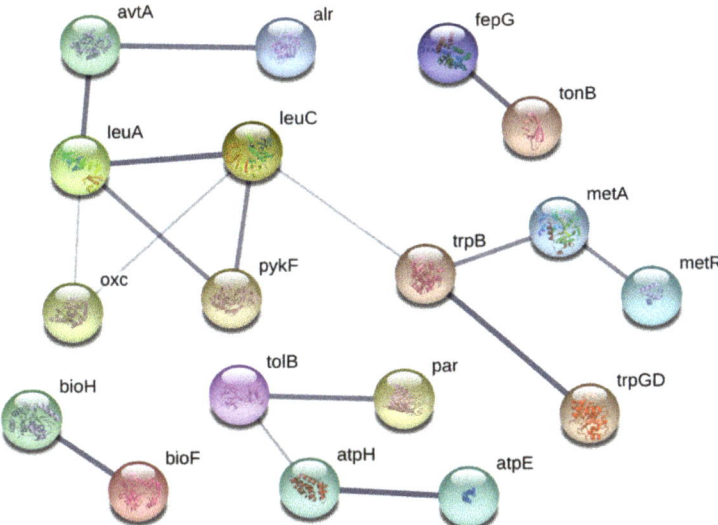

Figure 5. Connectivity map displaying the predicted functional associations between the Ga-resistant gene hits; disconnected gene hits not shown. The thicknesses of the lines indicate the degree of confidence prediction for the given interaction, based on fusion, curated database, experimental and co-expression evidence. Figure generated using STRING (version 10.5) and a medium confidence score of 0.4.

OmpC is a promiscuous porin that permits the transport of 30+ molecules, and is postulated to be a transporter of copper(I) and copper(II) [82] and potentially other metal species [83]. It has been hypothesized that Ga can cross the membrane of *E. coli* via porins [23]. While this hypothesis has not been demonstrated in *E. coli* directly, other works have confirmed findings in *P. aeruginosa* [9], *Mycobacterium smegmatis* [84] and *Francisella* strains [12]. Further evidence for the importance of OmpC in Ga resistance can be visualized using the STRING map (Figure 5). Here, OmpC is connected to two proteins that comprise the ATPase complex through the periplasmic protein TolB. TolB has been shown to physically interact with porins such as OmpC and is required for their assembly into the outer membrane of *E. coli* cells [85]. The resistance recovered in the ΔtolB strain may be due to a disruption in OmpC function, thereby hindering Ga import. In addition, CysU, which is involved in the uptake of sulfate and thiosulfate was also recovered as a resistant hit [86]. According to the hard-soft acid-base theory, Ga coordinates well with sulfate or thiosulfate [4]. A reduction in the uptake of these metabolites may prove useful against Ga stress due to decreased toxin import.

Genes involved in Fe import in other organisms have been shown to confer Ga resistance when deleted, or Ga sensitivity when overexpressed. A three-fold increase in Ga resistance was displayed upon the deletion of the gene *hitA*, which codes for a Fe-binding protein in *P. aeruginosa* [25]. The *Haemophilus influenzae* proteins FbpABC, which are involved in the delivery of Fe from the periplasm to the cytoplasm, were expressed in *E. coli* as a means of investigating their impact on Ga import, which increased in the presence of these genes [87]. Furthermore, earlier studies have examined the use of metal-chelators as antimicrobial enhancements. Although the majority of studies regarding Ga import have been performed in *P. aeruginosa*, some findings can be compared. For example, it has been demonstrated that the siderophore complex Ga-deferoxamine was slightly more effective at killing cells than Ga alone [10] and more promising results have been made with the complex Ga-protoporphyrin IX [88]. Altogether, these studies and our work suggest that Ga enters the cell via siderophore transport systems or Fe-binding transporters.

3.3.2. Amino Acid Biosynthesis

Ga resistant hits were functionally enriched for the synthesis of amino acids (Figure 3), classified in the subsystem, "Amino acid biosynthesis" (Table 2) and highly connected in the functional map (Figure 5). The genes recovered were found to be mainly involved in the biosynthesis of branched (*ilvB*, *ilvY*, *leuA* and *leuC*) and aromatic (*aroF*, *trpB*, and *trpD*) amino acids, methionine (*metA* and *metR*), and proline (*proA* and *proB*). The demand for NADPH in biosynthetic pathways of branched and aromatic amino acids, as well as methionine and proline, are among the highest [89]. It is plausible that a defect in the synthesis of these amino acids may increase levels of NADPH, which has been shown to neutralize the oxidative stress elicited from Ga exposure [7].

To further test this hypothesis, we exposed a number of the resistant hits mapped to branched amino acid biosynthesis to sublethal concentrations of Sulfometuron methyl (SMM), an inhibitor of acetolactate synthase [90], a key enzyme involved in the synthesis of branched amino acids. The resistance score of *ilvY* and *leuA* increased in the presence of SMM (Table 4). Sulfometuron methyl inhibits acetolactate synthase, which in turn may increase the liable NADPH pool. In fact, *ilvY* is a positive regulator of *ilvC* [91], which encodes a reductoisomerase and is the only enzyme in this pathway that directly uses NADPH. Here, *ilvB* and other genes involved in branched amino acid biosynthesis did not make the statistical cutoffs owing to large standard deviations. Finally, no changes in the sensitivity or resistance of *lacA* or *lacY*, negative controls in this work, were statistically identified.

Table 4. Sulfometuron methyl resistant gene hits, involved in the synthesis of amino acids, normalized to include only the effects of Ga exposure; only those with a score two deviations from the mean are included.

Gene	Score without SMM	Score with SMM [1,2]
leuA	0.302	0.341
ilvY	0.215	0.3

[1] Each individual score represents the mean of 9–12 trials. [2] Two-tailed *t*-test and significance was determined using the Benjamini–Hochberg; procedure; false discovery rate 10%. SMM: sulfometuron methyl

It has been postulated that the oxidation of amino acids is a common and damaging effect of metal-induced oxidative stress [92]. Certain side chains, such as Arg, Cys, His, Lys and Pro residues are major targets, leading to protein damage and intra/inter-crosslinking [92,93]. If Ga targets amino acids, both free and within proteins, a possible explanation for the recovery of amino acid gene resistant hits in this study may rest in the cell's requirement to repair or replace damaged amino acids. If these genes are absent fewer Ga targets remain and the cell expends less energy rebuilding these targeted biomolecules, while directing more energy elsewhere, such as scavenging and importing required metabolites. Furthermore, the oxidation of these amino acid side chains may lead to the propagation of ROS, and therefore a deficiency in amino acids may minimize damage by slowing the advancement of amino acid metal-induced oxidative stress.

3.3.3. Lipopolysaccharides and Peptidoglycan

The *E. coli* envelope is composed of lipopolysaccharides (LPS), which surround and protect the cytoplasm, and the cross-linked polymer peptidoglycan (PG), which is the primary stress-bearing biomolecule in the cell [94]. In this study, a number of genes involved in LPS or PG biosynthesis/maintenance were observed to cause Ga resistance when absent. These genes include *cpsG* and *rfaC* (LPS), and *alr*, *env* and *mrcB* (PG). Many of these genes are RpoS-regulated and participate in maintaining membrane integrity in response to pressure [95]. Loss of *mrcB*, which encodes for an inner membrane enzyme functioning in transglycosylation and transpeptidation of PG, has been shown to result in reduced surface PG density when absent [96]. The protein RfaC is essential in LPS production [97] and cells lacking this gene contain defects in the core heptose region [98]. The protein EnvC, which is a divisome-associated factor has been shown to have PG hydrolytic activity and result in decreased cell envelope integrity when deleted. Furthermore, the protein product of *tolB*, which plays a role in maintaining the structure of the cell envelope, was also a Ga-resistant hit. Cells deficient in *tolB* have been shown to release periplasmic proteins into the extracellular space [99]. An explanation for the appearance of *mrcB*, *envC* and *tolB* in this study may reside in the ability of PG to bind metals. Metal ions are known to bind the LPS or PG layer of Gram-negative and Gram-positive bacteria [100], and the presence of anionic groups such as carboxylic acids [101] and other hard acids within the cell envelope, provide suitable binding sights for free metal ions like Ga. Although the major ionic form of Ga is $Ga(OH)_4^-$, free Ga ions produced through equilibrium may be quickly bound by hard acids such as alcohols, carboxylates, and hydroxyls, which comprise the bulk of the PG. Despite their presence at low concentrations these species may further impede cell health and cause toxicity. However, if the LPS or PG layer is reduced, as would be the case in the absence of *mrcB*, *rfaC*, *envC* and *tolB*, then a reduction in Ga-cell envelope binding may occur. In the case of the Δ*tolB* strain, the potential release of periplasmic proteins with Ga-binding sites into the extracellular space may also provide protection via sequestration, which is a common bacterial resistance mechanism [24]. Another possible explanation for Ga resistance associated with LPS and PG genes may include the structural alteration of the cell envelope, which may disrupt Fe import systems. Inhibition of lipid biosynthesis prevents proper assembly and insertion of porins into the outer membrane since LPS-porin interaction sites have been shown to be important in their biogenesis [102,103]. Therefore, compromised function of siderophore receptors or porins in these mutants could decrease Ga import and mitigate toxicity.

4. Conclusions

In this study, the Keio collection was used as a means of drawing insight into the mechanisms of Ga toxicity and resistance in *E. coli* BW25113. In total, 3895 non-essential genes were screened and 3641 of these were normalized and scored. Genes demonstrating resistance or toxicity were mined to highlight processes and pathways affected by Ga exposure. Mutants demonstrating an increase in colony formation were considered resistant hits, in that the presence of the gene results in Ga sensitivity. In contrast, a decrease in colony size was regarded as a Ga-sensitive hit, consequently it was assumed that the presence of this gene would impart the resistant phenotype and mitigate the toxicity of prolonged Ga exposure.

Overall, comparable numbers of resistant and sensitive hits were mapped to each subsystem using Pathway Tools, which surveys against the EcoCyc Database. When examining the fold enrichment data, no biological process was enriched comparably between the two data sets. One general observation made evident from the latter conclusion is that distinct pathways are affected by Ga when comparing the mechanisms of toxicity and resistance since no overlap in functional enrichment was uncovered. Still, one significant exception was found: Fe-metabolism. Based on this study, and previous reports, there is a relationship between Ga and Fe-metabolism. The genes that code for TonB and FepG were two resistant hits highlighted in this work. On the contrary, Fdx, Bfr and LipA, proteins also involved in Fe-metabolism, gave rise to sensitivity when absent. Therefore, we propose that Fe-metabolism may serve as a mechanism of resistance and toxicity in *E. coli*. Here, the complexity of Ga exposure is made further apparent, fostering more questions regarding the interaction of this metal with microbes. What is clear however, is that the mechanism of Ga action is likely a result of a number of direct and indirect interactions, an observation made evident by the wide array of hits uncovered in this work.

Few studies have explored the mechanisms of adaptive resistance in *E. coli* under sub-lethal concentrations of Ga. In response, we have presented a number of genes that are implicated to be involved in adaptive survival. For example, genes involved in preventing oxidative damage and DNA repair were emphasized as sensitive hits, as such that their presence gives rise to resistance. In short, preventing and repairing DNA damage, a mechanism that has yet to be demonstrated in vivo, and redox maintenance may provide tools by which microbial organisms mitigate metal stress.

The use of Ga for the treatment of diseases and infections is gaining considerable attention. Still, to further the development of this metal as an antimicrobial agent it is imperative that we determine the associated mechanisms of toxicity and resistance. Further work must be completed to specifically test the various hypotheses we have presented here, such as determining the mode of Ga entry, the levels of ROS produced in the cell and the specific influence of Ga on Fe-metabolism. Nonetheless, this study provides a significant number of biomolecular mechanistic hypotheses to the community investigating the mechanisms of Ga action in *E. coli* and other microbes.

Supplementary Materials: The following are available online at http://www.mdpi.com/2073-4425/10/1/34/s1, Table S1: *Escherichia coli* gallium resistant and sensitive gene hits determined using a chemical genetic screen.

Author Contributions: Conceptualization, K.C.-R. and R.T.; Data curation, N.G. and K.C.-R.; Formal analysis, N.G. and K.C.-R.; Funding acquisition, R.T. and G.C.; Investigation, N.G.; Methodology, N.G.; Resources, G.C.; Supervision, R.T. and G.C.; Visualization, N.G., K.C.-R., R.T. and G.C.; Writing—original draft, N.G.; Writing—review & editing, N.G., K.C.-R., R.T., and G.C.

Funding: N.G. was funded by an Alexander Graham Bell Scholarship from the Natural Science and Engineering Research Council of Canada and an Eyes High Doctoral Scholarship from the University of Calgary. R.J.T. was funded by a project bridge grant from the Canadian Institutes of Health Research (CIHR bridge: PJT-149009). R.J.T. and G.C. were also funded by Discovery grants from the Natural Science and Engineering Research Council (NSERC DG: RGPIN/04811-2015) of Canada.

Acknowledgments: The authors of this paper would like to acknowledge Joe A. Lemire and Iain George for their assistance with experimental design and the BM3 robot from S&P Robotics Inc., and Ying Yan for his assistance with the statistical analyses.

Conflicts of Interest: The authors declare no conflict of interest.

Appendix A

Table A1. The Gallium(III) (Ga) resistant and sensitive hits were surveyed against the EcoCyc database permitting the clustering of the hits into systems, subsystems, component subsystems, and lastly into individual objects.

Systems	Subsystems [1]
Regulation	Signaling, Sigma factor regulon, Transcription factor, and Transcription factor regulons
Response to stimulus	Starvation, Heat, Cold, DNA damage, pH, Detoxification, Osmotic stress, and Other
Cellular processes	Cell cycle and division, Cell death, Genetic transfer, Biofilm formation, Quorum sensing, Adhesion, Locomotion, Viral response, Response to bacterium, Host interactions, Symbiosis, and Other proteins
Energy	Glycolysis, Pentose phosphate pathway, TCA cycle, Fermentation, Aerobic and anaerobic respiration, and Other proteins
Other pathways	Detoxification, Inorganic nutrient metabolism, Macromolecule modification, Activation/inactivation/interconversion, and Other enzymes
Degradation	Amino acids, Fatty acid/lipid, Nucleotide/nucleoside, Amine, Carbohydrate/carboxylate, Secondary metabolite, Alcohol, Polymer, Cell exterior and Other proteins
Biosynthesis	Amino acids, Nucleotide/nucleoside, Fatty acid/lipid, Amines, Carbohydrate/carboxylates, Cofactors, Secondary metabolites, Polymer, and Other proteins
Cell exterior	Transport, Cell wall biogenesis and organization, Lipopolysaccharide metabolism, Pilus, Flagellar, Outer membrane, Inner membrane, Periplasm, and Cell wall components
Central dogma	Transcription, Translation, DNA metabolism, RNA metabolism, Protein metabolism, and Protein folding, and secretion

[1] Genes can be found in multiple systems and subsystems.

References

1. Chitambar, C.R. The therapeutic potential of iron-targeting gallium compounds in human disease: From basic research to clinical application. *Pharmacol. Res.* **2017**, *115*, 56–64. [CrossRef]
2. Bonchi, C.; Imperi, F.; Minandri, F.; Visca, P.; Frangipani, E. Repurposing of gallium-based drugs for antibacterial therapy. *BioFactors* **2014**, *40*, 303–312. [CrossRef] [PubMed]
3. Bernstein, L.R. Mechanisms of therapeutic activity for gallium. *Pharmacol. Rev.* **1998**, *50*, 665–682. [PubMed]
4. Pearson, R.G. Hard and soft acids and bases. *J. Am. Chem. Soc.* **1963**, *85*, 3533–3539. [CrossRef]
5. Harris, W.R.; Pecoraro, V.L. Thermodynamic binding constants for gallium transferrin. *Biochemistry* **1983**, *22*, 292–299. [CrossRef] [PubMed]
6. Weiner, R.E.; Neumann, R.D.; Mulshine, J. Transferrin dependence of Ga (NO$_3$)$_3$ inhibition of growth in human-derived small cell lung cancer cells. *J. Cell. Biochem.* **1996**, *24*, 276–287. [CrossRef]
7. Beriault, R.; Hamel, R.; Chenier, D.; Mailloux, R.J.; Joly, H.; Appanna, V.D. The overexpression of NADPH-producing enzymes counters the oxidative stress evoked by gallium, an iron mimetic. *BioMetals* **2007**, *20*, 165–176. [CrossRef]
8. Al-Aoukaty, A.; Appanna, V.D.; Falter, H. Gallium toxicity and adaptation in *Pseudomonas fluorescens*. *FEMS Microbiol. Lett.* **1992**, *92*, 265–272. [CrossRef]
9. Kaneko, Y.; Thoendel, M.; Olakanmi, O.; Britigan, B.E.; Singh, P.K. The transition metal gallium disrupts *Pseudomonas aeruginosa* iron metabolism and has antimicrobial and antibiofilm activity. *J. Clin. Invest.* **2007**, *117*, 877–888. [CrossRef]

10. Banin, E.; Lozinski, A.; Brady, K.M.; Berenshtein, E.; Butterfield, P.W.; Moshe, M.; Chevion, M.; Greenberg, E.P.; Banin, E. The potential of desferrioxamine-gallium as an Anti-*Pseudomonas* therapeutic agent. *Proc. Natl. Acad. Sci. USA* **2008**, *105*, 16761–16766. [CrossRef]
11. Chitambar, C.R.; Narasimhan, J. Targeting iron-dependant DNA synthesis with gallium and transferrin-gallium. *Pathobiology* **1991**, *59*, 3–10. [CrossRef] [PubMed]
12. Olakanmi, O.; Gunn, J.S.; Su, S.; Soni, S.; Hassett, D.J.; Britigan, B.E. Gallium disrupts iron uptake by intracellular and extracellular *Francisella* strains and exhibits therapeutic efficacy in a murine pulmonary infection model. *Antimicrob. Agents Chemother.* **2010**, *54*, 244–253. [CrossRef] [PubMed]
13. Nikolova, V.; Angelova, S.; Markova, N.; Dudev, T. Gallium as a therapeutic agent: A thermodynamic evaluation of the competition between Ga^{3+} and Fe^{3+} ions in metalloproteins. *J. Phys. Chem. B* **2016**, *120*, 2241–2248. [CrossRef] [PubMed]
14. Kelson, A.B.; Carnevali, M.; Truong-Le, V. Gallium-based anti-infectives: Targeting microbial iron-uptake mechanisms. *Curr. Opin. Pharmacol.* **2013**, *13*, 707–716. [CrossRef] [PubMed]
15. Olakanmi, O.; Britigan, B.E.; Larry, S. Gallium disrupts iron metabolism of *Mycobacteria* residing within human macrophages. *Infect. Immun.* **2000**, *68*, 5619–5627. [CrossRef]
16. Olakanmi, O.; Kesavalu, B.; Pasula, R.; Abdalla, M.Y.; Schlesinger, L.S.; Britigan, B.E. Gallium nitrate is efficacious in murine models of tuberculosis and inhibits key bacterial Fe-dependent enzymes. *Antimicrob. Agents Chemother.* **2013**, *57*, 6074–6080. [CrossRef]
17. DeLeon, K.; Balldin, F.; Watters, C.; Hamood, A.; Griswold, J.; Sreedharan, S.; Rumbaugh, K.P. Gallium maltolate treatment eradicates *Pseudomonas aeruginosa* infection in thermally injured mice. *Antimicrob. Agents Chemother.* **2009**, *53*, 1331–1337. [CrossRef]
18. Arnold, C.E.; Bordin, A.; Lawhon, S.D.; Libal, M.C.; Bernstein, L.R.; Cohen, N.D. Antimicrobial activity of gallium maltolate against *Staphylococcus aureus* and methicillin-resistant *S. aureus* and *Staphylococcus pseudintermedius*: An in vitro study. *Vet. Microbiol.* **2012**, *155*, 389–394. [CrossRef]
19. Martens, R.J.; Miller, N.A.; Cohen, N.D.; Harrington, J.R.; Bernstein, L.R. Chemoprophylactic antimicrobial activity of gallium maltolate against intracellular *Rhodococcus equi*. *J. Equine Vet. Sci.* **2007**, *27*, 341–345. [CrossRef]
20. Antunes, L.C.S.; Imperi, F.; Minandri, F.; Visca, P. In vitro and in vivo antimicrobial activities of gallium nitrate against multidrug-resistant *Acinetobacter baumannii*. *Antimicrob. Agents Chemother.* **2012**, *56*, 5961–5970. [CrossRef]
21. Gugala, N.; Lemire, J.A.; Turner, R.J. The efficacy of different anti-microbial metals at preventing the formation of, and eradicating bacterial biofilms of pathogenic indicator strains. *J. Antibiot. (Tokyo)* **2017**, *70*, 775–780. [CrossRef]
22. Lemire, J.A.; Harrison, J.J.; Turner, R.J. Antimicrobial activity of metals: Mechanisms, molecular targets and applications. *Nat. Rev. Microbiol.* **2013**, *11*, 371–384. [CrossRef] [PubMed]
23. Minandri, F.; Bonchi, C.; Frangipani, E.; Imperi, F.; Visca, P. Promises and failures of gallium as an antibacterial agent. *Future Microbiol.* **2014**, *9*, 379–397. [CrossRef] [PubMed]
24. Harrison, J.J.; Ceri, H.; Turner, R.J. Multimetal resistance and tolerance in microbial biofilms. *Nat. Rev. Microbiol.* **2007**, *5*, 928–938. [CrossRef] [PubMed]
25. García-Contreras, R.; Lira-Silva, E.; Jasso-Chávez, R.; Hernández-González, I.L.; Maeda, T.; Hashimoto, T.; Boogerd, F.C.; Sheng, L.; Wood, T.K.; Moreno-Sánchez, R. Isolation and characterization of gallium resistant *Pseudomonas aeruginosa* mutants. *Int. J. Med. Microbiol.* **2013**, *303*, 574–582. [CrossRef] [PubMed]
26. Peeters, E.; Nelis, H.J.; Coenye, T. Resistance of planktonic and biofilm-grown *Burkholderia cepacia* complex isolates to the transition metal gallium. *J. Antimicrob. Chemother.* **2008**, *61*, 1062–1065. [CrossRef] [PubMed]
27. Mcquillan, J.S.; Shaw, A.M. Differential gene regulation in the Ag nanoparticle and Ag-Induced silver stress response in *Escherichia coli*: A full transcriptomic profile. *Nanotoxicology* **2014**, *8*, 177–184. [CrossRef]
28. Saulou-Bérion, C.; Gonzalez, I.; Enjalbert, B.; Audinot, J.N.; Fourquaux, I.; Jamme, F.; Cocaign-Bousquet, M.; Mercier-Bonin, M.; Girbal, L. *Escherichia coli* under ionic silver stress: An integrative approach to explore transcriptional, physiological and biochemical responses. *PLoS ONE* **2015**, *10*, 1–25. [CrossRef]
29. Ivask, A.; Elbadawy, A.; Kaweeteerawat, C.; Boren, D.; Fischer, H.; Ji, Z.; Chang, C.H.; Liu, R.; Tolaymat, T.; Telesca, D.; et al. Toxicity mechanisms in *Escherichia coli* vary for silver nanoparticles and differ from ionic silver. *ACS Nano* **2014**, *8*, 374–386. [CrossRef]

30. Gugala, N.; Lemire, J.; Chatfield-Reed, K.; Yan, Y.; Chua, G.; Turner, R. Using a chemical genetic screen to enhance our understanding of the antibacterial properties of silver. *Genes* **2018**, *9*, 344. [CrossRef]
31. Kershaw, C.J.; Brown, N.L.; Constantinidou, C.; Patel, M.D.; Hobman, J.L. The expression profile of *Escherichia coli* K-12 in response to minimal, optimal and excess copper concentrations. *Microbiology* **2005**, *151*, 1187–1198. [CrossRef] [PubMed]
32. Yamamoto, K.; Ishihama, A. Transcriptional response of *Escherichia coli* to external copper. *Mol. Microbiol.* **2005**, *56*, 215–227. [CrossRef] [PubMed]
33. Brocklehurst, K.R.; Morby, A.P. Metal-ion tolerance in *Escherichia coli*: Analysis of transcriptional profiles by gene-array technology. *Microbiology* **2000**, *146*, 2277–2282. [CrossRef] [PubMed]
34. Yamamoto, K. Transcriptional Response of *Escherichia coli* to External Zinc. *J. Bacteriol.* **2005**, *187*, 6333–6340. [CrossRef] [PubMed]
35. Baba, T.; Ara, T.; Hasegawa, M.; Takai, Y.; Okumura, Y.; Baba, M.; Datsenko, K.A.; Tomita, M.; Wanner, B.L.; Mori, H. Construction of *Escherichia coli* K-12 in-frame, single-gene knockout mutants: The Keio Collection. *Mol. Syst. Biol.* **2006**, *2*. [CrossRef] [PubMed]
36. Kuzminov, A. Recombinational repair of DNA damage in *Escherichia coli* and bacteriophage λ. *Microbiol. Mol. Biol. Rev.* **1999**, *63*, 751–813. [PubMed]
37. Hin, A.; Tong, Y.; Evangelista, M.; Parsons, A.B.; Xu, H.; Bader, G.D.; Page, N.; Robinson, M.; Raghibizadeh, S.; Hogue, C.W.V.; et al. Systematic genetic analysis with ordered arrays of yeast deletion mutants. *Science (80-.)* **2001**, *294*, 2364–2369. [CrossRef]
38. Wagih, O.; Usaj, M.; Baryshnikova, A.; VanderSluis, B.; Kuzmin, E.; Costanzo, M.; Myers, C.L.; Andrews, B.J.; Boone, C.M.; Parts, L. SGAtools: One-stop analysis and visualization of array-based genetic interaction screens. *Nucleic Acids Res.* **2013**, *41*, 591–596. [CrossRef] [PubMed]
39. Keseler, I.M.; Mackie, A.; Santos-Zavaleta, A.; Billington, R.; Bonavides-Martínez, C.; Caspi, R.; Fulcher, C.; Gama-Castro, S.; Kothari, A.; Krummenacker, M.; et al. The EcoCyc database: Reflecting new knowledge about *Escherichia coli* K-12. *Nucleic Acids Res.* **2017**, *45*, D543–D550. [CrossRef]
40. Bateman, A.; Martin, M.J.; O'Donovan, C.; Magrane, M.; Alpi, E.; Antunes, R.; Bely, B.; Bingley, M.; Bonilla, C.; Britto, R.; et al. UniProt: The universal protein knowledgebase. *Nucleic Acids Res.* **2017**, *45*, D158–D169. [CrossRef]
41. Huang, D.W.; Sherman, B.T.; Lempicki, R.A. Systematic and integrative analysis of large gene lists using DAVID bioinformatics resources. *Nat. Protoc.* **2009**, *4*, 44–57. [CrossRef] [PubMed]
42. Huang, D.W.; Sherman, B.T.; Lempicki, R.A. Bioinformatics enrichment tools: Paths toward the comprehensive functional analysis of large gene lists. *Nucleic Acids Res.* **2009**, *37*, 1–13. [CrossRef] [PubMed]
43. Szklarczyk, D.; Morris, J.H.; Cook, H.; Kuhn, M.; Wyder, S.; Simonovic, M.; Santos, A.; Doncheva, N.T.; Roth, A.; Bork, P.; et al. The STRING database in 2017: Quality-controlled protein-protein association networks, made broadly accessible. *Nucleic Acids Res.* **2017**, *45*, D362–D368. [CrossRef]
44. Patrick, M.W.; Quandt, M.E.; Swartzlander, B.D.; Matsumura, I. Multicopy suppression underpins metabolic evolvability. *Mol. Microbiol. Evol.* **2007**, *24*, 2716–2722. [CrossRef] [PubMed]
45. Joyce, A.R.; Reed, J.L.; White, A.; Edwards, R.; Osterman, A.; Baba, T.; Mori, H.; Lesely, S.A.; Palsson, B.; Agarwalla, S. Experimental and computational assessment of conditionally essential genes in *Escherichia coli*. *J. Bacteriol.* **2006**, *188*, 8259–8271. [CrossRef]
46. Mchugh, J.P.; Rodríguez-Quiñones, F.; Abdul-Tehrani, H.; Svistunenko, D.A.; Poole, R.K.; Cooper, C.E.; Andrews, S.C. Global iron-dependent gene regulation in *Escherichia coli*. *J. Biol. Chem.* **2003**, *278*, 29478–29486. [CrossRef] [PubMed]
47. Kakuta, Y.; Horio, T.; Takahashi, Y.; Fukuyama, K. Crystal structure of *Escherichia coli* Fdx, an adrenodoxin-type ferredoxin involved in the assembly of iron-sulfur clusters. *Biochemistry* **2001**, *40*, 11007–11012. [CrossRef]
48. Cicchillo, R.M.; Lee, K.H.; Baleanu-Gogonea, C.; Nesbitt, N.M.; Krebs, C.; Booker, S.J. *Escherichia coli* lipoyl synthase binds two distinct [4Fe-4S] clusters per polypeptide. *Biochemistry* **2004**, *43*, 11770–11781. [CrossRef]
49. Waller, J.C.; Alvarez, S.; Naponelli, V.; Lara-Nunez, A.; Blaby, I.K.; Da Silva, V.; Ziemak, M.J.; Vickers, T.J.; Beverley, S.M.; Edison, A.S.; et al. A role for tetrahydrofolates in the metabolism of iron-sulfur clusters in all domains of life. *Proc. Natl. Acad. Sci. USA* **2010**, *107*, 10412–10417. [CrossRef]
50. Jang, S.; Imlay, J.A. Hydrogen peroxide inactivates the *Escherichia coli* Isc iron-sulfur assembly system, and OxyR induces the suf system to compensate. *Mol. Microbiol.* **2010**, *78*, 1448–1467. [CrossRef]

51. Laxman, S.; Sutter, B.M.; Wu, X.; Kumar, S.; Guo, X.; David, C.; Mirzaei, H.; Tu, B.P. Sulfur amino acids regulate translational capacity and metabolic homeostasis through modulation of tRNA thiolation. *Cell* **2014**, *154*, 416–429. [CrossRef]
52. Miethke, M.; Marahiel, M.A. Siderophore-based iron acquisition and pathogen control. *Microbiol. Mol. Biol. Rev.* **2007**, *71*, 413–451. [CrossRef] [PubMed]
53. Raymond, K.N.; Dertz, E.A.; Kim, S.S. Enterobactin: An archetype for microbial iron transport. *Proc. Natl. Acad. Sci. USA* **2003**, *100*, 3584–3588. [CrossRef] [PubMed]
54. Chitambar, C.R. Gallium and its competing roles with iron in biological systems. *Biochim. Biophys. Acta* **2016**, *1863*, 2044–2053. [CrossRef] [PubMed]
55. Koronakis, V. TolC-the bacterial exit duct for proteins and drugs. *FEBS Lett.* **2003**, *555*, 66–71. [CrossRef]
56. Komendarczyk, R.; Pullen, J. Transcriptional regulation of drug efflux genes by EvgAS, a two-component system in *Escherichia coli*. *Microbiology* **2003**, *149*, 2819–2828. [CrossRef]
57. Bou-Abdallah, F.; Lewin, A.C.; Le Brun, N.E.; Moore, G.R.; Dennis Chasteen, N. Iron detoxification properties of *Escherichia coli* Bacterioferritin. Attenuation of Oxyradical Chemistry. *J. Biol. Chem.* **2002**, *277*, 37064–37069. [CrossRef] [PubMed]
58. Imlay, J.A. Pathways of oxidative damage. *Annu. Rev. Microbiol.* **2003**, *57*, 395–418. [CrossRef]
59. Imlay, J.A. Cellular defenses against superoxide and hydrogen peroxide. *Annu. Rev. Biochem.* **2008**, *77*, 755–776. [CrossRef]
60. Latinwo, L.M.; Donald, C.; Ikediobi, C.; Silver, S. Effects of intracellular glutathione on sensitivity of *Escherichia coli* to mercury and arsenite. *Biochem. Biophys. Res. Commun.* **1998**, *242*, 67–70. [CrossRef]
61. Ozyamak, E.; Black, S.S.; Walker, C.A.; MacLean, M.J.; Bartlett, W.; Miller, S.; Booth, I.R. The critical role of S-Lactoylglutathione formation during methylglyoxal detoxification in *Escherichia coli*. *Mol. Microbiol.* **2010**, *78*, 1577–1590. [CrossRef] [PubMed]
62. Fernandes, A.P.; Fladvad, M.; Berndt, C.; Andrésen, C.; Lillig, C.H.; Neubauer, P.; Sunnerhagen, M.; Holmgren, A.; Vlamis-Gardikas, A. A novel monothiol glutaredoxin (Grx4) from Escherichia coli can serve as a substrate for thioredoxin reductase. *J. Biol. Chem.* **2005**, *280*, 24544–24552. [CrossRef] [PubMed]
63. Yeung, N.; Gold, B.; Liu, N.L.; Prathapam, R.; Sterling, H.J.; Willams, E.R.; Butland, G. The *E. coli* monothiol glutaredoxin GrxD forms homodimeric and heterodimeric FeS cluster containing complexes. *Biochemistry* **2011**, *50*, 8957–8969. [CrossRef] [PubMed]
64. Meganathan, R. Ubiquinone biosynthesis in microorganisms. *FEMS Microbiol. Lett.* **2001**, *203*, 131–139. [CrossRef] [PubMed]
65. Søballe, B.; Poole, R.K. Ubiquinone limits oxidative stress in *Escherichia coli*. *Microbiology* **2000**, *146*, 787–796. [CrossRef] [PubMed]
66. Nocentini, G. Ribonucleotide reductase inhibitors: New strategies for cancer chemotherapy. *Crit. Rev. Oncol. Hematol.* **1996**, *22*, 89–126, 1040. [CrossRef]
67. Chitambar, C.R.; Seligman, P.A. Effects of different transferrin forms on transferrin receptor expression, iron uptake, and cellular proliferation of human leukemic HL60 cells. Mechanisms Responsible for the Specific Cytotoxicity of Transferrin-Gallium. *J. Clin. Invest.* **1986**, *78*, 1538–1546. [CrossRef]
68. Chitambar, C.R.; Narasimhan, J.; Guy, J.; Sem, D.S.; Brien, W.J.O. Inhibition of ribonucleotide reductase by gallium in murine. *Cancer Res.* **1991**, *51*, 6199–6202.
69. Smith, J.M.; Daum, H.A. Identification and nucleotide sequence of a gene encoding 5′-phosphoribosylglycinamide transformylase in *Escherichia coli* K12. *J. Biol. Chem.* **1987**, *262*, 10565–10569.
70. Zhu, M.; Dai, X.; Guo, Z.; Yang, M.; Wang, H.; Wang, Y.-P. Manipulating the bacterial cell cycle and cell size by titrating the expression of ribonucleotide reductase. *MBio* **2017**, *8*, 6–11. [CrossRef]
71. Salguero, I.; Guarino, E.; Guzmàn, E.C. RecA-dependent replication in the NrdA101(Ts) mutant of *Escherichia coli* under restrictive conditions. *J. Bacteriol.* **2011**, *193*, 2851–2860. [CrossRef] [PubMed]
72. Guarino, E.; Jiménez-Sánchez, A.; Guzmán, E.C. Defective ribonucleoside diphosphate reductase impairs replication fork progression in *Escherichia coli*. *J. Bacteriol.* **2007**, *189*, 3496–3501. [CrossRef] [PubMed]
73. Roux, B.; Walsh, C.T. P-Aminobenzoate Synthesis in *Escherichia coli*: Kinetic and mechanistic characterization of the amidotransferase PabA. *Biochemistry* **1992**, *31*, 6904–6910. [CrossRef] [PubMed]
74. Bermingham, A.; Derrick, J.P. The folic acid biosynthesis pathway in bacteria: Evaluation of potential for antibacterial drug discovery. *BioEssays* **2002**, *24*, 637–648. [CrossRef] [PubMed]

75. Henderson, G.B.; Huennekens, F.M. Membrane-associated folate transport proteins. *Methods Enzymol.* **1986**, *122*, 260–269. [PubMed]
76. Krakoff, I.H.; Brown, N.C.; Reichard, P. Inhibition reductase of ribonucleoside by hydroxyurea diphosphate. *Cancer Res.* **1968**, *28*, 1559–1565. [PubMed]
77. Sinha, N.K.; Snustad, D.P. Mechanism of inhibition of deoxyribonucleic acid synthesis in *Escherichia coli* by hydroxyurea. *J. Bacteriol.* **1972**, *112*, 1321–1324.
78. Rice, D.W.; Rafferty, J.B.; Artymiuk, P.J.; Lloyd, R.G. Insights into the mechanisms of homologous recombination from the structure of RuvA. *Curr. Opin. Struct. Biol.* **1997**, *7*, 798–803. [CrossRef]
79. Chenault, S.S.; Earhart, C.F. Organization of genes encoding membrane proteins of the *Escherichia coli* ferrienterobactin permease. *Mol. Microbiol.* **1991**, *5*, 1405–1413. [CrossRef]
80. Cartron, M.L.; Maddocks, S.; Gillingham, P.; Craven, C.J.; Andrews, S.C. Feo-transport of ferrous iron into bacteria. *BioMetals* **2006**, *19*, 143–157. [CrossRef]
81. Templin, M.F.; Ursinus, A.; Höltje, J.V. A defect in cell wall recycling triggers autolysis during the stationary growth phase of *Escherichia coli*. *EMBO J.* **1999**, *18*, 4108–4117. [CrossRef] [PubMed]
82. Egler, M.; Grosse, C.; Grass, G.; Nies, D.H. Role of the extracytoplasmic function protein family σ factor RpoE in metal resistance of *Eschenchia coli*. *J. Bacteriol.* **2005**, *187*, 2297–2307. [CrossRef] [PubMed]
83. Nies, D.H. Bacterial tranistion metal homeostasis. In *Molecular Microbiology of Heavy Metals*; Springer: Berlin/Heidelberg, Germany, 2007; pp. 118–142.
84. Jones, C.M.; Niederweis, M. Role of porins in iron uptake by *Mycobacterium smegmatis*. *J. Bacteriol.* **2010**, *192*, 6411–6417. [CrossRef] [PubMed]
85. Rigal, A.; Bouveret, E.; Lloubes, R.; Lazdunski, C.; Benedetti, H. The TolB protein interacts with the porins of *Escherichia coli*. *J. Bacteriol.* **1997**, *179*, 7274–7279. [CrossRef] [PubMed]
86. Sirko, A.; Hryniewicz, M.; Hulanicka, D.; Böck, A. Sulfate and thiosulfate transport in *Escherichia coli* K-12: Nucleotide sequence and expression of the CysTWAM gene cluster. *J. Bacteriol.* **1990**, *172*, 3351–3357. [CrossRef] [PubMed]
87. Anderson, D.S.; Adhikari, P.; Nowalk, A.J.; Chen, C.Y.; Mietzner, T.A. The HFbpABC transporter from *Haemophilus influenzae* Functions as a Binding-Protein-Dependent ABC transporter with high specificity and affinity for ferric iron. *J. Bacteriol.* **2004**, *186*, 6220–6229. [CrossRef] [PubMed]
88. Stojiljkovic, I.; Kumar, V.; Srinivasan, N. Non-iron metalloporphyrins: Potent antibacterial compounds that exploit Haem/Hb uptake systems of pathogenic bacteria. *Mol. Microbiol.* **1999**, *31*, 429–442. [CrossRef]
89. Xu, J.Z.; Yang, H.K.; Zhang, W.G. NADPH Metabolism: A survey of its theoretical characteristics and manipulation strategies in amino acid biosynthesis. *Crit. Rev. Biotechnol.* **2018**, *38*, 1061–1076. [CrossRef]
90. LaRossa, R.A.; Schloss, J.V. The sulfonylurea herbicide sulfometuron methyl is an extremely potent and selective inhibitor of acetolactate synthase in *Salmonella typhimurium*. *J. Biol. Chem.* **1984**, *259*, 8753–8757. [CrossRef]
91. Watson, M.D.; Wild, J.; Umbarger, H.E. Positive control of IlvC expression in *Escherichia coli* K-12; Identification and Mapping of Regulatory Gene IlvY. *J. Bacteriol.* **1979**, *139*, 1014–1020.
92. Stohs, S.J.; Bagchi, D. Oxidative mechanisms in the toxicity of metal ions. *Free Radic. Biol. Med.* **1995**, *18*, 321–336. [CrossRef]
93. Ercal, N.; Gurer-Orhan, H.; Aykin-Burns, N. Toxic metals and oxidative stress part I: Mechanisms involved in metal induced oxidative damage. *Curr. Top. Med. Chem.* **2001**, *1*, 529–539. [CrossRef] [PubMed]
94. Cabeen, M.T.; Jacobs-Wagner, C. Bacterial cell shape. *Nat. Rev. Microbiol.* **2005**, *3*, 601–610. [CrossRef]
95. De La Fuente-Núñez, C.; Korolik, V.; Bains, M.; Nguyen, U.; Breidenstein, E.B.M.; Horsman, S.; Lewenza, S.; Burrows, L.; Hancock, R.E.W. Inhibition of bacterial biofilm formation and swarming motility by a small synthetic cationic peptide. *Antimicrob. Agents Chemother.* **2012**, *56*, 2696–2704. [CrossRef] [PubMed]
96. Caparrós, M.; Quintela, J.C.; de Pedro, M.A. Variability of peptidoglycan surface density in *Escherichia coli*. *FEMS Microbiol. Lett.* **1994**, *121*, 71–76. [CrossRef]
97. Gronow, S.; Brabetz, W.; Brade, H. Comparative functional characterization in vitro of heptosyltransferase I (WaaC) and II (WaaF) from *Escherichia coli*. *Eur. J. Biochem.* **2000**, *267*, 6602–6611. [CrossRef] [PubMed]
98. Beher, M.; Schnaitman, C. Regulation of the OmpA outer membrane protein of *Escherichia coli*. *J. Bacteriol.* **1981**, *147*, 972–985.
99. Lazzaroni, J.C.; Portalier, R.C. Genetic and biochemical characterization of periplasmic leaky mutants of *Escherichia coli* K-12. *J. Bacteriol.* **1981**, *145*, 1351–1358. [CrossRef]

100. Beveridge, T.J.; Koval, S.F. Binding of Metals to Cell Envelopes of *Escherichia coli* binding of metals to cell envelopes of *Escherichia coli* K-12. *Appl. Environ. Microbiol.* **1981**, *42*, 325–335.
101. Hoyle, B.D.; Beveridge, T.J. Metal binding by the peptidoglycan sacculus of *Escherichia coli* K-12. *Can. J. Microbiol.* **1984**, *30*, 204–211. [CrossRef]
102. Arunmanee, W.; Pathania, M.; Solovyova, A.S.; Le Brun, A.P.; Ridley, H.; Baslé, A.; van den Berg, B.; Lakey, J.H. Gram-negative Trimeric Porins Have Specific LPS binding sites that are essential for porin biogenesis. *Proc. Natl. Acad. Sci. USA* **2016**, *113*, 5034–5043. [CrossRef] [PubMed]
103. Bolla, J.M.; Lazdunski, C.; Pagès, J.M. The assembly of the major outer membrane protein OmpF of *Escherichia coli* depends on lipid synthesis. *EMBO J.* **1988**, *7*, 3595–3599. [CrossRef] [PubMed]

© 2019 by the authors. Licensee MDPI, Basel, Switzerland. This article is an open access article distributed under the terms and conditions of the Creative Commons Attribution (CC BY) license (http://creativecommons.org/licenses/by/4.0/).

Article

Integrase-Controlled Excision of Metal-Resistance Genomic Islands in *Acinetobacter baumannii*

Zaaima AL-Jabri [1,2,3], Roxana Zamudio [1], Eva Horvath-Papp [3], Joseph D. Ralph [1], Zakariya AL-Muharrami [2], Kumar Rajakumar [3] and Marco R. Oggioni [1,*]

1. Department of Genetics and Genome Biology, University of Leicester, Leicester LE1 7RH, UK; zaljabri01@gmail.com (Z.A.-J.); rzz1@leicester.ac.uk (R.Z.); jdwr2@leicester.ac.uk (J.D.R.)
2. Department of Microbiology and Immunology, Sultan Qaboos University, Muscat 123, Oman; muharrmi@squ.edu.om
3. Department of Infection Immunity and Inflammation, University of Leicester, Leicester LE1 7RH, UK; ehp5@student.le.ac.uk (E.H.-P.); kr46@le.ac.uk (K.R.)
* Correspondence: mro5@leicester.ac.uk; Tel.: +44-116-252-2261

Received: 5 June 2018; Accepted: 16 July 2018; Published: 20 July 2018

Abstract: Genomic islands (GIs) are discrete gene clusters encoding for a variety of functions including antibiotic and heavy metal resistance, some of which are tightly associated to lineages of the core genome phylogenetic tree. We have investigated the functions of two distinct integrase genes in the mobilization of two metal resistant GIs, G08 and G62, of *Acinetobacter baumannii*. Real-time PCR demonstrated integrase-dependent GI excision, utilizing isopropyl β-D-1-thiogalactopyranoside IPTG-inducible integrase genes in plasmid-based mini-GIs in *Escherichia coli*. In *A. baumannii*, integrase-dependent excision of the original chromosomal GIs could be observed after mitomycin C induction. In both *E. coli* plasmids and *A. baumannii* chromosome, the rate of excision and circularization was found to be dependent on the expression level of the integrases. Susceptibility testing in *A. baumannii* strain ATCC 17978, A424, and their respective ΔG62 and ΔG08 mutants confirmed the contribution of the GI-encoded efflux transporters to heavy metal decreased susceptibility. In summary, the data evidenced the functionality of two integrases in the excision and circularization of the two *Acinetobacter* heavy-metal resistance GIs, G08 and G62, in *E. coli*, as well as when chromosomally located in their natural host. These recombination events occur at different frequencies resulting in genome plasticity and may participate in the spread of resistance determinants in *A. baumannii*.

Keywords: copper resistance; genomic island; integrase; *Acinetobacter baumannii*; mobile genetic element

1. Introduction

Genomic islands (GIs) are discrete gene clusters most of which are found as DNA segments within the chromosome. The GIs were originally known as pathogenicity islands (PAIs) by Hacker et al. in late 1980s, when they were examining the genetic virulence mechanisms in *Escherichia coli* [1]. Genomic islands are variable in size, ranging from 10 to 200 kb, and are usually detected during comparative genomic analysis of different closely related strains [2]. Genomic islands generally harbor genes coding for an integrase or recombinase, but could also carry insertion sequences or transposons within the element contributing to their movement [3,4]. Genomic islands are by definition part of the accessory gene pool of a species and allow for mobilization of whole pathways and gene clusters, thus acting as reservoirs for genetic diversity [5].

In the last few decades, *Acinetobacter baumannii* has been recognized as a multi-drug resistant opportunistic pathogen as a result of being burdened with massive use of broad-spectrum antibiotics [6]. The presence of *A. baumannii* with closely related Gram-negative bacteria in clinical settings has contributed to the development of new resistance mechanisms on top of their own intrinsic factors [7].

Acinetobacter baumannii and the other Gram-negative pathogens are known for their genome plasticity and capability of evolving mainly due to acquiring new virulence determinants carried on mobile genetic elements [8]. This issue has rendered antibiotic therapy ineffective in life-threatening *A. baumannii* infections, and in some cases even with the last line of combination therapy of high doses of antibiotics [9].

In *A. baumannii*, the first GI identified was a large GI of 86 kbp in size, during the sequencing of the epidemic strain AYE, and was named "AbaR1" harboring 45 resistance determinants [10]. Similar GIs were later recognized in *A. baumannii* and named from AbaR0 to AbaR27 [10–12]. Despite being similar in the backbone structure, AbaR-like GIs are often variable in terms of size and genetic composition. For example, the multiple-antibiotic resistance region (MARR) of the AbaR GI harbor a set of genes encoding antibiotic, heavy metal, and antiseptic resistance and efflux determinants [10]. This diversity in AbaR GIs compositions were a result of several events of recombination like integration, excision, and rearrangements [13].

Comparative genomic analysis showed extensive synteny throughout the genome and identified 63 DNA regions, ranging in size from 4–126 kb, all exhibiting certain features of GIs including a group of resistance GIs with different genes encoding resistance to antibiotics and heavy metals which are grouped in clusters [14]. For example, the *aadA1* (streptomycin-resistance encoding) gene, flanked by *satR* (streptothricin-resistance encoding) and *dhfr* (trimethoprim-encoding resistance) genes were found in GIs in clusters. Moreover, genes involved in mercury resistance (*merRCAD* cluster) were found to be located in a separate cluster, and a 4.5 kb DNA segment containing *feoAB* (ferrous iron transport operon), *czc* (tricomponent proton/cation antiporter efflux system), and *ars* (arsenite transporters) genes were co-existing as a group, next to the *cus* (copper resistance) genes conserved in the same chromosomal locations of certain GIs [14]. However, these genes differ in sequence and the overall arrangement from other homologous GIs in *A. baumannii*. This supports the notion that the set of accessory genes had been independently acquired by the different strains.

The two GIs of interest in this study, G08 and G62, harbor a set of putative heavy metal resistance conferring genes which are identical (Figure 1A) [14]. Only a few studies have described the G62 island [14–16], for example the presence of a similar resistance island has been shown in an *A. baumannii* hyper-virulent and outbreak-associated isolate, LAC-4 in China [15]. In LAC-4 clinical isolate, the G62 harbors the exact set of resistance genes and was referred to as a "copper resistance gene cluster"; however, in that strain, G62 was found to be sandwiched between two copies of IS*Aba26* element [15]. The ATCC17978 genome has been extensively analyzed in previous studies [17] and 13 putative zinc/copper resistance efflux pumps have been identified, including the efflux pumps present in G08 and G62 [16] (Figure 1A). The chromosomal region harboring zinc and/or copper efflux genes are likely to have been acquired laterally on mobile genetic elements, with the G62 of ATCC 17978 and LAC-4 being the largest of these elements [16]. Comparative analysis of putative zinc and/or copper efflux systems in *A. baumannii* and *A. baylyi* (strain ADP1) other than the ones in G08 and G62, identified a number of genes ranging between eight (strain SDF) and 18 (strain AB6870155) in each of the strains examined, all of which were chromosomally located [14,18]. Further BLAST search revealed that five strains harbored more than ten genes encoding putative zinc and/or copper efflux components, including ATCC 17978, ATCC 19606T, AB0057, AB6870155, and ACICU. On the other hand, the G08 island is found more frequently in *A. baumannii* including in strains AB0057, AB6870155, and AYE, where the element is inserted into the *dusA* locus encoding for the enzyme tRNA-dihydrouridinesynthase A, catalyzing the post-transcriptional reduction of uridine to dihydrouridine in tRNA [19].

The analysis of the distribution of genomic islands and other genes belonging to the accessory genome has shown in many species that many so-called mobile elements or accessory genes cluster tightly with specific lineages on a phylogenetic core genome tree [5,10,20]. This raises questions on this association which could be due to some positive selection or more likely due to loss of function of the mobile elements. In this work, we aim to test the hypothesis that the two metal-resistant related genomic islands G08 and G62 of *A. baumannii* are still functional.

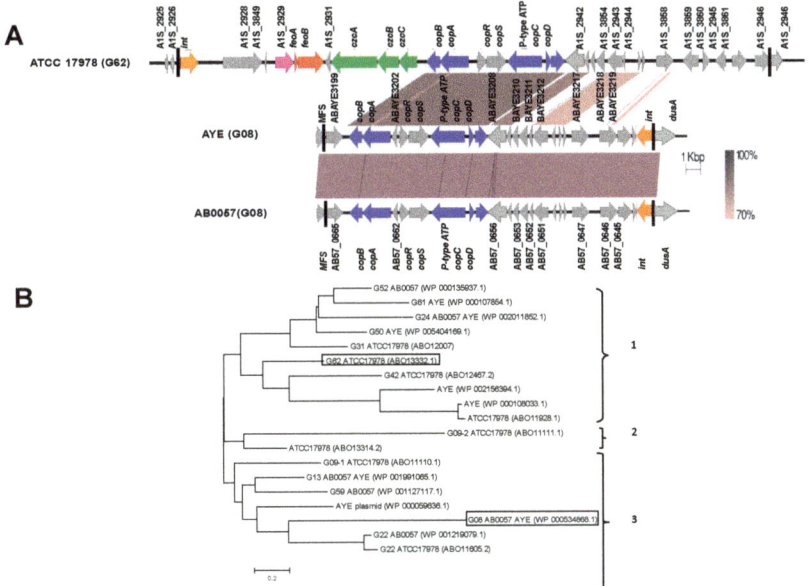

Figure 1. Schematic map of the *Acinetobacter baumannii* strain genomic islands G08 and G62 and phylogenetic tree of genomic island (GI) integrases. (**A**) G62 island is taken from strain ATCC 17978 (accession CP000521; updated refseq NC_009085.1) and G08 from both strains AB0057 (NC_011586.2/CP001182.2) and AYE (NC_010410.1/CU459141.1). The annotated sequences were aligned and visualized by the Easyfig tool [21]. The genes involved in encoding copper efflux systems are shown in blue, all *czc*-like genes *czcA*, *czcB*, and *czcC* encoding cadmium, zinc, and cobalt resistance are represented in green color, genes encoding ferrous iron transport proteins are shown in red (*feoA* and *feoB*), and a putative further heavy metal efflux system A1S_2929 is shown in pink. The integration sites (*att* sites) for both GIs are shown as black vertical lines. The genes encoding the integrases are colored in orange and all other genes in grey. Locus tags of relevant genes are shown above or below the genes. The image is drawn in scale and the percentage of DNA identity between various regions is shown by gradient shading. (**B**) Phylogenetic tree of tyrosine recombinases from strains ATCC 17978, AB0057, and AYE. The evolutionary relationship between phage integrase family proteins (NCBI Reference Sequence) detected in three strains of *A. baumannii* was inferred using the Neighbor-Joining method [22]. Integrases were labelled according to the genomic island number, strain name, followed by Refseq accession numbers. In certain instances, some GIs have not been given a specific number, but instead were indicated by their accession. When two strains are mentioned next to a GI, this means that the same GI is present in both strains. Identical proteins in different genomes which yield identical Refseq numbers are shown only once. The three major branches are indicated by numbers on the right side of the figure. G08 and G62 are indicated by black boxes.

2. Materials and Methods

2.1. Bacterial Strains and Cultivation

A set of *A. baumannii* strains from different geographical origins were used. Hundred strains were collected from clinical samples at Sultan Qaboos University Hospital (SQUH, Oman) between 2012 to 2013. These strains were collected from various body sites of patients admitted in the internal medical wards in SQUH. The rest of the strains were from the collection at the Department of Infection, Immunity and Inflammation of the University of Leicester (Table 1). All strains were stored at −80 °C in 30% glycerol. Strains were re-streaked in LB agar (BD) or broth for liquid cultures.

Table 1. List of *A. baumannii* strains.

Strain	ST	Relevant Characteristics	Reference
A424	1	Clinical isolate from Croatia	[11]
A424 ΔG08	1	G08::aacC1	This study
A424 pWSK129-WHG08	1	Complemented ΔG08 strain	This study
AYE	1	Epidemic MDR type strain, France	[10]
AB0057	1	MDR type strain	[23]
ATCC 17978	437	Reference strain	[17]
ATCC 17978 ΔG62	437	G62::aacC1	This study
ATCC 17978 pWSK129-WHG62	437	Complemented ΔG62 strain	This study
KR3831	1	Clinical isolate from SQUH, Oman	This study

ST = Sequence type [24].

2.2. Genome Analysis

The whole genome phylogenetic single nucleotide polymorphism (SNP) tree was build using the FFP (Feature frequency profile) version 3.19 suite of programs (http://sourceforge.net/projects/ffp-phylogeny/) [25]. As input, the 101 complete *A. baumannii* genomes deposited in GenBank (8 July 2018) were used with the addition of our own two strains A424 (GCA_003185755.1) and KR3831 (GCF_003185745.1/GCA_003185745.1). The matrix of integrase presence in the genomes was generated using as a query the integrase genes present in AB0057, AYE, and ATCC 17978 (shown also in Figure 1B). The matrix was generated using command line BLAST (90% identity; 95% coverage) and the R platform to generate the output.

2.3. Colony Genotpying

The polymerase chain reaction was performed on crude cell extracts from serially diluted suspensions of *E. coli* and *A. baumannii* cells. Thirty µL aliquot of the dilution was boiled for 5 min, and 2 µL of supernatant was used as a polymerase chain reaction (PCR) template. PCR was conducted in a 20 µL reaction volume, containing 1 µL DNA (50 ng/µL), 5 µL 10× reaction buffer (Promega, Madison, WI, USA), 1 µL dNTPs (10 mM), 1 µL of each 10 mM primer, and 0.2 µL Go*Taq* DNA polymerase (Promega). Amplification was performed in a thermal cycler (Mastercycler gradient, Eppendorf, Hamburg, Germany) with an initial denaturation at 95 °C for 2 min, followed by 25 cycles of 95 °C for 1 min, 57 °C for 1 min, and 72 °C for 2 min, and a final extension at 72 °C for 10 min. The PCR products were imaged from a 1% TAE-agarose gel.

2.4. Real-Time PCR

The real-time PCR reactions had a total volume of 20 µL containing 5 µL of template DNA, 10 µL of the SensiMixPlus SYBR Green mastermix (Bioline, London, UK), and 0.5 µL of each 15 µM primer (F-G08-exc and R-G08-circ) or (F-G62-exc and R-G62-circ). Since there was no positive control used, every run included a negative control without target DNA, and all reactions were performed in triplicate. The reactions were performed in an Applied Biosystems Prism (Foster City, CA, USA) model 7500HT Sequence Detection System with the following settings: 40 cycles of 20 s at 95 °C and 1 min at 60 °C. Determinations of cycle threshold (Ct), or the PCR cycle where fluorescence first occurred, were performed automatically by the Sequence Detection Systems software of the instrument (version 2.3; Applied Biosystems, Foster City, CA, USA).

2.5. RNA Extraction and Retrotranscription

Following induction, 5 mL of cells were harvested at predetermined time points by centrifugation for 5 min at 4000× *g*, and resuspended in 1 mL RNALater (Invitrogen, Carlsbad, CA, USA), and stored at 4 °C. The total RNA was extracted using the Geneflow total RNA purification kit protocol (Norgen, Thorold, ON, Canada). The RNA was eluted from the column into a 1.5 mL microcentrifuge tube by addition of 30 µL RNase-free water and stored at −20 °C. Total RNA was quantified by spectrophotometry at A260 (Nanodrop 2000; Fisher ThermoScientific, Waltham, MA, USA), and cDNA was created by taking 1 µg of RNA per each reverse transcription reaction which was 20 µL, and the procedure was completed according to the High Capacity RNA-to-cDNA kit (Applied Biosystems).

2.6. Construction of Plasmids with G08 and G62 Mini-GIs

To test the functionality of the integrase genes of G08 and G62 in excising their respective GIs, two plasmid constructs were created. Mini-islands were generated by creating smaller circular molecules with precise site-specific excision via the attachment sites *attL/attR* included within the left and right flanking regions, and integrase coding gene cloned in a plasmid under an inducible promoter. The new fragments generated were later cloned into pUC18 vector. The primer pairs F-LF-08/R-LF-08 and F-RF-08/R-RF-08 (Table S1) were used to amplify the left and right G08 flanking regions including the *att* sites from strain A424 (Genbank accession: GCA_003185755.1). Similarly, the left- and right-flanking regions including the *att* sites of G62 GI were amplified from the strain ATCC 17978 using the primer pairs F-LF-62/R-LF-62 and F-RF-62 and R-RF-62 (Table S1), respectively. The two integrase genes G08*int* (A424_1287 from A424) and G62*int* (A1S-2927 from ATCC 17978) were separately amplified by PCR using primer pairs F-G08int/R-G08int and F-G62int/R-G62int, respectively. Amplicons containing the integrases were ligated in the HindIII within the multiple cloning site (MCS) to be expressed under the *lacZ* promoter in the final recipient vector. The three PCR fragments (LF, RF, and integrase) were finally joined by fusion PCR resulting in a recombinant DNA product.

2.7. Construction of Inducible Plasmids for A. baumannii

The vectors carrying the min-islands of G08 and G62 were sub-cloned into pWSK129, a low-copy-number plasmid carrying aminoglycoside 3′-phosphotransferase (*aphA1*) gene conferring kanamycin-resistance (Km^R) [26]. As this plasmid turned out to be non-functional in *A. baumannii*, we amplified the origin of transfer from pWH1277, a cryptic plasmid from an *A. lwoffii* strain fragment of pWH1266 (kindly donated by Philip Rather, Emory University, USA), using the primer pair PR3136 and PR3137 (Table S1). These pWSK129-WH plasmids were successfully transferred into competent *A. baumannii* knock-out strains (A424 and ATCC 17978) by conjugation.

2.8. Suicide Vector-Based Allelic Exchange for Mutant Construction in A. baumannii

Deletion mutants of the GIs were constructed in *A. baumannii* using the suicide vector pJTOOL-3 [27], containing 500 bp long fragments of each of the borders of either G08 or G62. For transformation, *E. coli* CC118λ*pir* and S17.1λ*pir* were used as a host for replication and as a conjugative strain, respectively. Plasmid single cross-over insertion into *Acinetobacter* was selected by gentamicin and the double cross over by plating on 6% sucrose containing to check for the loss of the levansucrase *sacB* gene of pJTOOL-3. The expected genotype was obtained in all three randomly selected colonies that possessed the expected chloramphenicol-sensitive and gentamicin-resistant phenotype.

2.9. IPTG and Mitomycin C Induction

For isopropyl β-D-1-thiogalactopyranoside (IPTG), 5 mL overnight culture of *A. baumannii* or *E. coli* were diluted at 1:100 into fresh LB and then incubated at 37 °C in the shaking incubator at $200\times g$, until OD$_{600nm}$ = 0.2 is reached. IPTG was added at concentration of 1.0 mM, and the cultures were then incubated, with 500 µL of the culture removed at time points 0, 4, 8, and 24 h for DNA preparation and qPCR analysis. In *Staphylococcus*, mitomycin was shown to induce excision of genomic islands [28]. For mitomycin C, induction 5 mL of overnight cultures of *A. baumannii* strains were treated with sub-lethal concentrations of mitomycin C MIC (0.5 times the MIC) for 2 h. Mitomycin C MIC of AYE, AB0057, A424, KR3831, and ATCC 17978 was found to range from 32–64 µg/mL. Non-induced cultures were run alongside in each occasion under identical conditions.

2.10. Metal Susceptibility Testing

For testing of susceptibility to the heavy-metal salts, analytical-grade salts of $CdCl_2 \cdot H_2O$, $CoCl_2 \cdot 6H_2O$, $NiSO_4 \cdot 6H_2O$, and $ZnSO_4 \cdot 7H_2O$, $CuSO_4 \cdot 5H_2O$, $FeSO_4 \cdot 7H_2O$, $MnSO_4 \cdot H_2O$. and As_2SO_3 (Sigma–Aldrich, Gillingham, UK) were used to prepare 1.0 M stock solutions, which were dissolved in ultrapure distilled water and later filter-sterilized and added to the medium at final concentrations of 1 mM. MIC and MBC assays to heavy metals was performed as described by the Clinical and Laboratory Standards Institute (CLSI) guidelines using a broth microdilution method [29]. Briefly, starting inocula of 1×10^5 CFU/mL of all *A. baumannii* strains were aliquoted in 96-well plates containing serial dilutions of each metal compound in the range 0.02–10 mM using MHB (Oxoid Ltd., Basingstoke, UK).

3. Results

The distribution of integrases, as proxies of their genomic islands, varies widely in different lineages of a whole genome phylogenetic SNP tree constructed on all complete deposited *A. baumannii* genomes (Figure 2). One of the integrases not associated to genomic islands (*int1* ABAYE_RS10930) present in almost all isolates, some others such as G08, G13, G16, and G42 are detected in many lineages and may be present only in a subgroup of isolates of a given ST. Other integrases like G09, G31 or G62 are present only in single or very few STs. The integrase of the metal resistance associated genomic islands G08 and G62 are representatives of this latter groups being G08 present in ST1, 25, 26, 52, 79, 81, 126, 138, 229, 422, and 638m while G62 only in ST10 and 437 (Figure 2).

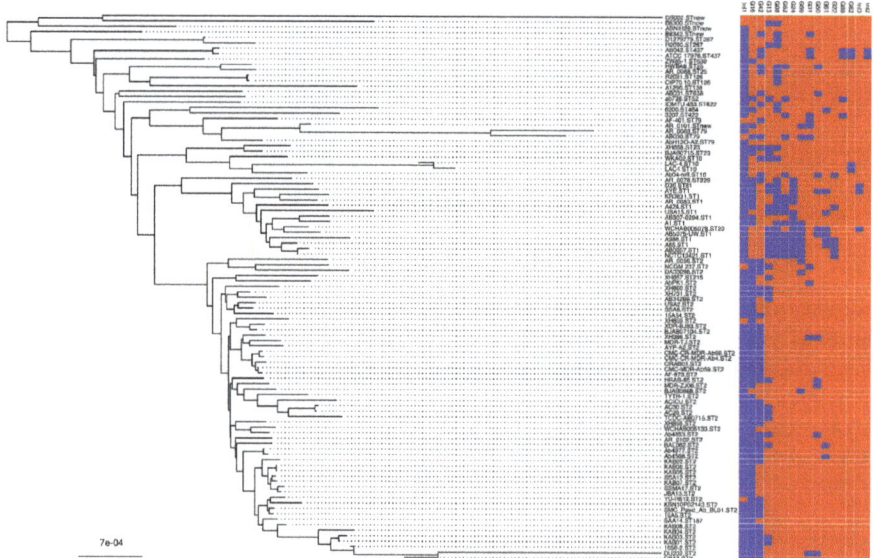

Figure 2. Integrase distribution on an *A. baumannii* phylogenetic tree. The whole genome phylogenetic SNP tree was built using the FFP and input from the 105 complete *A. baumannii* genomes deposited in GenBank (08/07/2018) with the addition of strains A424 and KR3831. Strain names include the MLST "Pasteur" sequence type. The matrix of 16 integrases on the right part of the figure includes all integrase genes present in AB0057, AYE, and ATCC 17978 (red absent; blue present). The integrases of the GI are numbered according to Di Nocera et al. [14]. Three further integrases present in the three stains but not associated to genomic islands were included in the analysis and named *int1* (ABAYE_RS10930), *int2* (AUO97_RS03560), and the p3ABAYE integrase *int3* (ABAYE_RS00155).

In order to test the hypothesis that these islands, even if present only in defined lineages are still mobile, we selected the genomic islands G08 and G62 respectively in strains AB0057 and ATCC 17978 [14] (Figure 1A). Phylogenetic analysis of the respective GI integrases confirmed that the G08*int* and G62*int* are not related, and each belong to a separate clade within the *Acinetobacter* GI-related tyrosine recombinases (Figure 1B). Both GIs carry the *copABCD* and *copRS* copper resistance genes [16], and G62 carries in addition the cadmium, zinc, iron, and cobalt resistance genes (Figure 1A). To check the distribution of G08 and G62, we screened by PCR a 100 sample collection of *A. baumannii* clinical isolates obtained from SQUH, Oman from 2012 to 2013 using primers for conserved sequences flanking the target region. The PCR screening analysis yielded a possibly occupied G08 in only one clinical isolate, and the presence of G08 was confirmed by WGS (strain KR3831, accession GCF_003185745.1, GCA_003185745.1). To test the contribution of G08 and G62 to metal susceptibility phenotypes, we constructed deletion mutants respectively in strain A424 and ATCC 17978 (Table 2). In the G08 knock out mutant, only the minimal inhibitory concentration (MIC) for manganese ($MnSO_4$) decreased form 1 µg/mL to 0.5 µg/mL, while the copper MIC remained unchanged. Deletion of G62 in ATCC 17978 resulted in a decrease of the MIC of zinc ($ZnSO_4$ from 4 to 2 µg/mL), cobalt ($CoCl_2$ 4 to 2 µg/mL), cadmium ($CdCl_2$ 4 to 2 µg/mL), and nickel ($NiSO_4$ 4 to 2 µg/mL), and again no decrease in the MIC of copper was detected ($CuSO_4$ 8 µg/mL). No differences were observed in susceptibility to iron ($FeSO_4$) and arsenic (As_2O_3). Four independent ko mutants were assayed against the wild type in all tests and the difference in MICs found to be statically relevant.

Table 2. Metal susceptibility testing wild type (wt) and GI mutants.

	ATCC 17978	ATCC 17978 ΔG62	A424	A424 ΔG08
$ZnSO_4$	4 *	2	0.5	0.5
$CuSO_4$	8	8	8	8
$CdCl_2$	4	2	2	2
$MnSO_4$	10	10	1	0.5
$FeSO_4$	4	4	4	4
$CoBr_2$	4	2	4	4
$NiSO_4$	10	5	10	10
As_2O_3	4	4	4	4

* MIC in µg/mL.

To test the functionality of integrase genes of the G08 and G62 islands in excising their respective GIs in a heterologous *E. coli* background, mini-islands were generated. Mini-GIs were obtained by cloning the integrase coding genes G08*int* (locus_tag A424_1287, strain A424, NC_011586.2/CP001182.2) and G62*int* (locus_tag A1S-2927, strain ATCC 17978, NC_009085.1/CP000521) under control of the *lacZ* promoter and flanked by *attL* and *attR* sites [30] (Figure 3). The resultant plasmids pUC18-G08*int* and pUC18-G62*int* carrying the mini-islands of G08 and G62 were complemented in their respective knock out strains to be tested for integrase dependent excision using sets of divergent primers. The *dusA*–associated integrases have been shown to excise as circular elements with the restoration of the junction [19]. Our data showed IPTG-dependent mini-GI excision, by amplification of both the reconstituted target site and the junction of the circular form, for both the G08 and G62 constructs over the whole growth phase in liquid medium (Figure 4A,B). No or only marginal excision of the mini-GIs was detected without IPTG induction of the plasmid-carried mini-GIs (Figure 4A,B). To test for the excision of mini-GIs in *A. baumannii* background, the constructs were transferred on pWH1266 and pWK129 shuttle vectors. The IPTG-induced excision of the mini-GIs in *A. baumannii* was detected by amplification of the circular intermediates of the G08 and G62 mini-GIs using the same primer sets as in *E. coli* (Figure 4C).

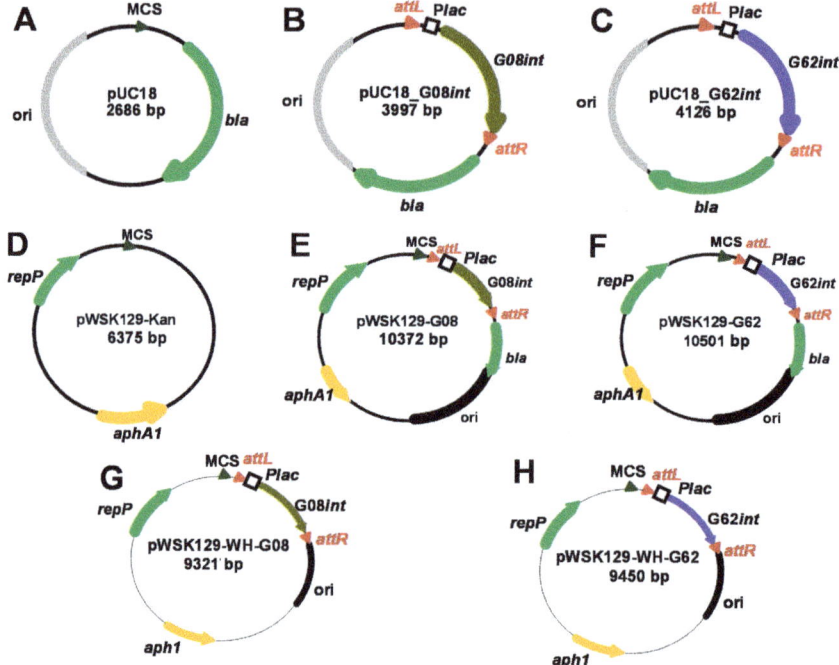

Figure 3. G08 and G62 mini-GI vectors for *E. coli* and *A. baumannii*. pUC18 (**A**) used as a vector harboring *bla* gene conferring ampicillin resistance (green), *E. coli* origin of replication *ori* (grey), multiple cloning site MCS shown as green triangle, to clone mini-GIs containing upstream and downstream flanking borders with *attL/attR* sites (red) of the respective islands as well as integrase genes (**B**) G08*int* (olive) and (**C**) G62*int* (blue). The IPTG-inducible P*lac* promoter was fused by PCR in front of the integrase genes during construction of the plasmids (**B,C**). Plasmids used to assess integrase activity in *A. baumannii* (**D–F**) were constructed by fusing pUC18-based G08 and G62 mini-island vectors with pWSK129 carrying the *aphA1* conferring kanamycin resistance (orange) and pLG339 replication initiation protein (*repP*, light green). To allow for stable transfer of the constructs to *Acinetobacter* pWSK129-WH-G08 (**G**) and pWSK129-WH-G62 (**H**) were constructed. pWSK129-kan plasmids were used to clone the *Acinetobacter* origin of Replication (*oriR*) from pWH1266 (black color) resulting in pWSK129-WH-derived new constructs compatible with *A. baumannii* strains.

To test the dynamics of excision and reconstitution of the chromosomal target site of the G08 and G62 islands in *Acinetobacter*, we amplified the circular intermediates and targets in our four G08-positive AB0057, AYE, A424, and KR3831 strains, as well as the G62-positive strain ATCC 17978. To test GI excision, bacteria were grown to mid log phase and either tested directly or after exposure to 38 μg/mL of mitomycin C for 2 h. The junction and the circular forms were sequenced by Sanger sequencing to map the *att* sites of G08 and G62. In ATCC 17978, the *att* sites of G62 were identical at both ends with the consensus "AATAACTTTAAAGATTAA" [14]. However, our data show that in all examined strains, G08 was flanked by two 17 bp semiconserved attachment sequences [19], which showed variation in the *attR/attL* and *attP/attB* of a single nucleotide (SNP) in strains AYE and KR3831, and of two SNPs in strain AB0057 compared to ATCC 17978 (Figure 5). To examine the reason for these differences in the *att* sites among G08-harbouring strains, the database was searched for *att* sites of strains devoid of G08 and showed that two variable alleles of the *attB* sites in *dusA* gene exist in two G08-negative strains ATCC 17978 and AB307-0294. The ATCC 17978 had the more frequently occurring allele with a single SNP, whereas AB307-0294 had two SNPs similar to those seen in AB0057. This could mean that either of the *att* sites could be recognized by the integrases as preferable integration/excision sites during mobilization of the G08.

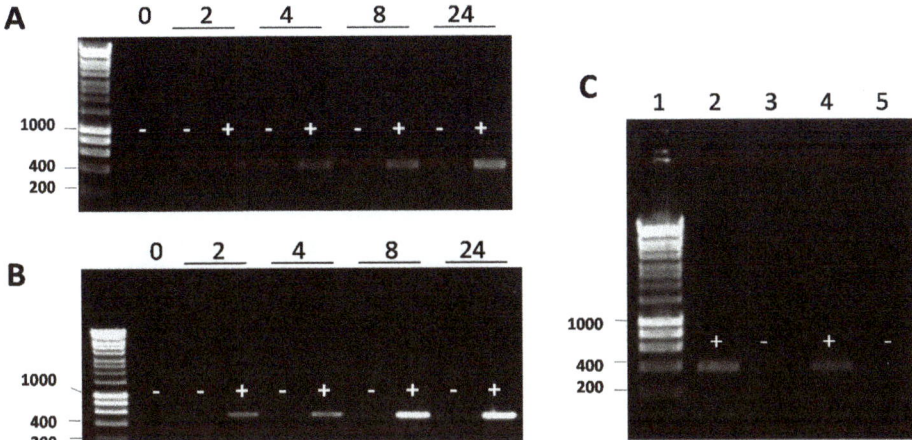

Figure 4. Integrase induction drives excision of G08 and G62 mini-islands in *E. coli* and *A. baumannii*. Gel images of circularized G08 (from AB0057) (**A**) and G62 (from ATCC 17978) (**B**) mini-islands respectively, with (+) and without (−) IPTG induction. Mini-GI excision was tested by amplification of circular intermediates using outwards facing primer sets yielding 567 bp product for G08 (**A**) and a 489 bp product for G62 excision (**B**). Different time points (in hours) after IPTG induction in early exponential phase are shown (legend above gel). Integrase dependent excision of G08 and G62 mini-islands in *A. baumannii* is shown in panel C. Gel images of circularized G08 and G62 mini-islands 8 h after IPTG induction with the same primers as in *E. coli* (panel **A** and **B**). The non-induced samples in lane 3 for G8 and in lane 5 for G62 in lanes 4 and 5 of panel C. The marker is Gene Ruler 1 kb and the size of some bands in bp is given on the left of each gel.

Figure 5. Allelic variation in the *attB* site of *A. baumannii* strains. (**A**) Shows the variability of *att* sequences in AB0057 and AYE and KR3831 compared to ATCC 17978. The *att* sites sequenced in this study are abbreviated as (seq). (**B**) Shows two variable alleles of *attB* sequences in *dusA* in two *A. baumannii* strains ATCC 17978 and AB307-0294 lacking the G08 island. The *attB* site is underlined. The variable alleles differing from the consensus sequence by two or three SNPs are shown in red.

Real-time PCR was performed to quantify the circular forms after excision and variation in number of excised elements between samples was corrected by arbitrarily setting the values of strain AB0057 to 1 and expressing the data in the other strains and after mitomycin C treatment at fold change. The primers' efficiencies were checked by performing serial dilutions (Figure S1). Without any induction, the excision of the elements was low and no strong baseline variation between the G08 carrying strains AB0057, AYE, A424, and KR3831 were seen for both the detection of the reconstituted target site and the circular intermediates (Figure 6A). After exposure of *A. baumannii* cells to mitomycin C, the detection of G08 circular intermediates increased in all strains significantly 4- to 8-fold (Figure 6B). Similarly, when testing excision of the G62 element in strain ATCC 17978, we detected a significant increase of about 4-fold in the formation of circular intermediates after exposure to mitomycin C (Figure 6C).

To test whether the increased excision of the G08 and G62 elements after mitomycin C treatment in *A. baumannii* was integrase mediated, we tested the expression of the integrases G08*int* (A424_1287 from A424) and G62*int* (A1S-2927 from ATCC 17978). This was done by real-time PCR with primers internally to G08*int* and G62*int*. Data show significant upregulation of integrase expression after mitomycin C exposure of about 5-fold for G08*int* and 6-fold for G62*int* (Figure 6D).

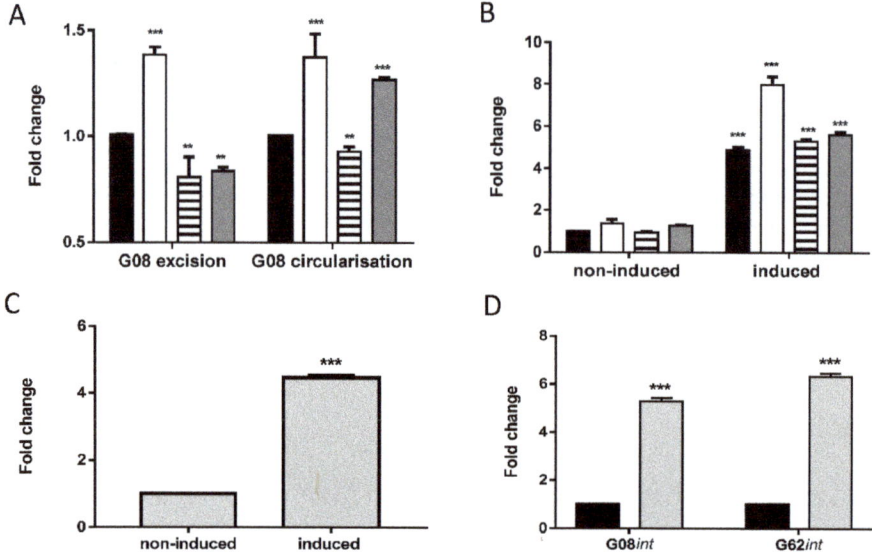

Figure 6. Quantification of G08 and G62 island excision and integrase expression in *A. baumannii* strains with and without mitomycin C induction. The data in (**A**) represent the real-time PCR detection of reconstituted target sites after excision of G08 (excision) and of the circular intermediates (circularization) without any induction. The G08 positive strains are AB0057 strain (black, accession NC_011586.2/CP001182.2). AYE (white, accession NC_010410.1/CU459141.1), A424 (striped, accession GCA_003185755.1), KR3831 (grey, accession: GCF_003185745.1. GCA_003185745.1), and data are normalized to AB0057. (**B**) Reports the variation in G08 excision by detecting circular intermediates with and without mitomycin C induction (0.75 × MIC) for 2 h. The asterisks represent the significance of change in each strain when compared to AB0057. Error bars represent SEM of three independent replicates. (**C**) Repost real-time PCR quantification of G62 circular intermediates with and without mitomycin C induction. (**D**) Shows the expression G08*int* (A424_1287 from strain A424, protein ID: PRJNA473420:DMB35_05960, Genbank accession: GCA_003185755.1) and G62*int* (A1S-2927 from ATCC 17978) measured by real-time PCR without or with mitomycin C exposure (grey). Error bars indicate the SEM of three independent replicates in each experiment as analyzed by two-way ANOVA test. ** $p < 0.01$, *** $p < 0.001$.

4. Discussion

Previous comparative genomic analysis of *A. baumannii* explored the chromosomal loci of 63 GIs including the two GIs objects of this study, G08 and G62, within seven strains belonging to different genotypes ST1, ST2, ST25, ST77, and ST78 [14]. The genomic alignment revealed GIs of various functions such as those encoding for surface components and transport systems, as well as resistance to drugs and heavy metals. More recently, data of a pan-genome analysis of 50 *A. baumannii* isolates and 249 previously sequenced *A. baumannii* strains were compiled [31], and their dataset confirmed the diversity of gene pools found within the GIs identified as an adaptive response of the *A. baumannii* strains to facilitate their survival in a nutrient-deficient environment. In many instances, the integrases of these GIs were found to be non-functional in various species due to frameshift and nonsense-mutations [32–34]. This resulted in most of these GIs being permanently positioned in their chromosomal location [35]. Therefore, this work aimed to check whether the G08 and G62 integrases are functional and contribute to excision of the islands by generating mini-GIs carrying the essential components for mobilization of integrases and *att* sites. In addition, direct excision from the chromosomal host was tested via mitomycin C induction. This approach has been previously employed in other *A. baumannii* studies [36] as well as other GIs circularization and have successfully demonstrated excision after the use of modified protocols [37–39]. The response to mitomycin C that was studied and research showed that pathogenicity island excision was facilitated by mitomycin C which induces an SOS response [28,40]. The data presented here have confirmed that the use of mitomycin C can effectively induce the excision of the GIs G08 and G62 via the visualization of the bands in gel electrophoresis. Similar observations were reported on other *dusA/dusB* associated integrases [19]. Real-time PCR data supported our observation and even demonstrated excision and circular events occurring at later cycles without induction. Sequencing of the circular intermediates and the chromosomal junctions after excision showed the possibility of having multiple *att* sites in AB0057, which probably can lead to having multiple insertions occurring at different frequencies depending on the most prevalent or preferable sites.

It could also be argued that the importance of such variability in attachment sites is probably minor, due to the low excision frequencies under laboratory conditions. Studies in which integrase activity was assessed in *A. baumannii* background are limited, most of which were performed in integron studies in the closely related non-pathogenic species *Acinetobacter baylyi* due to the ease of genetic modification and transformations [41–43]. In this context, our study demonstrated integrase-dependent excision and circularization in both tested metal-resistance GIs that were hypothesized to be non-mobile in the majority of the cases.

Moreover, the contribution of these two GIs towards metal susceptibility phenotypes was addressed by generating deletion mutants. Susceptibility data of both G08 and G62 mutants respectively of ATCC 17978 and A424 showed a significant, but minor decrease in the MIC for zinc ($ZnSO_4$), cobalt ($CoCl_2$), cadmium ($CdCl_2$), and nickel ($NiSO_4$), whereas the MIC remained unchanged for copper ($CuSO_4$), iron ($FeSO4$), and arsenic (As_2O_3). These GI deletion mutants showed slight phenotypic changes as compared to their wild type counterparts, and their tolerance to the rest of the metals could be attributed the presence of other chromosomal efflux transports. Putative efflux pumps for copper and zinc have been previously analyzed in *A. baumannii* ATCC 17978 by TransAAP [44]. Thirteen efflux systems were identified that belong to either the CDF family, P-type ATPase family, CorA metal ion transporter family, HME family of RND transporters, or CopB-type family of Cu exporters. Transcriptional data by qPCR revealed that some of these putative efflux genes were induced by addition of either or both zinc and copper [16]. The MIC/MBC data of broth microdilution were identical in both A424 and ATCC 17978 wild-type and mutants for copper, cobalt, iron, and nickel. This observation was partially explained by the presence of *cnrCBA* mediating resistance to cobalt-nickel, as well as *czcCBA* genes which are cobalt-zinc-cadmium resistance determinants in this bacterial strain [45,46]. Similar iron susceptibility data in both wild-type and mutants of ATCC 17978 and A424 could be due to the presence of putatively non-functional FeoB in

ATCC 17978, and the tolerance could be attributed to another iron efflux transporter. The additive value of metal resistance carried on mobile elements, for example the copper resistance conferred by the *Staphylococcus aureus* COMER element in USA3000, still confers the strain's increased resistance to copper-related macrophage killing, and showed significant higher virulence, despite the weak phenotypes detected in vitro [47].

Collectively, this work reveals that metal resistance GIs in *A. baumannii* are of clinical significance as they confer metal resistance phenotypes, and their mobility could be demonstrated by the integrase assays. This issue can raise concern as these metal-resistance GIs could be readily transferred among strains (and patients) in clinical settings, and could be viewed as vehicles disseminating resistance as well as other potential virulence genes. The use of sub-lethal doses of antimicrobials and metal-containing compounds not only accelerate their resistance, but could also potentiate their virulence and spread in hospital environments.

Supplementary Materials: The following are available online at http://www.mdpi.com/2073-4425/9/7/366/s1, Table S1: List of primers, Figure S1: Standard curves for the qPCR.

Author Contributions: Z.A.-J. performed the experiments and wrote the manuscript, R.Z., performed genome analysis and contributed to revision of the manuscript, E.H.-P. contributed to the constructions of the knock out strains J.D.R. contributed to the genome analysis and the manuscript Z.A.-M. was responsible for the strain collection, K.R. designed and supervised the initial phases of the project, M.R.O. supervised the work and wrote the manuscript.

Funding: Z.A.-J. was funded by a PhD fellowship of the Sultan Qaboos University, Oman. This research received no further external funding.

Acknowledgments: Authors thank Philip Rather, Emory University, USA for plasmid pWH1266.

Conflicts of Interest: The authors declare no conflict of interest.

References

1. Hacker, J.; Blum-Oehler, G.; Muhldorfer, I.; Tschape, H. Pathogenicity islands of virulent bacteria: Structure, function and impact on microbial evolution. *Mol. Microbiol.* **1997**, *23*, 1089–1097. [CrossRef] [PubMed]
2. Juhas, M.; van der Meer, J.R.; Gaillard, M.; Harding, R.M.; Hood, D.W.; Crook, D.W. Genomic islands: Tools of bacterial horizontal gene transfer and evolution. *FEMS Microbiol. Rev.* **2009**, *33*, 376–393. [CrossRef] [PubMed]
3. Buchrieser, C.; Brosch, R.; Bach, S.; Guiyoule, A.; Carniel, E. The high-pathogenicity island of *Yersinia pseudotuberculosis* can be inserted into any of the three chromosomal *asn tRNA* genes. *Mol. Microbiol.* **1998**, *30*, 965–978. [CrossRef] [PubMed]
4. Gal-Mor, O.; Finlay, B.B. Pathogenicity islands: A molecular toolbox for bacterial virulence. *Cell. Microbiol.* **2006**, *8*, 1707–1719. [CrossRef] [PubMed]
5. Hsiao, W.W.L.; Ung, K.; Aeschliman, D.; Bryan, J.; Finlay, B.B.; Brinkman, F.S.L. Evidence of a large novel gene pool associated with prokaryotic genomic islands. *PLoS Genet.* **2005**, *1*, e62. [CrossRef] [PubMed]
6. Peleg, A.Y.; de Breij, A.; Adams, M.D.; Cerqueira, G.M.; Mocali, S.; Galardini, M.; Nibbering, P.H.; Earl, A.M.; Ward, D.V.; Paterson, D.L.; et al. The success of *Acinetobacter* species; genetic, metabolic and virulence attributes. *PLoS ONE* **2012**, *7*, e46984. [CrossRef] [PubMed]
7. Camp, C.; Tatum, O.L. A Review of *Acinetobacter baumannii* as a highly successful pathogen in times of war. *Lab. Med.* **2010**, *41*, 649–657. [CrossRef]
8. Bergogne-Bérézin, E.; Friedman, H.; Bendinelli, M. (Eds.) *Acinetobacter Biology and Pathogenesis*; Springer: New York, NY, USA, 2008; 220p.
9. Amat, T.; Gutiérrez-Pizarraya, A.; Machuca, I.; Gracia-Ahufinger, I.; Pérez-Nadales, E.; Torre-Giménez, A.; Garnacho-Montero, J.; Cisneros, J.M.; Torre-Cisneros, J. The combined use of tigecycline with high-dose colistin might not be associated with higher survival in critically ill patients with bacteraemia due to carbapenem-resistant *Acinetobacter baumannii*. *Clin. Microbiol. Infect.* **2018**, *24*, 630–634. [CrossRef] [PubMed]
10. Fournier, P.E.; Vallenet, D.; Barbe, V.; Audic, S.; Ogata, H.; Poirel, L.; Richet, H.; Robert, C.; Mangenot, S.; Abergel, C.; et al. Comparative genomics of multidrug resistance in *Acinetobacter baumannii*. *PLoS Genet.* **2006**, *2*, e7. [CrossRef] [PubMed]

11. Kochar, M.; Crosatti, M.; Harrison, E.M.; Rieck, B.; Chan, J.; Constantinidou, C.; Pallen, M.; Ou, H.Y.; Rajakumar, K. Deletion of Tn*AbaR23*rResults in both expected and unexpected antibiogram changes in a multidrug-resistant *Acinetobacter baumannii* strain. *Antimicrob. Agents Chemother.* **2012**, *56*, 1845–1853. [CrossRef] [PubMed]
12. Zhu, L.; Yan, Z.; Zhang, Z.; Zhou, Q.; Zhou, J.; Wakeland, E.K.; Fang, X.; Xuan, Z.; Shen, D.; Li, Q.-Z. Complete genome analysis of three *Acinetobacter baumannii* clinical isolates in China for insight into the diversification of drug resistance elements. *PLoS ONE* **2013**, *8*, e66584. [CrossRef] [PubMed]
13. Post, V.; Hall, R.M. AbaR5, a large multiple-antibiotic resistance region found in *Acinetobacter baumannii*. *Antimicrob. Agents Chemother.* **2009**, *53*, 2667–2671. [CrossRef] [PubMed]
14. Di Nocera, P.P.; Rocco, F.; Giannouli, M.; Triassi, M.; Zarrilli, R. Genome organization of epidemic *Acinetobacter baumannii* strains. *BMC Microbiol.* **2011**, *11*, 224. [CrossRef] [PubMed]
15. Ou, H.Y.; Kuang, S.N.; He, X.; Molgora, B.M.; Ewing, P.J.; Deng, Z.; Osby, M.; Chen, W.; Xu, H.H. Complete genome sequence of hypervirulent and outbreak-associated *Acinetobacter baumannii* strain LAC-4: Epidemiology, resistance genetic determinants and potential virulence factors. *Sci. Rep.* **2015**, *5*, 8643. [CrossRef] [PubMed]
16. Hassan, K.A.; Pederick, V.G.; Elbourne, L.D.; Paulsen, I.T.; Paton, J.C.; McDevitt, C.A.; Eijkelkamp, B.A. Zinc stress induces copper depletion in *Acinetobacter baumannii*. *BMC Microbiol.* **2017**, *17*, 59. [CrossRef] [PubMed]
17. Smith, M.G.; Gianoulis, T.A.; Pukatzki, S.; Mekalanos, J.J.; Ornston, L.N.; Gerstein, M.; Snyder, M. New insights into *Acinetobacter baumannii* pathogenesis revealed by high-density pyrosequencing and transposon mutagenesis. *Genes Dev.* **2007**, *21*, 601–614. [CrossRef] [PubMed]
18. Eijkelkamp, B.; Stroeher, U.; Hassan, K.; Paulsen, I.; Brown, M. Comparative analysis of surface-exposed virulence factors of *Acinetobacter baumannii*. *BMC Genom.* **2014**, *15*, 1020. [CrossRef] [PubMed]
19. Farrugia, D.N.; Elbourne, L.D.; Mabbutt, B.C.; Paulsen, I.T. A novel family of integrases associated with prophages and genomic islands integrated within the tRNA-dihydrouridine synthase A (*dusA*) gene. *Nucleic Acids Res.* **2015**, *43*, 4547–4557. [CrossRef] [PubMed]
20. Croucher, N.J.; Coupland, P.G.; Stevenson, A.E.; Callendrello, A.; Bentley, S.D.; Hanage, W.P. Diversification of bacterial genome content through distinct mechanisms over different timescales. *Nat. Commun.* **2014**, *5*, 5471. [CrossRef] [PubMed]
21. Sullivan, M.J.; Petty, N.K.; Beatson, S.A. Easyfig: A genome comparison visualizer. *Bioinformatics* **2011**, *27*, 1009–1010. [CrossRef] [PubMed]
22. Saitou, N.; Nie, M. The neighbor-joining method: A new method for reconstructing phylogenetic trees. *Mol. Biol. Evol.* **1987**, *4*, 406–425. [PubMed]
23. Adams, M.D.; Goglin, K.; Molyneaux, N.; Hujer, K.M.; Lavender, H.; Jamison, J.J.; MacDonald, I.J.; Martin, K.M.; Russo, T.; Campagnari, A.A.; et al. Comparative genome sequence analysis of multidrug-resistant *Acinetobacter baumannii*. *J. Bacteriol.* **2008**, *190*, 8053–8064. [CrossRef] [PubMed]
24. Diancourt, L.; Passet, V.; Nemec, A.; Dijkshoorn, L.; Brisse, S. The population structure of *Acinetobacter baumannii*: expanding multiresistant clones from an ancestral susceptible genetic pool. *PLoS ONE* **2010**, *5*, e10034. [CrossRef] [PubMed]
25. van Vliet, A.H.; Kusters, J.G. Use of alignment-free phylogenetics for rapid genome sequence-based typing of *Helicobacter pylori* virulence markers and antibiotic susceptibility. *J. Clin. Microbiol.* **2015**, *53*, 2877–2888. [CrossRef] [PubMed]
26. Wang, R.F.; Kushner, S.R. Construction of versatile low-copy number vectors for cloning, sequencing and gene expression in *Escherichia coli*. *Gene* **1991**, *100*, 195–199. [CrossRef]
27. van Aartsen, J.J.; Rajakumar, K. An optimized method for suicide vector-based allelic exchange in *Klebsiella pneumoniae*. *J. Microbiol. Methods* **2011**, *86*, 313–319. [CrossRef] [PubMed]
28. Ubeda, C.; Maiques, E.; Knecht, E.; Lasa, I.; Novick, R.P.; Penades, J.R. Antibiotic-induced SOS response promotes horizontal dissemination of pathogenicity island-encoded virulence factors in staphylococci. *Mol. Microbiol.* **2005**, *56*, 836–844. [CrossRef] [PubMed]
29. Barry, A.L.; Craig, W.A.; Nadler, H.; Reller, L.B.; Sanders, C.C.; Swenson, J.M. Methods for determining bactericidal activity of antimicrobial agents; approved guideline. *NCCLS* **1999**, *19*, M26-A.
30. Yanisch-Perron, C.; Vieira, J.; Messing, J. Improved M13 phage cloning vectors and host strains: Nucleotide sequences of the M13mpI8 and pUC19 vectors. *Gene* **1985**, *33*, 103–119. [CrossRef]

31. Chan, A.P.; Sutton, G.; DePew, J.; Krishnakumar, R.; Choi, Y.; Huang, X.Z.; Beck, E.; Harkins, D.M.; Kim, M.; Lesho, E.P.; et al. A novel method of consensus pan-chromosome assembly and large-scale comparative analysis reveal the highly flexible pan-genome of *Acinetobacter baumannii*. *Genome Biol.* **2015**, *16*, 143. [CrossRef] [PubMed]
32. Nemergut, D.R.; Robeson, M.S.; Kysela, R.F.; Martin, A.P.; Schmidt, S.K.; Knight, R. Insights and inferences about integron evolution from genomic data. *BMC Genom.* **2008**, *9*, 261. [CrossRef] [PubMed]
33. Gillings, M.R.; Holley, M.P.; Stokes, H.W.; Holmes, A.J. Integrons in *Xanthomonas*: A source of species genome diversity. *Proc. Natl. Acad. Sci. USA* **2005**, *102*, 4419–4424. [CrossRef] [PubMed]
34. Cambray, G.; Sanchez-Alberola, N.; Campoy, S.; Guerin, É.; Da Re, S.; González-Zorn, B.; Ploy, M.C.; Barbé, J.; Mazel, D.; Erill, I. Prevalence of SOS-mediated control of integron integrase expression as an adaptive trait of chromosomal and mobile integrons. *Mob. DNA* **2011**, *2*, 6. [CrossRef] [PubMed]
35. Osborn, A.M.; Boltner, D. When phage, plasmids, and transposons collide: Genomic islands, and conjugative- and mobilizable-transposons as a mosaic continuum. *Plasmid* **2002**, *48*, 202–212. [CrossRef]
36. Rose, A. Tn*AbaR1*: A novel Tn7-related transposon in *Acinetobacter baumannii* that contributes to the accumulation and dissemination of large repertoires of resistance genes. *Biosci. Horiz.* **2010**, *3*, 40–48. [CrossRef]
37. Doublet, B.; Boyd, D.; Mulvey, M.R.; Cloeckaert, A. The *Salmonella* genomic island 1 is an integrative mobilizable element. *Mol. Microbiol.* **2005**, *55*, 1911–1924. [CrossRef] [PubMed]
38. Rameckers, J.; Hummel, S.; Herrmann, B. How many cycles does a PCR need? Determinations of cycle numbers depending on the number of targets and the reaction efficiency factor. *Naturwissenschaften* **1997**, *84*, 259–262. [CrossRef] [PubMed]
39. Dominguez, N.M.; Hackett, K.T.; Dillard, J.P. XerCD-mediated site-specific recombination leads to loss of the 57-kilobase gonococcal genetic island. *J. Bacteriol.* **2011**, *193*, 377–388. [CrossRef] [PubMed]
40. Aranda, J.; Poza, M.; Shingu-Vazquez, M.; Cortes, P.; Boyce, J.D.; Adler, B.; Barbe, J.; Bou, G. Identification of a DNA-damage-inducible regulon in *Acinetobacter baumannii*. *J. Bacteriol.* **2013**, *195*, 5577–5582. [CrossRef] [PubMed]
41. Starikova, I.; Harms, K.; Haugen, P.; Lunde, T.T.; Primicerio, R.; Samuelsen, O.; Nielsen, K.M.; Johnsen, P.J. A trade-off between the fitness cost of functional integrases and long-term stability of integrons. *PLoS Pathog.* **2012**, *8*, e1003043. [CrossRef] [PubMed]
42. Domingues, S.; da Silva, G.J.; Nielsen, K.M. Integrons: Vehicles and pathways for horizontal dissemination in bacteria. *Mob. Genet. Elements* **2012**, *2*, 211–223. [CrossRef] [PubMed]
43. Vaneechoutte, M.; Young, D.M.; Ornston, L.N.; De Baere, T.; Nemec, A.; Van Der Reijden, T.; Carr, E.; Tjernberg, I.; Dijkshoorn, L. Naturally transformable *Acinetobacter* sp. strain ADP1 belongs to the newly described species *Acinetobacter baylyi*. *Appl. Environ. Microbiol.* **2006**, *72*, 932–936. [CrossRef] [PubMed]
44. Elbourne, L.D.; Tetu, S.G.; Hassan, K.A.; Paulsen, I.T. TransportDB 2.0: A database for exploring membrane transporters in sequenced genomes from all domains of life. *Nucleic Acids Res.* **2017**, *45*, D320–D324. [CrossRef] [PubMed]
45. Legatzki, A.; Grass, G.; Anton, A.; Rensing, C.; Nies, D.H. Interplay of the Czc system and two P-type ATPases in conferring metal resistance to *Ralstonia metallidurans*. *J. Bacteriol.* **2003**, *185*, 4354–4361. [CrossRef] [PubMed]
46. Nies, D.H. Efflux-mediated heavy metal resistance in prokaryotes. *FEMS Microbiol. Rev.* **2003**, *27*, 313–339. [CrossRef]
47. Purves, J.; Thomas, J.; Riboldi, G.P.; Zapotoczna, M.; Tarrant, E.; Andrew, P.W.; London, A.; Planet, P.J.; Geoghegan, J.A.; Waldron, K.J.; et al. A horizontally gene transferred copper resistance locus confers hyper-resistance to antibacterial copper toxicity and enables survival of community acquired methicillin resistant *Staphylococcus aureus* USA300 in macrophages. *Environ. Microbiol.* **2018**, *40*, 1576–1589. [CrossRef] [PubMed]

© 2018 by the authors. Licensee MDPI, Basel, Switzerland. This article is an open access article distributed under the terms and conditions of the Creative Commons Attribution (CC BY) license (http://creativecommons.org/licenses/by/4.0/).

Article

Genomic Islands Confer Heavy Metal Resistance in *Mucilaginibacter kameinonensis* and *Mucilaginibacter rubeus* Isolated from a Gold/Copper Mine

Yuan Ping Li [1,†], Nicolas Carraro [2,†], Nan Yang [1], Bixiu Liu [1], Xian Xia [3], Renwei Feng [1,*], Quaiser Saquib [4], Hend A Al-Wathnani [5], Jan Roelof van der Meer [2] and Christopher Rensing [1,6,*]

1. Institute of Environmental Microbiology, Fujian Agriculture and Forestry University, Fuzhou 350002, China; li343000@126.com (Y.P.L.); cindyyn1118@163.com (N.Y.); liu_bixiu@163.com (B.L.)
2. Department of Fundamental Microbiology, University of Lausanne, Lausanne 1015, Switzerland; nicolas.carraro@unil.ch (N.C.); janroelof.vandermeer@unil.ch (J.R.v.d.M.)
3. State Key Laboratory of Agricultural Microbiology, College of Life Science and Technology, Huazhong Agricultural University, Wuhan 430070, China; xiaxian@webmail.hzau.edu.cn
4. Zoology Department, College of Sciences, King Saud University, P.O. Box 2455, Riyadh 11451, Saudi Arabia; qsaquib@ksu.edu.sa
5. Department of Botany & Microbiology, College of Sciences, P.O. Box 2455, Riyadh 11451, Saudi Arabia; wathnani@ksu.edu.sa
6. Key Laboratory of Urban Environment and Health, Institute of Urban Environment, Chinese Academic of Sciences, 361021 Xiamen, China
* Correspondence: frwzym@aliyun.com (R.F.); rensing@fafu.edu.cn (C.R.)
† These authors contributed equally to this work.

Received: 22 October 2018; Accepted: 19 November 2018; Published: 23 November 2018

Abstract: Heavy metals (HMs) are compounds that can be hazardous and impair growth of living organisms. Bacteria have evolved the capability not only to cope with heavy metals but also to detoxify polluted environments. Three heavy metal-resistant strains of *Mucilaginibacer rubeus* and one of *Mucilaginibacter kameinonensis* were isolated from the gold/copper Zijin mining site, Longyan, Fujian, China. These strains were shown to exhibit high resistance to heavy metals with minimal inhibitory concentration reaching up to 3.5 mM $Cu^{(II)}$, 21 mM $Zn^{(II)}$, 1.2 mM $Cd^{(II)}$, and 10.0 mM $As^{(III)}$. Genomes of the four strains were sequenced by Illumina. Sequence analyses revealed the presence of a high abundance of heavy metal resistance (HMR) determinants. One of the strain, *M. rubeus* P2, carried genes encoding 6 putative P_{IB-1}-ATPase, 5 putative P_{IB-3}-ATPase, 4 putative $Zn^{(II)}/Cd^{(II)}$ P_{IB-4} type ATPase, and 16 putative resistance-nodulation-division (RND)-type metal transporter systems. Moreover, the four genomes contained a high abundance of genes coding for putative metal binding chaperones. Analysis of the close vicinity of these HMR determinants uncovered the presence of clusters of genes potentially associated with mobile genetic elements. These loci included genes coding for tyrosine recombinases (integrases) and subunits of mating pore (type 4 secretion system), respectively allowing integration/excision and conjugative transfer of numerous genomic islands. Further in silico analyses revealed that their genetic organization and gene products resemble the *Bacteroides* integrative and conjugative element CTnDOT. These results highlight the pivotal role of genomic islands in the acquisition and dissemination of adaptive traits, allowing for rapid adaption of bacteria and colonization of hostile environments.

Keywords: *Mucilaginibacer rubeus*; *Mucilaginibacter kameinonensis*; genomic island; evolution; heavy metal resistance; draft genome sequence; CTnDOT

1. Introduction

Heavy metals (HMs) have a dualistic impact on living organisms. On the one hand, metal ions are essential for numerous biological processes mandatory for cellular activity, including homeostasis, enzyme activity, and protein functionality [1]. On the other hand, when present in excess in the environment, HM can have toxic effect hindering diverse cellular processes and thus cellular life.

Heavy metal pollution has been part of Earth's history as it can originate from natural processes such as volcanic eruption. Recent (over)industrialization and exploitation of Earth resources worldwide has accelerated HM release into the environment and led to high levels of water, air, and soil pollution. Especially, mine exploitation for metal extraction is one of the most important sources of heavy metal pollution [2]. This comes not only from excavating deep-buried HMs to be exposed to the surface, but also from extraction protocols that often rely on the use of other contaminants, including HMs [2].

Beyond its effects on people, HM toxicity was shown to have profound impacts on microbial communities, including fungi and bacteria [2]. Heavy metals were shown to have critical consequences on bacterial viability due to their pleiotropic effect on cellular processes. Excess of HM can disrupt the cell membrane, damage nucleic acids and proteins, impair enzymatic activities, and inhibit key processes such as transcription [1]. The presence of HM pollution exerts a high selective pressure on microbial communities, reducing their diversity, biomass, and activity, thus strongly impacting the biological activity of polluted environments [3].

In order to cope with the presence of elevated concentration of HMs, a myriad of bacterial genetic programs has been selected encoding functions that allow efflux and/or sequestration of HMs, and modification to inactivate or reduce reactivity of certain metal ions. The main mechanism to resist toxicity of HMs is efflux [1]. Important classes of HM transporters include P_{IB}-type ATPases and cation diffusion facilitators (CDF). Both types of transporters translocate HM ions from the cytoplasm across the cytoplasmic membrane into the periplasm [4]. In the context described here with microbes having to handle very high external concentrations of HMs, P-type ATPases are much more relevant since they are much more powerful using ATP to pump HMs against their concentration gradient out of the cytoplasm [5]. In addition, HMs are translocated from the periplasm across the outer membrane into the extracellular space by resistance-nodulation-division (RND)-type transport systems. These multicomponent transporters of the RND type contain 3 RND transport proteins, 6 membrane fusion proteins (MFPs), and 3 outer membrane factor (OMF) proteins. The fascinating transport mechanism of the RND-type transport complex has been described in detail [4]. P_{IB}-type ATPases and RND-type transport systems were described as being the most important systems to confer a high HM resistance (HMR).

Bacteria also show an astonishing capability to spread HMR genes within bacterial communities via horizontal gene transfer. Dissemination of genetic material conferring HMR is frequently associated with conjugative plasmids, genomic islands, and transposons [6]. Conjugative plasmids are extrachromosomal replicative entities able to transfer from a donor cell toward a recipient cell by conjugation [7]. Conjugative plasmids have been recognized as major contributors for the spread of adaptive traits such as antibiotic resistance, new metabolic capacities, and HMR [8]. Conjugative plasmid-borne HMR is associated with occurrence of large clusters of HMR genes that can span over several kb [9–15]. Portions of genomic DNA called genomic islands (GIs) were also shown to play a pivotal role into the horizontal dissemination of genetic material [16]. Although the mechanisms underlying the mobility of some GIs remain obscure, current knowledge describes different strategies that ultimately rely on conjugative transfer [17,18]. GI-associated HMR was described in *Enterobacteriaceae* and *Shewanellaceae* [19], *Listeria monocytogenes* [20], and *Acinetobacter baumannii* [21]. Also, HMR was shown to be conferred by an IncC-dependent mobilizable genomic island SGI1 variant called SGI1-K in *Salmonella enterica* [22–24]. Transposons are genetic entities able to move intra-molecularly (on the same replicon) or inter-molecularly (between different replicons) [25]. Most transposons can hitchhike by integrating into a conjugative plasmid or a GI for intercellular mobility. Transposons conferring HMR were described to be in association with other mobile genetic elements [13,22].

In this study, we describe the isolation and characterization of four heavy metal-resistant *Mucilaginibacter* strains isolated from a gold/copper mine in China. Genomes of these strains were sequenced and further in silico analysis revealed a high number of heavy metal resistance determinants. Moreover, at least part of these HMR gene clusters were shown to be potentially mobile as they are in the close vicinity of the core region of putative integrative and conjugative elements (ICEs).

2. Materials and Methods

2.1. Bacterial Isolation

Strains *Mucilaginibacter rubeus* P1, P2, and P3 were isolated from samples collected at 5–10 cm below the surface of a soil located near a waste water treatment dam of a copper-gold mine, and *Mucilaginibacter kameinonensis* P4, was isolated from a hillside with little human activity within the gold and copper mine (Zijin mining) in Longyan city of Fujian province, China (Table 1). After serial dilutions with 0.85% NaCl, the soil sample was spread on R2A (DSM medium 830) agar plates containing 2 mM $CuSO_4 \cdot 5H_2O$. After incubation at 28 °C for 1 week, the strains were isolated and later stored at −80 °C in 20% glycerol (w/v).

Table 1. Characteristics of the heavy metal (HM)-contaminated soil from where the strains were isolated.

	Mucilaginibacter kameinonensis P4	*Mucilaginibacter rubeus* P3	*M. rubeus* P2	*M. rubeus* P1
Altitude (m)	216	192	192	192
Longitude	N25°09.719′	N25°09.724′	N25°09.724′	N25°09.724′
Latitude	E116°23.258′	E116°23.258′	E116°23.258′	E116°23.258′
pH	6.64	5.52	6.32	6.32
Water content	9.38%	6.41%	7.05%	7.05%

2.2. Taxonomic Analysis

Strains were incubated at 28 °C for 24 h on R2A agar plates. As described in Brosius et al. [26], the universal primer pair 27F/1492R was used to amplify 16S sequences and the amplified PCR product was subsequently sequenced [26]. PCR products were sequenced by Biosune Company (Fuzhou, China) using the Sanger method. Based on the EzTaxon database (http://eztaxon-e.ezbiocloud.net) [27], pairwise sequence similarity and phylogenetic neighbors of the sequences of each individual strain (1382–1432 bp) were obtained through BLAST searches. In total, 19 *Mucilaginibacter* strains with publicly available 16S ribosomal RNA (rRNA) gene sequences were selected, with *Pedobacter africanus* DSM 12126T (AJ438171) as an out-group, to do the alignment via Mega 7.0 software [28]. A Neighbor-joining (NJ) tree was generated and the Kimura's two-parameter model was used to calculate evolutionary distances [29], and bootstrap analysis with 1000 replications was conducted to obtain confidence levels of the branches [30].

2.3. Determination of the Minimal Inhibitory Concentration

To determine the level of resistance to various metals of all strains, *M. rubeus* P1, P2, and P3 and *M. kameinonensis* P4 were grown on Cu, As, Cd, and Zn agar plates containing different $Cu^{(II)}$, $Zn^{(II)}$, $As^{(III)}$, and $Cd^{(II)}$ concentrations to determine the minimal inhibitory concentration (MIC). The different R2A plates contained 0–10.0 mM of copper or arsenic, with 0.5 mM increments, 0–30.0 mM with 1.5 mM increments in case of zinc, and 0–2.0 mM cadmium with the increments being 0.2 mM. 1M $CuSO_4 \cdot 5H_2O$, $ZnCl_2$, $NaAsO_2$, and $CdCl_2 \cdot 5H_2O$ stock solutions were prepared and stored after filtration through a 0.22 μm filter.

2.4. Cell Morphology and Flagella Observation

Overnight cultures of strains *M. rubeus* P1, P2, and P3 and *M. kameinonensis* P4 were inoculated into 50 mL of R2A medium at 28 °C with 180 rpm shaking. After 24 h of growth with shaking, cells were centrifuged (1000× g, 10 min, 4 °C) and observed under scanning electron microscopy (SEM). Cells were harvested and washed three times with cold (4 °C) phosphate buffered saline (0.2 M PBS, pH 7.2). Fixation was performed with 2.5% glutaraldehyde (24 h, 4 °C). Fixed cells were dehydrated through a series of alcohol dehydration steps (30%, 50%, 70%, 85%, 95%, and 100%) and finally freeze dried and sputter coated. The samples were then viewed using a scanning electron microscope JSM-6390 SEM (JEOL, Tokyo, Japan).

2.5. Growth Conditions Optimization

To optimize NaCl concentration and pH of the medium for growth of the *Mucilaginibacter* strains, 50 µL precultures were added to 5 mL R2A liquid medium supplemented with 0–3% NaCl at pH 7, or to R2A without any NaCl and with pH set to the range between pH 2–11. Cultures were incubated at 28 °C for 7 days, after which culture turbidities optical density (OD) at 600nm were evaluated. Anaerobic growth was tested by incubating R2A plates in an anaerobic chamber at 28 °C for 1 week. Optimal growth temperature was tested in the incubator on R2A agar plates at temperatures between 4 to 40 °C for 1 week.

2.6. Genomic DNA Extraction

Genomic DNA (gDNA) was extracted by using a TIANamp Bacteria DNA Kit (Tiangen Biotech, Beijing, China) from cultures grown on R2A. The quantity and purity of gDNA were assessed using an UV spectrophotometry (Nanodrop ND-1000, J & H Technology Co., Ltd. Wilmington, USA). Genomic DNA with OD260/280 value higher than 1.80 was selected and examined on agarose gel electrophoresis (0.8%). Samples containing more than 25 µg of intact gDNA (fragment size > 20 kb) were sent out for whole-genome sequencing.

2.7. Whole-Genome Sequencing

Whole-genome shotgun sequencing was preformed using an Illumina HiSeq X Ten System provided by Vazyme Biotech Co., Ltd. (Nanjing, China). The DNA library was constructed using the Illumina V3 VAHTS Universal DNA Library Prep Kit according to the VAHTS Universal DNA sample preparation protocol (Illumina, Santiago, USA). The insert size was 300 bp for all strains, and 16,980,768, 18,531,104, 18,306,636, and 20,005,292 read-pairs and 2.86, 3.12, 3.09, and 3.37 Gb of raw data were obtained for strains *M. kameinonensis* P4 and *M. rubeus* P1, P2, and P3, respectively.

2.8. De novo Genome Assembly and Annotation

Illumina reads were quality-filtered, trimmed, and de novo assembled with default settings using CLC Genomic Workbench 11.0 (QIAGEN, Hilden, Germany). The draft genome sequences were annotated by NCBI PGAP, and are accessible under GenBank numbers QEYR0000000, QFKW0000000, QFKV0000000, and QFKU0000000 for *M. kameinonensis* P4 and *M. rubeus* P1, P2, and P3, respectively. *M. kameinonensis* P4 generated 78 contigs with an n50 value of 350.607 bp. *M rubeus* P1 generated 158 contigs with an n50 value of 139.339 bp. *M rubeus* P2 generated 118 contigs with an n50 value of 132.524 bp. *M rubeus* P3 generated 107 contigs with an n50 value of 148.541 bp.

2.9. TraG Proteins Phylogenetic Analyses

Molecular phylogenetic analysis of TraG proteins was performed using MEGA6 [28]. The 807- to 850-amino acid sequences of TraG proteins were recovered from genome sequences of *Mucilaginibacter* isolated in this study. The corresponding sequence in CTnDOT (TraG$_{DOT}$ accession number: AAG17832.1) was added to the dataset as an outgroup. Analyses were computed using an amino acid alignment generated by MUSCLE [31]. The evolutionary history was inferred by using the Maximum Likelihood method based on the Jone, Taylor and Thornton (JTT) matrix-based model [32]. Initial tree(s) for the heuristic search were obtained by applying the NJ method to a matrix of pairwise distances estimated using a JTT model. A discrete Gamma distribution was used to model evolutionary rate differences among sites (five categories (+G, parameter = 2.9848)). The analysis involved 18 amino acid sequences. All positions with less than 95% site coverage were eliminated, providing a total of 716 positions in the final dataset.

3. Results and Discussion

3.1. Isolation of Four Heavy Metal-Resistant Mucilaginibacter

We intended to isolate heavy metal resistant strains from the ZiJin copper-gold mine to gain insights into how bacterial strains adapt to high concentrations of HMs. We recovered four strains that were morphologically similar with a high tolerance to a number of HMs.

Based on phylogenetic analysis (NJ) of the 16S rRNA gene three strains (P1, P2, and P3) were closely related to *M. rubeus* EF23^T (98.34–99.93%) and *M. gossypiicola* Gh-67^T (98.12–99.01 %). The fourth strain (P4) grouped closely with *M. kameinonensis* SCK^T (98.8 %) (Figure 1). All strains belonged to the *Sphingobacteriaceae* family in the class *Sphingobacteriia*.

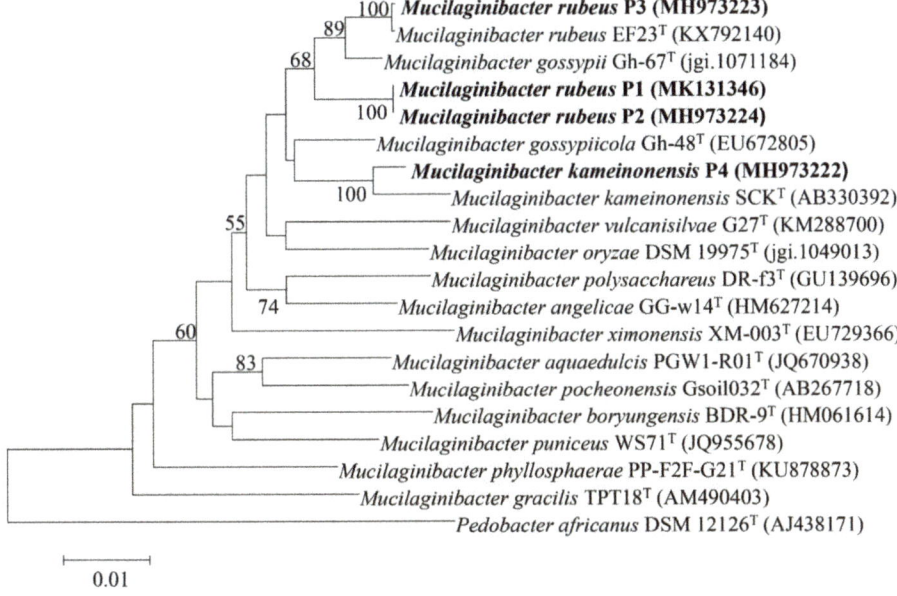

Figure 1. Neighbour-joining phylogenetic tree constructed based on the 16S ribosomal RNA (rRNA) gene sequences from the draft genome sequence showing the phylogenetic relationships between strains *Mucilaginibacter rubeus* P1, P2 and P3 and *Mucilaginibacter kameinonensis* P4 and other species in the genus *Mucilaginibacter*. Values indicate percentages of identical branching in 1000 bootstrappings. The sequence of *Pedobacter koreensis* WPCB189T was used as an out-group. Bar, 0.01 substitutions per nucleotide position.

3.2. Phenotypic Characterization of Mucilaginibacter Strains Uncovered Multiple Heavy Metal Resistances

The HM concentration of the soil is extremely high, even the lowest concentration of total Zn, As, Cd, and Cu was found to be 49.27, 1.43, 1.19, and 18.37 mg·kg^{-1}, respectively. The MICs of the four strains reached up to 3.5 mM Cu$^{(II)}$, 21 mM Zn$^{(II)}$, 1.2 mM Cd$^{(II)}$, and 10.0 mM As$^{(III)}$.

Strain *M. kameinonensis* P4 displayed higher Cd resistance compared to strains *M. rubeus* P1, P2, and P3 (Table 2). Related, not heavy metal resistant *Mucilaginibacter pedocola* sp. TBZ30^T, cultured under similar conditions displayed MICs of 0.4 mM Cu$^{(II)}$, 3 mM Zn$^{(II)}$, 0.2 mM Cd$^{(II)}$, and 0.2 mM As$^{(III)}$ [33]. Such high resistance to multiple HMs as reported here has therefore not been observed before in the genus *Mucilaginibacter* [33].

Table 2. Minimal inhibitory concentration (MIC) of strains to Zn$^{(II)}$, As$^{(III)}$, Cd$^{(II)}$, and Cu$^{(II)}$ and respective concentrations of the HM in the soil where the strains were isolated from.

Metals	*M. kameinonensis* P4	*M. rubeus* P3	*M. rubeus* P2	*M. rubeus* P1	*M. pedocola* sp. TBZ30^T
Zn$^{(II)}$/mM	10.5	21.0	10.5	21.0	3.0
As$^{(III)}$/mM	3.5	4.5	9.0	10.0	0.2
Cd$^{(II)}$/mM	1.2	0.2	0.4	0.4	0.2
Cu$^{(II)}$/mM	3.5	3.5	3.5	3.5	0.4
Zn/mg·kg^{-1}	49.27	176.79	96.56	96.56	ND
Cd/mg·kg^{-1}	1.21	1.19	2.26	2.26	ND
As/mg·kg^{-1}	55.89	51.99	1.43	1.43	ND
Cu/mg·kg^{-1}	365.10	1067.82	18.37	18.37	ND

Note. ND means not found.

Strains *M. rubeus* P1, P2, and P3 and *M. kameinonensis* P4 formed a light orange or pink, moist, circular, and convex colony with smooth margins on R2A agar plates. All strains were Gram-negative and aerobic. Growth of strains was observed at 4–30 °C (optimum, 28 °C). Optimal growth occurred in absence of further NaCl, but the strains could still grow in R2A with up to 1.5% NaCl added. These characteristics are consistent with description of the genus *Mucilaginibacter* [34–36]. Medium pH for optimal growth (~pH 5.0) and pH tolerance (pH 5.0–9.0) varied slightly between the four strains (Table 3).

Table 3. General features of strains *M. rubeus* P1, P2, and P3 and *M. kameinonensis* P4.

Property	*M. kameinonensis* P4	*M. rubeus* P3	*M. rubeus* P2	*M. rubeus* P1	*M. pedocola* sp. TBZ30^T
Gram strain	Negative	Negative	Negative	Negative	Negative
Cell shape	Rod-shaped	Rod-shaped	Rod-shaped	Rod-shaped	Rod-shaped
Colony colour	Light-yellow	Pink	Pink	Pink	Pink
pH	5.0–7.0 (5.0)	5.0–9.0 (5.0)	5.0–8.0 (5.0)	5.0–8.0 (6.0)	5.0–8.5 (7.0)
Temperature range (°C)	4–37 (28)	4–37 (28)	4–37 (28)	4–37 (28)	4–28 (25)
Oxygen requirement	Aerobic	Aerobic	Aerobic	Aerobic	Aerobic
Salinity (%)	0–1.5 (0)	0–1.0 (0)	0–1.5 (0)	0–1.0 (0)	0–1.0 (0)

3.3. Mucilaginobacter Strains Exhibit an Arsenal of Genetic Determinants to Deal with High Concentrations of Heavy Metals

To gain insight in the genetic basis of how the four strains were able to deal with these high HM concentrations, we determined draft genome sequences. Draft genomes were automatically annotated through RAST (Rapid Annotation using Subsystem Technology) database (http://rast.nmpdr.org/). Based on inferred protein homologies, between 6 and 16 putative P$_{1B}$ type-ATPase [37] were encoded in the four genomes (Table 4). All strains further encoded a variety of putative RND type metal transporter systems of the CzcCBAD type. Three strains further encoded putative CusCBA Cu$^{(I)}$ translocating RND-type transport systems, except *M. kameinonensis* P4 (Table 3). Multiple genes for putative multicopper oxidases were found on the different genomes, which may constitute the basis for the observed copper resistance (Table 3). Genes for putative multicopper oxidases were only taken into account if they were located adjacent to genes encoding P$_{1B}$ type Cu$^{(I)}$ translocating P-type ATPase. Finally, between 2 and 4 putative *ars* operons (*arsNCR*, *acr3*, *arsMCR*) were among the *Mucilaginibacter* genomes (Table 3). The higher number of *ars* operons in strain P1 and P2 genomes correlated to their high MICs on As$^{(III)}$ (10.0 and 9.0 mM, respectively).

The number of HMR determinants in *Mucilaginibacter* genomes was unusually high, even in comparison to the well-known HM-resistant strain *Cupriavidus metallidurans* CH34 [5,38,39] (Table 4), suggesting a strong selection for HM resistance in their natural living environment. The HM resistance determinants are often clustered together, and often located adjacent to *tra* genes. They could be identified on many different contigs.

Table 4. Heavy metal related genes in the analyzed *Mucilaginibacter* strains in comparison to *Cupriavidus metallidurans* CH34.

Genes Encoding Heavy Metal Resistance Determinants		*M. kameinonensis* P4	*M. rubeus* P3	*M. rubeus* P2	*M. rubeus* P1	*C. metallidurans* CH34
P$_{1B}$-type-ATPase	P$_{IB-1}$ type-ATPase	3	6	6	6	7
	P$_{IB-3}$ type-ATPase	1	4	5	4	0
	P$_{IB-4}$ type-ATPase	2	2	4	4	1
	Mg$^{(II)}$	0	1	1	1	0
RND type metal transport systems	CzcCBAD	8	10	10	11	9
	CusCBA	0	2	4	3	2
	NccCBA	0	0	2	2	1
ars operons		2	3	4	4	1
Multicopper oxidases		2	6	5	6	2

RND: Resistance-nodulation-division.

3.4. Heavy Metal Resistance Is Associated with CTnDOT-Related Genomics Islands

Tolerance to HMs is frequently acquired by horizontal transmission among and between bacterial populations. Given the important size of HMR clusters identified in *Mucilaginibacter* genomes (up to 150 kb), we wondered whether some might be encompassed by GIs (Figure 2). We examined the close vicinity of the HMR clusters for the hallmark of conjugative systems, also known as Type 4 secretion systems (T4SSs) [40]. T4SSs have been classified based on their VirB4 protein, a ubiquitous constituent of conjugative systems [41]. A total of 17 genes encoding VirB4 proteins (*traG*) exhibiting sizes between 807 to 850 amino acids were identified on the 4 genomes, most of them in the close proximity of HMR clusters: 6 in *M. rubeus* P1 (contigs 1, 4, 24, 29, 42, 55), 6 in *M. rubeus* P2 (contigs 24, 26, 29, 32, 34, 42), 2 in *M. rubeus* P3 (contigs 5 and 11), and 3 in *M. kameinonensis* P4 (contigs 6, 9, and 35).

TraG proteins of the putative conjugative GIs identified in *Mucilaginibacter* genomes were compared to the 838-amino acid TraG of CTnDOT (TraG$_{DOT}$) and showed 32 to 53% of identity over 94 to 98% of their amino acid sequence. The evolutionary history of TraG proteins was inferred using TraG$_{DOT}$ as an outgroup (Figure 3). Two strongly supported clades were delineated, suggesting that they belong to two distinct lineages (Figure 3, green and red boxes). As expected, each one of the TraG proteins of strain *M. rubeus* P1 grouped with one TraG protein from strain *M. rubeus* P2, confirming that these identical strains contain the same 6 elements. More interestingly, TraG$_{P1-1}$, TraG$_{P2-26}$, TraG$_{P3-5}$, and TraG$_{P4-9}$ grouped together, and their gene sequences were identical.

Closer analysis revealed the presence of genes coding for other T4SS subunits adjacent to each one of the *traG* genes. This grouping of *tra* genes may be regarded as a conjugation module, i.e., genes and sequences implicated in the same biological process [42]. In particular the regions including *traG*$_{P1-1}$, *traG*$_{P2-26}$, *traG*$_{P3-5}$, and *traG*$_{P4-9}$ were 100% identical across a circa 150-kb region, including an about 75-kb cluster coding for multiple HMR. The organization of these putative conjugation modules resembles the one encoded in CTnDOT, a protypical ICE of *Bacteroides* [41,43]. As in CTnDOT, putative conjugative modules encoded by GIs of *Mucilaginibacter* thus belong to the mating pair formation (MPF) category B (MPFB) [41]. Also, most of the putative conjugative modules identified in *Mucilaginibacter* strains are in the proximity of genes encoding a putative tyrosine recombinase related to IntDOT, the integrase of CTnDOT [44]. Moreover, the *Mucilaginibacter* GIs carried a gene predicted to encode a RteC-like protein reminiscent of the CTnDOT regulation system [45,46]. *Mucilaginibacter*

GIs are thus likely to be ICEs, whose maintenance relies on integration into the chromosome and dissemination depends on its excision from the chromosome as a circular element that would transfer by conjugation [17,47,48]. The presence of at least one identical contiguous region over 150 kb (represented by the $traG_{P1-1}$ gene) in the four different *Mucilaginibacter* recovered strains suggest active mobility and recent transfer of this GI.

Figure 2. Representative putative genomic island carrying genes encoding HM determinants in contig 26 of *M. Rubeus* P2. Genomic analysis was performed via RAST (http://rast.nmpdr.org/). Genes encoding determinants related to metalloid (Arsenical and Antimony) resistance are highlighted in maroon, metal (Copper/Cobalt/Cadmium/Zinc/Lead/Mercury) resistance in blue, and the genes encoding putative transfer functions in red. Genes encoding hypothetical proteins and unknown functions are highlighted in gray.

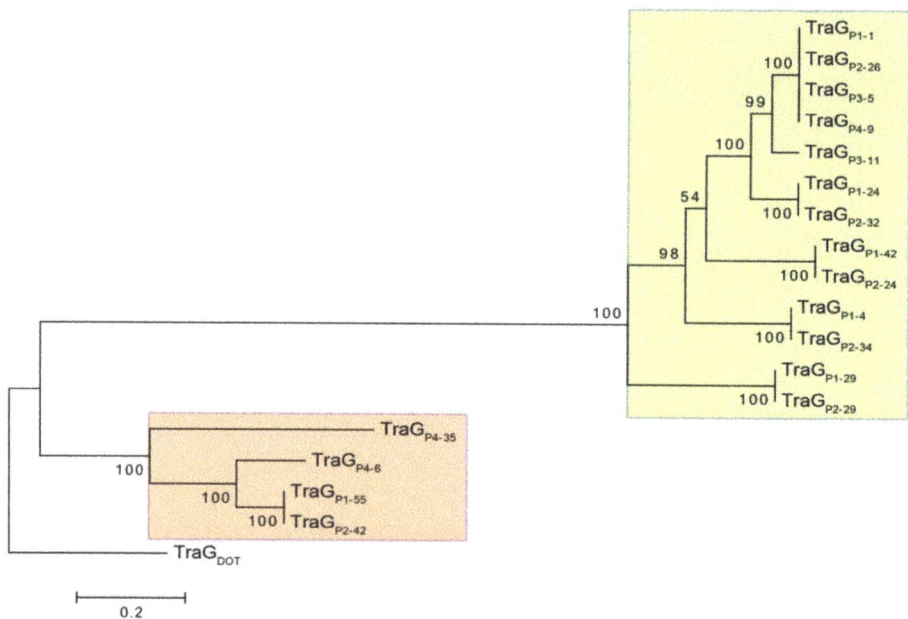

Figure 3. Molecular phylogenetic analysis of TraG proteins of putative conjugative genomic islands (GIs) of *Mucilaginibacter*. The evolutionary history was inferred by using the Maximum Likelihood method based on the JTT matrix-based model [32]. The percentage of trees in which the associated taxa clustered together is shown next to the branches. The tree is drawn to scale, with branch lengths measured in the number of substitutions per site. Evolutionary analyses were conducted in MEGA6 [28]. Initial alignment of sequences was performed using Muscle for the presented tree. An identical tree with minor changes in bootstrap values was obtained using ClustalW for alignment. The VirB4 subunit of MPFB T4SS is named TraG [41]. For convenience and consistence in TraG protein identification, nomenclature is as follows: TraGPX-Y, where X is the strain number and Y the contig carrying the gene coding for TraG. TraG CTnDOT (TraGDOT) accession number: AAG17832.1.

Since the draft genomes were not completely curated to a single contiguous scaffold, we could not confidently delimit the boundaries of the putative conjugative GIs. As a matter of fact, a single GI might be spread over multiple contigs, or could be a defective element lacking flanking or internal parts of the original GI. Also, IntDOT was reported to not require strict homology between the recombining sites in contrast to the majority of tyrosine recombinases [46,49]. The integration/excision is, in that case, site-selective rather than site-specific, strongly impairing the precise identification of the right and left attachment sites (*attR* and *attL*, respectively).

4. Concluding Remark

This work allowed the isolation and characterization of four heavy metal-resistant *Mucilaginibacter* strains recovered from polluted soil of gold mines. Sequencing of genomic content allowed inspection of HMR loci into the chromosome of these strains and their close association with loci coding for conjugation of CTnDOT-related GIs. Further genome closure and experimental investigation should allow testing the functionality of such putative ICEs found in the *Mucilaginibacter* strains. Notably, phenotypes such as their ability to excise from the chromosome and their capability to transfer toward a new host by conjugation will be monitored. In particular the 150-kb (at least) conserved putative ICE present in the four strains is an interesting candidate, likely to be functional given its complete identity among all four strains. Protein BLAST using TraG of this conserved element and search

Mucilaginibacter genomes did not give other perfect hits, suggesting that presence of this GI is restricted to sampling locations of this study (Table 1). Increasing availability of fully sequenced genomes should allow further data meaning in order to evaluate the abundance of specific GIs in *Mucilaginibacter* and predict their functionality.

The presence of putative CTnDOT-related ICEs into genomes of *Mucilaginibacter* strains does not seem uncommon. Protein BLAST analysis using TraG$_{DOT}$ as query and searching *Mucilaginibacter* genomes gave multiple hits with identity ranging from 70% over 99% of aa sequence, down to 30% over 79% of aa sequence (considering 96 hits with more than 75% of coverage). This observation highlights the presence of multiple putative CTnDOT-related GIs populating *Mucilaginibacter* genomes, most likely playing a key role in their genome evolution and adaptation. One can speculate that such GIs may be involved in conferring specific capabilities such as the ability to degrade pectin, xylan, and laminarin of *M. paludis* and *M. gracilis* [50] plant growth promotion capacity conferred by *M. gossypii* and *M. gossypiicola* [51], or yet to be discovered adaptive functions that may be conferred by such GIs. Further *in silico* analyses may reveal interesting features of such putative ICEs considering accessory functions they could confer, or functionality of their recombination, conjugation, or regulatory systems.

This exploratory work on HMR GIs of *Mucilaginibacter* together with other research done on *Mucilaginibacter* species so far constitutes a solid ground for future experimental research aiming at developing molecular tools. Such tools would greatly facilitate further investigation of *M. rubeus* and *M. kameinonensis* biology and could likely be extended to other *Mucilaginibacter* species.

Author Contributions: Y.P.L. isolated strains from mine and generated the phylogenetic tree of 16SrDNA, annotated the genome, mapped the gene cluster, and partially drafted the manuscript. N.C. contributed to the general concept of the study, performed gene sequence prediction, molecular phylogenetic analysis of TraG proteins, and partially drafted the manuscript. N.Y. contributed to the MICs determination and partially drafted the manuscript. B.L. performed growth conditions optimization and genomic DNA extraction. X.X. assembled the genomes. Q.S. and H.A.Al-W. contributed to data analysis. R.F. contributed to the general concept of the study and manuscript verification. J.R.v.d.M. contributed to the general concept of the study and manuscript verification. C.R. contributed to the general concept of the study, design of experiments and manuscript preparation.

Funding: This research was funded by National Natural Science Foundation of China (NSFC) grant number 31770123, State administration of Foreign Expert Project numbers GDT20173600005 and 110000214620170006, and Swiss National Science Foundation (SNSF) grant number 31003A_175638, Twasol Research Excellence Program (TRE), King Saud University, Saudi Arabia for support.

Acknowledgments: We like to thank engineers Huaiguo Huang, Hongwen Li, and Xianzheng Chen of Zijin Mining Group Co., Ltd. for the technical support in the sampling.

Conflicts of Interest: The authors declare no conflicts of interest.

References

1. Chandrangsu, P.; Rensing, C.; Helmann, J.D. Metal homeostasis and resistance in bacteria. *Nat. Rev. Microbiol.* **2017**, *15*, 338–350. [CrossRef] [PubMed]
2. Nguyen-Viet, H.; Gilbert, D.; Mitchell, E.A.D.; Badot, P.M.; Bernard, N. Effects of experimental lead pollution on the microbial communities associated with *Sphagnum fallax* (Bryophyta). *Microb. Ecol.* **2007**, *54*, 232–241. [CrossRef] [PubMed]
3. Huang, L.N.; Kuang, J.L.; Shu, W.S. Microbial ecology and evolution in the acid mine drainage model system. *Trends Microbiol.* **2016**, *24*, 581–593. [CrossRef] [PubMed]
4. Nies, D.H. Efflux-mediated heavy metal resistance in prokaryotes. *FEMS Microbiol. Rev.* **2003**, *27*, 313–339. [CrossRef]
5. Nies, D.H. The biological chemistry of the transition metal "transportome" of *Cupriavidus metallidurans*. *Metallomics* **2016**, *8*, 481–507. [CrossRef] [PubMed]
6. Reva, O.; Bezuidt, O. Distribution of horizontally transferred heavy metal resistance operons in recent outbreak bacteria. *Mob. Genet. Elem.* **2012**, *2*, 96–100. [CrossRef] [PubMed]
7. Llosa, M.; Gomis-Rüth, F.X.; Coll, M.; De la Cruz, F. Bacterial conjugation: A two-step mechanism for DNA transport. *Mol. Microbiol.* **2010**, *45*, 1–8. [CrossRef]

8. Cabezón, E.; Ripoll-Rozada, J.; Peña, A.; De la Cruz, F.; Arechaga, I. Towards an integrated model of bacterial conjugation. *FEMS Microbiol. Rev.* **2014**, *39*, 81–95. [CrossRef] [PubMed]
9. Monchy, S.; Benotmane, M.A.; Janssen, P.; Vallaeys, T.; Taghavi, S.; Van der Lelie, D.; Mergeay, M. Plasmids pMOL28 and pMOL30 of *Cupriavidus metallidurans* are specialized in the maximal viable response to heavy metals. *J. Bacteriol.* **2007**, *189*, 7417–7425. [CrossRef] [PubMed]
10. Mergeay, M.; Monchy, S.; Vallaeys, T.; Auquier, V.; Benotmane, A.; Bertin, P.; Taghavi, S.; Dunn, J.; van der Lelie, D.; Wattiez, R. *Ralstonia metallidurans*, a bacterium specifically adapted to toxic metals: Towards a catalogue of metal-responsive genes. *FEMS Microbiol. Rev.* **2010**, *27*, 385–410. [CrossRef]
11. Hernández-Ramírez, K.C.; Reyes-Gallegos, R.I.; Chávez-Jacobo, V.M.; Díaz-Magaña, A.; Meza-Carmen, V.; Ramírez-Díaz, M.I. A plasmid-encoded mobile genetic element from *Pseudomonas aeruginosa* that confers heavy metal resistance and virulence. *Plasmid* **2018**, *98*, 15–21. [CrossRef] [PubMed]
12. Bezuidt, O.; Pierneef, R.; Mncube, K.; Limamendez, G.; Reva, O.N. Mainstreams of horizontal gene exchange in enterobacteria: Consideration of the outbreak of enterohemorrhagic *E. coli O104:H4 in Germany in 2011*. *PLoS ONE* **2012**, *2*, 96–100.
13. Schneiker, S.; Keller, M.; Dröge, M.; Lanka, E.; Pühler, A.; Selbitschka, W. The genetic organization and evolution of the broad host range mercury resistance plasmid pSB102 isolated from a microbial population residing in the rhizosphere of alfalfa. *Nucl. Acids Res.* **2001**, *29*, 5169–5181. [CrossRef] [PubMed]
14. Gilmour, M.W.; Thomson, N.R.; Sanders, M.; Parkhill, J.; Taylor, D.E. The complete nucleotide sequence of the resistance plasmid R478: Defining the backbone components of incompatibility group H conjugative plasmids through comparative genomics. *Plasmid* **2004**, *52*, 182–202. [CrossRef] [PubMed]
15. Kamachi, K.; Sota, M.; Tamai, Y.; Nagata, N.; Konda, T.; Inoue, T.; Top, E.M.; Arakawa, Y. Plasmid pBP136 from *Bordetella pertussis* represents an ancestral form of IncP-1β plasmids without accessory mobile elements. *Microbiology* **2006**, *152*, 3477–3484. [CrossRef] [PubMed]
16. Guglielmini, J.; Quintais, L.; Garcillán-Barcia, M.P.; De la Cruz, F.; Rocha, E.P.C. The repertoire of ICE in prokaryotes underscores the unity, diversity, and ubiquity of conjugation. *PLoS Genet.* **2011**, *7*, e1002222. [CrossRef] [PubMed]
17. Carraro, N.; Burrus, V. The dualistic nature of integrative and conjugative elements. *Mob. Genet. Elem.* **2015**, *5*, 98–102. [CrossRef] [PubMed]
18. Carraro, N.; Rivard, N.; Burrus, V.; Ceccarelli, D. Mobilizable genomic islands, different strategies for the dissemination of multidrug resistance and other adaptive traits. *Mob. Genet. Elem.* **2017**, *7*, 1–6. [CrossRef] [PubMed]
19. Staehlin, B.M.; Gibbons, J.G.; Rokas, A.; O'Halloran, T.V.; Slot, J.C. Evolution of a heavy metal homeostasis/resistance island reflects increasing copper stress in enterobacteria. *Genome Boil. Evol.* **2016**, *8*, 811–826. [CrossRef] [PubMed]
20. Lee, S.; Ward, T.J.; Jima, D.D.; Parsons, C.; Kathariou, S. The arsenic resistance *Listeria* genomic island LGI2 exhibits sequence and integration site diversity and propensity for three Listeria monocytogenes clones with enhanced virulence. *Appl. Environ. Microbiol.* **2017**, *83*, 1189–11117. [CrossRef] [PubMed]
21. Al-Jabri, Z.; Zamudio, R.; Horvath-Papp, E.; Ralph, J.; Al-Muharrami, Z.; Rajakumar, K.; Oggioni, M. Integrase-controlled excision of metal-resistance genomic islands in *Acinetobacter baumannii*. *Genes* **2018**, *9*. [CrossRef] [PubMed]
22. Levings, R.S.; Partridge, S.R.; Djordjevic, S.P.; Hall, R.M. SGI1-K, a variant of the SGI1 genomic island carrying a mercury resistance region, in *Salmonella enterica* serovar Kentucky. *Antimicrob. Agents Chemother.* **2007**, *51*, 317–323. [CrossRef] [PubMed]
23. Carraro, N.; Matteau, D.; Luo, P.; Rodrigue, S.; Burrus, V. The master activator of IncA/C conjugative plasmids stimulates genomic islands and multidrug resistance dissemination. *PLoS Genet.* **2014**, *10*, e1004714. [CrossRef] [PubMed]
24. Carraro, N.; Durand, R.; Rivard, N.; Anquetil, C.; Barrette, C.; Humbert, M.; Burrus, V. *Salmonella* genomic island 1 (SGI1) reshapes the mating apparatus of IncC conjugative plasmids to promote self-propagation. *PLoS Genet.* **2017**, *13*, e1006705. [CrossRef] [PubMed]
25. Roberts, A.P.; Chandler, M.; Courvalin, P.; Guédon, G.; Mullany, P.; Pembroke, T.; Rood, J.I.; Smith, C.J.; Summers, A.O.; Tsuda, M.; et al. Revised nomenclature for transposable genetic elements. *Plasmid* **2008**, *60*, 167–173. [CrossRef] [PubMed]

26. Brosius, J.; Arfsten, U. Primary structure of protein L19 from the large subunit of *Escherichia coli* ribosomes. *Biochemistry* **1978**, *17*, 508–516. [CrossRef] [PubMed]
27. Chun, J.; Lee, J.H.; Jung, Y.; Kim, M.; Kim, S.; Kim, B.K.; Lim, Y.W. EzTaxon: A web-based tool for the identification of prokaryotes based on 16S ribosomal RNA gene sequences. *Int. J. Syst. Evol. Microbiol.* **2007**, *57*, 2259–2261. [CrossRef] [PubMed]
28. Tamura, K.; Stecher, G.; Peterson, D.; Filipski, A.; Kumar, S. MEGA6: Molecular evolutionary genetics analysis version 6.0. *Mol. Biol. Evol.* **2013**, *30*, 2725–2729. [CrossRef] [PubMed]
29. Kimura, M. A simple method for estimating evolutionary rates of base substitutions through comparative studies of nucleotide sequences. *J. Mol. Evol.* **1980**, *16*, 111–120. [CrossRef] [PubMed]
30. Felsenstein, J. Confidence limit on phylogenies: An approach using the bootstrap. *Evolution* **1985**, *39*, 783–791. [CrossRef] [PubMed]
31. Edgar, R.C. Muscle: Multiple sequence alignment with high accuracy and high throughput. *Nucl. Acids Res.* **2004**, *32*, 1792–1797. [PubMed]
32. Jones, D.T.; Taylor, W.R.; Thornton, J.M. The rapid generation of mutation data matrices from protein sequences. *Bioinformatics* **1992**, *8*, 275–282. [CrossRef]
33. Tang, J. Identification and Genome Analysis of *Mucilaginibacter pedocola* sp. nov. Ph.D. Thesis, Huazhong Agriculture University, Wuhan, China, 2017.
34. Liu, Q.; Siddiqi, M.Z.; Kim, M.S.; Kim, S.Y.; Im, W.T. *Mucilaginibacter hankyongensis*, sp. nov. isolated from soil of ginseng field Baekdu mountain. *J. Microbiol.* **2017**, *55*, 525–530. [CrossRef] [PubMed]
35. Kim, M.M.; Siddiqi, M.Z.; Im, W.T. *Mucilaginibacter ginsenosidivorans*, sp. nov. isolated from soil of ginseng field. *Curr. Microbiol.* **2017**, *74*, 1–7. [CrossRef] [PubMed]
36. Tang, J.; Huang, J.; Qiao, Z.; Wang, R.; Wang, G. *Mucilaginibacter pedocola* sp. nov. isolated from a heavy-metal-contaminated paddy field. *Int. J. Syst. Evol. Microbiol.* **2016**, *66*, 4033–4038. [PubMed]
37. Purohit, R.; Ross, M.O.; Batelu, S.; Kusowski, A.; Stemmler, T.L.; Hoffman, B.M.; Rosenzweig, A.C. Cu^+-specific CopB transporter: Revising P_{1B}-type ATPase classification. *Proc. Natl. Acad. Sci. USA* **2018**, *115*, 2108–2113. [CrossRef] [PubMed]
38. Mergeay, M.; Nies, D.; Schlegel, H.G.; Gerits, J.; Charles, P.; Van Gijsegem, F. *Alcaligenes eutrophus* CH34 is a facultative chemolithotroph with plasmid-bound resistance to heavy metals. *J. Bacteriol.* **1985**, *162*, 328–334. [PubMed]
39. Janssen, P.J.; Van Houdt, R.; Moors, H.; Monsieurs, P.; Morin, N.; Michaux, A.; Benotmane, M.A.; Leys, N.; Vallaeys, T.; Lapidus, A.; et al. The complete genome sequence of *Cupriavidus metallidurans* strain CH34, a master survivalist in harsh and anthropogenic environments. *PLoS ONE* **2010**, *5*, e10433. [CrossRef] [PubMed]
40. Christie, P.J. The mosaic type IV secretion systems. *EcoSal Plus* **2016**, *7*, 1–22.
41. Guglielmini, J.; Néron, B.; Abby, S.S.; Garcillánbarcia, M.P.; Cruz, F.D.L.; Rocha, E.P.C. Key components of the eight classes of type IV secretion systems involved in bacterial conjugation or protein secretion. *Nucl. Acids Res.* **2014**, *42*, 5715–5727. [CrossRef] [PubMed]
42. Toussaint, A.; Merlin, C. Mobile elements as a combination of functional modules. *Plasmid* **2002**, *47*, 26–35. [CrossRef] [PubMed]
43. Johnson, C.M.; Grossman, A.D. Integrative and conjugative elements (ICEs): What they do and how they work. *Annu. Rev. Genet.* **2015**, *49*, 577–601. [CrossRef] [PubMed]
44. Shoemaker, N.B.; Wang, G.R.; Salyers, A.A. NBU1, a mobilizable site-specific integrated element from *Bacteroides* spp. can integrate non-specifically in *Escherichia coli*. *J. Bacteriol.* **1996**, *178*, 3601–3607. [CrossRef] [PubMed]
45. Park, J.; Salyers, A.A. Characterization of the *Bacteroides* CTnDOT regulatory protein RteC. *J. Bacteriol.* **2011**, *193*, 91–97. [CrossRef] [PubMed]
46. Cheng, Q.; Sutanto, Y.; Shoemaker, N.B.; Gardner, J.F.; Salyers, A.A. Identification of genes required for excision of CTnDOT, a *Bacteroides* conjugative transposon. *Mol. Microbiol.* **2001**, *41*, 625–632. [CrossRef] [PubMed]
47. Delavat, F.; Miyazaki, R.; Carraro, N.; Pradervand, N.; Van der Meer, J.R. The hidden life of integrative and conjugative elements. *FEMS Microbiol. Rev.* **2017**, *41*, 512–537. [CrossRef] [PubMed]
48. Carraro, N.; Burrus, V. Biology of three ICE families: SXT/R391, ICE*Bs1*, and ICE*St1*/ICE*St3*. *Microbiol. Spectr.* **2014**, *2*. [CrossRef]

49. Wood, M.M.; Gardner, J.F. The integration and excision of CTnDOT. *Microbiol. Spectr.* **2015**, *3*. [CrossRef]
50. Pankratov, T.A.; Tindall, B.W.; Dedysh, S.N. *Mucilaginibacter paludis* gen. nov. sp. nov. and *Mucilaginibacter gracilis* sp. nov. pectin-, xylan- and laminarin-degrading members of the family *Sphingobacteriaceae* from acidic *Sphagnum* peat bog. *Int. J. Syst. Evol. Microbiol.* **2007**, *57*, 2349–2354. [CrossRef] [PubMed]
51. Madhaiyan, M.; Poonguzhali, S.; Jungsook, L.; Senthilkumar, M.; Lee, K.C.; Sundaram, S. *Mucilaginibacter gossypii* sp. nov. and *Mucilaginibacter gossypiicola* sp. nov. plant-growth-promoting bacteria isolated from cotton rhizosphere soils. *Int. J. Syst. Evol. Microbiol.* **2010**, *60*, 2451–2457. [CrossRef] [PubMed]

© 2018 by the authors. Licensee MDPI, Basel, Switzerland. This article is an open access article distributed under the terms and conditions of the Creative Commons Attribution (CC BY) license (http://creativecommons.org/licenses/by/4.0/).

Review

Heavy Metal Resistance Determinants of the Foodborne Pathogen *Listeria monocytogenes*

Cameron Parsons [1,*,†], Sangmi Lee [2,†] and Sophia Kathariou [1]

1 Department of Food, Bioprocessing and Nutrition Sciences, North Carolina State University, Raleigh, NC 27695-7624, USA; skathar@ncsu.edu
2 Seoul National University, Seoul 08826, Korea; slee19@ncsu.edu
* Correspondence: ctparson@ncsu.edu
† Authors contributed equally to this manuscript.

Received: 16 November 2016; Accepted: 18 December 2018; Published: 24 December 2018

Abstract: *Listeria monocytogenes* is ubiquitous in the environment and causes the disease listeriosis. Metal homeostasis is one of the key processes utilized by *L. monocytogenes* in its role as either a saprophyte or pathogen. In the environment, as well as within an animal host, *L. monocytogenes* needs to both acquire essential metals and mitigate toxic levels of metals. While the mechanisms associated with acquisition and detoxification of essential metals such as copper, iron, and zinc have been extensively studied and recently reviewed, a review of the mechanisms associated with non-essential heavy metals such as arsenic and cadmium is lacking. Resistance to both cadmium and arsenic is frequently encountered in *L. monocytogenes*, including isolates from human listeriosis. In addition, a growing body of work indicates the association of these determinants with other cellular functions such as virulence, suggesting the importance of further study in this area.

Keywords: *Listeria monocytogenes*; heavy metal resistance; mobile genetic element; cadmium; arsenic

1. Introduction

Listeria monocytogenes is a Gram-positive facultative intracellular pathogen and the causative agent of the disease listeriosis. In healthy individuals, listeriosis can manifest as gastroenteritis; however, in at-risk individuals such as the elderly, pregnant women, or immunocompromised patients, listeriosis can result in severe symptoms, including septicemia, meningitis, stillbirths and even death [1–3]. Listeriosis is responsible for approximately 1455 hospitalizations and 255 deaths in the United States annually [4]. *L. monocytogenes* is found ubiquitously in the environment, is capable of growing in the cold, and can persistently colonize food production facilities [2,5]. This, along with the severe outcomes and life-threatening potential of listeriosis, makes *L. monocytogenes* a major cause for food safety and public health concern.

L. monocytogenes is well-adapted to survive both in the environment as well as within the body of humans and other animals [6,7]. One of the key adaptations for these dual survival modalities is metal homeostasis. Certain metals such as copper, zinc, and iron are required for essential cellular functions but become toxic at higher concentrations. In contrast, metals such as arsenic and cadmium appear to serve no cellular function and are considered toxic at any concentration [8]. In the environment, metals are typically found at low levels, but their concentrations can increase due to various anthropogenic interventions, including industrial pollution or agricultural practices [9,10]. In an animal host, metal concentrations are dependent on various factors, such as diet and tissue type [11–13]. The immune system can utilize metals in response to pathogens, either by restricting metal availability or by accumulating metals, to exert toxic effects on pathogens in the course of an infection [14–16]. For these reasons, the ability to import or export metals as needed is essential for *L. monocytogenes* to survive in its diverse environmental niches. Here we update the information currently available for functions related to the essential metals copper, iron, and zinc in *L. monocytogenes*, while providing the first comprehensive review of the widely-distributed resistances to toxic heavy metals, specifically cadmium and arsenic.

2. Essential Yet Potentially Toxic Metals

Metals such as copper, iron, and zinc are cofactors for essential enzymes, and insufficient amounts of these metals can result in cellular death [8]. However, at excessive concentrations these metals become toxic to the cells, disrupting membrane potential, interfering with enzyme function, and creating reactive oxygen species [8,17,18]. *L. monocytogenes* has several determinants to acquire these metals at highly regulated levels, and to expel, sequester, or convert and detoxify these metals when they are in excess [8,18–20]. Both conditions can occur in the animal host. Substantial work has been done to elucidate these processes for essential metals in *L. monocytogenes*, culminating in several reviews [8,18,19]. In relation to iron, recent findings have clarified the role of FrvA, which is implicated in haem toxicity and pathogenicity and is a high-affinity Fe(II)-exporting P-type ATPase with specificity for elemental iron [21,22]. Additionally, a recent study by Yousuf, Ahire, and Dicks elucidated the likely mechanism underlying copper toxicity in *L. monocytogenes* in which copper disrupts the cell membrane through lipid peroxidation and protein oxidation, as shown in other organisms as well [23–25]. Additionally of note from this study was the finding that *L. monocytogenes* had the most pronounced resistance to copper of all the Gram-positive organisms tested (*L. monocytogenes*, *Streptococcus* spp., *Enterococcus* spp. and *Bacillus cereus*), which was considered by the authors to be worthy of further investigation [25]. Recent work identified a dual role for the penicillin-binding protein encoded by *pbp4* (*lmo 2229* homolog), both in tolerance of *L. monocytogenes* to β-lactam antibiotics and in copper homeostasis [26].

3. Cadmium and Arsenic: Non-Essential Toxic Metals

In contrast to multiple reviews of *L. monocytogenes* determinants mediating homeostasis for essential metals, no comprehensive reviews are available on this pathogen's resistance to non-essential toxic metals, such as arsenic and cadmium, even though resistance to these agents has been one of the earliest-documented phenotypes of *L. monocytogenes* [27,28]. Such resistance was encountered frequently enough to be utilized as a subtyping tool before the advent of higher-resolution techniques such as ribotyping, pulsed-field gel electrophoresis and multilocus sequencing, and was often associated with epidemic-associated clones [29,30]. Determinants mediating resistance to these heavy metals are widely distributed within *L. monocytogenes*, both on the chromosome and on plasmids [31–36]; all such determinants described here are summarized in Table 1.

Table 1. Heavy metal resistance-associated determinants in *Listeria monocytogenes*.

Metal Resistance-Associated Determinant	Annotation
arsA	Arsenic efflux ATPase [37]
arsB	Membrane transporter [37]
arsC	Arsenate reductase [37]
arsD	Transcriptional regulator [37]
arsR	Transcriptional regulator [37]
cadA [1]	Cadmium efflux ATPase [38]
cadC [1]	Transcriptional regulator [38]

[1] Cadmium resistance determinants can exhibit sequence divergence sufficient to be considered as different alleles of *cadA* and *cadC*. As discussed in this review, these have been designated with numbers, e.g., *cadA1 cadC1*, *cadA2 cadC2*, etc., based on the order in which they were identified or characterized.

4. Arsenic Resistance

Arsenic resistance has been primarily associated with serotype 4b, which is over-represented among clinical isolates in comparison to those from foods and food processing environments [29,39,40]. Further studies on the distribution of arsenic-resistant isolates within serotype 4b revealed that arsenic resistance was most frequently encountered among clones associated with outbreaks [30,36,39]. In particular, clonal complex (CC) 2 (formerly epidemic clone (EC) Ia) displayed the highest prevalence of arsenic-resistant isolates, and CC1 (previously ECI) was the second highest in the percentage of arsenic-resistant isolates, with resistance also encountered in several other clones, including the hypervirulent serotype 4b clone CC4 and isolates of CC315 and CC9 [30,36,39]. Interestingly, however, no CC6 (former ECII) isolates were found to be resistant to arsenic [30,36,39].

Albeit infrequent, arsenic resistance can also be found in serotypes 1/2a, 1/2b and 1/2c [39]. In serotype 1/2a, for instance, approximately 2% of the isolates tested were resistant to arsenic [39]. Interestingly, non-pathogenic *Listeria* species seem to largely lack arsenic resistance, with one of the exceptions being the reference strain *Listeria innocua* CLIP 11262, which harbors arsenic resistance genes on its plasmid pLI100 [31]. These findings suggest that arsenic resistance is primarily encountered in *L. monocytogenes*, especially in serotype 4b.

Analysis of strains that were persistently isolated from a rabbit meat processing facility in Italy revealed that approximately 90% of the isolates of clone CC14 (serotype 1/2a) exhibited arsenic resistance [41]. The extent to which arsenic resistance may contribute to persistence of this or other *L. monocytogenes* clones in food-processing facilities remains to be elucidated.

5. Arsenic Resistance Determinants

Typically, arsenic resistance cassettes are comprised of three (*arsRBC*) to five (*arsRDABC*) genes that are transcribed into a single polycistronic mRNA [37,42,43]. The genes *arsA* and *arsB* encode an ATPase and a membrane transporter, respectively, which form an ATP-dependent anion pump that exports arsenite from the cells [37,44]. The *arsA* gene product can function independently as a passive transporter of arsenite [37,42]. The *arsC* gene encodes a reductase that performs the conversion of arsenate to arsenite, which is then extruded by ArsA or the ArsA/ArsB complex [37,42,45,46]. Thus, deletion of *arsC* impairs resistance to arsenate but does not influence resistance to arsenite [45]. The *arsA*, *arsB*, and *arsC* genes are regulated by two regulatory proteins encoded by *arsR* and *arsD* [37,42,45,47]. The *arsR* gene product is a repressor that binds to the operator of the *ars* cassette in the absence of the inducer (arsenate and arsenite) but dissociates from the operator upon interaction with the inducer [37,45]. In other words, ArsR determines the basal expression level of the arsenic resistance cassette [37,45]. Meanwhile, the *arsD* gene product is not affected by the inducer and controls the maximal level of the *ars* operon, preventing the deleterious effects of *arsB* overexpression, such as hypersensitivity to arsenite [37,42,47].

Whole genome sequencing of *L. monocytogenes* has revealed three operons putatively associated with arsenic detoxification in *Listeria* spp. [31,34,35]. The first putative arsenic resistance operon (*arsR1D2R2A2B1B2*) was reported on plasmid pLI100, which is harbored in *L. innocua* CLIP 11262 [31,34]. As indicated above, no plasmid-borne arsenic resistance determinants have been reported in *L. monocytogenes*, which is consistent with earlier findings that arsenic resistance is chromosomally mediated in this species [29]. The other two putative arsenic resistance determinants are both located on the chromosome and were identified only in *L. monocytogenes*, each with a tendency to contribute to arsenic resistance in different serotypes. The first consists of the arsenic resistance cassette identified in pLI100 (*arsR1D2R2A2B1B2*) and two additional upstream genes, *arsD1* and *arsA1* [48]. The second consists of the arsenic resistance cassette *arsCBADR* harbored on a Tn554-like element [35].

The *arsR1D2R2A2B1B2* cassette, together with the upstream genes *arsD1* and *arsA1*, were initially identified on a 35-kb chromosomal island, termed *Listeria* genomic island 2 (LGI2), harbored by the CC2 strain Scott A [48]. Downstream of the arsenic resistance cassette, LGI2 also harbored the novel cadmium resistance determinant *cadA4* [35,36,48,49]. Therefore, all tested serotype 4b arsenic-resistant isolates were also resistant to cadmium [30,36].

Further studies using the LGI2 genes as genetic markers and whole genome sequence analysis showed that LGI2 was present in all tested arsenic-resistant isolates of serotype 4b, including isolates belonging to clones CC1 and CC2, and the hypervirulent clone CC4 [36,39,50]. Interestingly, the entire island was markedly diversified in a majority of arsenic-resistant CC1 strains; this diversified derivative was termed LGI2-1 and was inserted at the same chromosomal locus in all CC1 isolates that harbored it [36,39,50,51]. These findings suggest that arsenic resistance of serotype 4b can be attributable to arsenic resistance genes harbored on LGI2. Furthermore, regardless of the diversification, all serotype 4b arsenic-resistant isolates harboring LGI2 displayed tolerance to higher concentrations of arsenic (arsenite minimum inhibitory concentration (MIC) of 1.250 to 2.500 µg/mL) compared with susceptible strains (arsenite MIC of 250 to 500 µg/mL) [36,39]. However, direct experimental evidence is still warranted to explicitly demonstrate the involvement of arsenic resistance genes on LGI2 in arsenic detoxification.

LGI2 genes were rarely encountered among arsenic-resistant isolates that belong to serotypes other than 4b [39]. Even if strains were positive for LGI2 genes, both PCR typing and whole genome sequence analysis suggested sequence divergence from either LGI2 or LGI2-1, except in the case of a serotype 1/2a belonging to CC14, which harbored LGI2 that was highly conserved with that in the serotype 4b strain Scott A [39,41]. As mentioned earlier, CC14 isolates harboring this highly-conserved LGI2 were also found to be persistent in a rabbit meat processing plant [40].

While the genetic content of LGI2 is highly conserved in *L. monocytogenes*, genome analysis of arsenic-resistant isolates harboring LGI2 revealed this island to be inserted in at least eight different locations, primarily within open reading frames [39]. The GC content of LGI2 is lower than average (34% versus the *L. monocytogenes* average of 38%), and LGI2 also harbors a putative phage integrase gene. Such findings suggest that LGI2 was acquired via horizontal gene transfer from other bacterial genomes [36,39]. However, likely donors for LGI2 or LGI2-1 in *L. monocytogenes* remain unidentified.

The chromosomally-encoded arsenic resistance cassette (*arsCBADR*) that is associated with a Tn*554*-like transposon was first identified via whole genome sequencing of the serotype 1/2c strain SLCC 2372 [35]. When *arsA* associated with this Tn*554*-like element was used as a genetic marker, it was exclusively found among arsenic-resistant isolates of serotypes other than 4b, and approximately 90% of these isolates were negative for any LGI2-associated arsenic resistance genes, while positive for *arsA* harbored on the Tn*554*-like transposon [39]. These observations suggest that, in contrast to LGI2 which is predominantly found among serotype 4b isolates, the arsenic resistance cassette harbored on the Tn*554*-like transposon is responsible for arsenic resistance of strains of *L. monocytogenes* of other serotypes. Further experimental and *in-silico* evidence can deepen our understanding of the evolution and function of arsenic resistance associated with the Tn*554*-like element.

6. Cadmium Resistance

Several studies examined prevalence of cadmium resistance in strains isolated from food and food processing facilities in different regions [52–55]. Prevalence ranged from 50 to 66%, suggesting that cadmium resistance is globally widespread and highly prevalent in food-associated isolates. One study also noted that isolates repeatedly isolated from milk and dairy foods in Northern Ireland were more likely to be cadmium-resistant than those that were only sporadically recovered [56], suggesting that food or food processing facilities may provide unique pressures that select for cadmium resistance in *L. monocytogenes*.

In contrast to arsenic resistance often associated with serotype 4b, cadmium resistance was frequently encountered among isolates of serotype 1/2a, which are over-represented among food isolates compared with those of clinical origin [29,40]. In congruence with this association, cadmium resistance was generally much more prevalent than arsenic resistance among *L. monocytogenes* from foods and food processing environments [29,52–55]. Even in serotype 4b, the prevalence of cadmium-resistant isolates surpassed that of arsenic-resistant isolates; however, approximately 50% of serotype 4b cadmium-resistant isolates were also resistant to arsenic due to LGI2, as discussed above. While genes putatively associated with both cadmium and arsenic detoxification are co-localized on

LGI2, multiple other cadmium resistance determinants have been identified in *L. monocytogenes* and are harbored both chromosomally and on plasmids, as will be discussed in the following section.

7. Plasmid-Associated Cadmium Resistance Determinants

In a survey of *L. monocytogenes* plasmids from strains of diverse origins (food, environmental, and clinical), an estimated 28% of the isolates were found to harbor plasmids, and most (95%) of these plasmid-harboring strains were found to be cadmium-resistant [28]. This was consistent with a later survey of plasmids in the genus *Listeria* that spanned multiple species, serogroups, and origins which found that, besides the origin of replication, the most common plasmid-borne elements were cadmium resistance cassettes encountered in all plasmids that were analyzed [34]. The cadmium resistance cassettes on the plasmids encoded a cadmium efflux P-type ATPase (*cadA*) and its putative repressor *cadC*. Multiple *cadA* determinants have been identified in *L. monocytogenes* and have been serially numbered, e.g., *cadA1*, *cadA2*, *cadA3*, etc., in the order in which they were discovered. Many *cadA*-harboring plasmids also harbored putative copper-resistance determinants [34]. Intriguingly, plasmid-encoded *cadA* in conjunction with a cassette of genes for arsenic detoxification has only been encountered once, in the aforementioned pLI100 of *L. innocua* CLIP 11262 [31,35]. The evolutionary and ecological mechanisms mediating the scarcity of pLI100-like plasmids with genes for both cadmium and arsenic resistance remain unclear.

The first *cadA* determinant to be identified in *L. monocytogenes* was *cadA1*, which was genetically similar to the *cadA* characterized in *Staphylococcus aureus* [38]. While the *cadA1* in *S. aureus* conferred resistance to both cadmium and zinc, the plasmid-harbored *cadA1* in *L. monocytogenes* was specific to cadmium [38]. CadA1 was harbored on the mobile genetic element Tn5422 [57]. Interestingly, Tn5422 was never detected chromosomally but appeared to integrate extensively into plasmids, leading Lebrun et al. to speculate that this element was responsible for much of the size variation of plasmids in *L. monocytogenes* [38]. Tn5422-associated *cadA1* has been subsequently identified on numerous other plasmids of *L. monocytogenes* (Figure 1) [57,58].

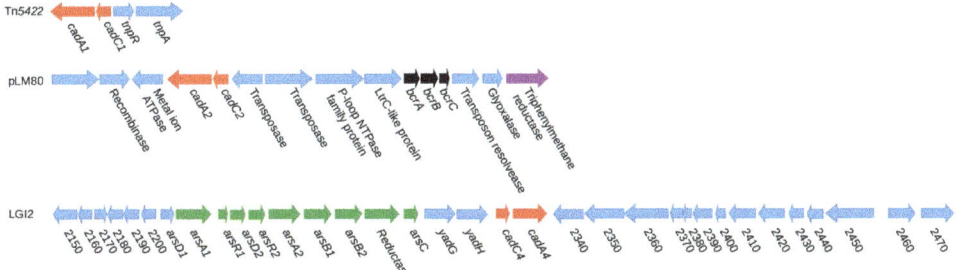

Figure 1. Resistance gene distribution across mobile genetic elements for *cadA1*, *cadA2*, and *cadA4*. Cad family members, benzalkonium chloride resistance determinants, toxic triphenylmethane dye resistance determinants, and putative arsenic detoxification determinants are in red, black, purple, and green, respectively.

A second putative *cadA* (*cadA2*) was first identified on the large plasmid pLI100 of *L. innocua* CLIP 11262, followed by its discovery on the approximately 80 kb plasmid pLM80 of *L. monocytogenes* H7858, a strain implicated in a large, multistate outbreak in the U.S. in 1998–1999, which involved contaminated hotdogs [31,33] (Figure 1). The latter plasmid was later experimentally confirmed to confer not only cadmium resistance but also resistance to the quaternary ammonium compound (QAC) benzalkonium chloride, via the *bcrABC* efflux cassette that mediated enhanced tolerance to benzalkonium chloride and other QACs [32,59]. In addition, pLM80 conferred resistance to toxic triphenylmethane dyes such as crystal violet and malachite green via *tmr*, a determinant that detoxifies these dyes and appears to have been acquired from Gram-negative bacteria [32,60] (Figure 1).

Plasmids harboring *cadA1* and *cadA2* have been observed in both *L. monocytogenes* and other, non-pathogenic *Listeria* spp. [34,61]. Several studies have suggested that in *L. monocytogenes*, *cadA1* is more common than *cadA2* [53,54,62]. Data also suggest that *cadA1* may have been predominant earlier in *L. monocytogenes* [28,38,57], since *cadA2* was not identified until the characterization of strains implicated in the 1998–1999 hot dog outbreak [32,33]. It is thought-provoking that pLM80-like plasmids were not identified previously in *L. monocytogenes* and that *cadA2* determinants were not detected among the cadmium-resistant plasmid-harboring isolates in earlier studies [28,57]. It is tempting to speculate that the reasons are related to co-selection of *cadA2*-harboring plasmids for resistance to QACs, which were not recognized as disinfectants until 1934 [63] and which have been routinely and extensively employed for sanitation of food processing facilities only in the past few decades. A survey of *L. monocytogenes* isolates of primarily serotype 1/2a or 1/2b from turkey processing plants in the U.S. revealed that cadmium-resistant isolates that also exhibited enhanced QAC tolerance were more likely to harbor *cadA2*, either alone or together with *cadA1*, than isolates that were cadmium resistant but without enhanced tolerance to QACs [62].

Strains harboring both *cadA1* and *cadA2* were encountered in some surveys [54,62], while other studies reported little or no co-occurrence of *cadA1* and *cadA2* in the same strains [36,53]. Such findings suggest that certain environments may be more conducive to the co-occurrence of these two *cadA* determinants. Analysis of *Listeria* plasmid sequences failed to reveal plasmids that harbor both *cadA1* and *cadA2* [34], suggesting that strains positive for both *cadA1* and *cadA2* harbored these determinants on different plasmids.

Overall, *cadA1* and *cadA2* were far more prevalent in serotype 1/2a and 1/2b strains from food and food processing plants in comparison to serotype 4b [53,62]. This could potentially be explained by the overall greater prevalence of plasmids in serogroup 1 than in serogroup 4 [34]. Strains that harbored only *cadA1* were significantly more likely to be serotype 1/2a, while those harboring only *cadA2* were significantly more likely to be serotype 1/2b [62]. An unexpected finding has been the lack of detection of *cadA1* among serotype 4b clinical isolates belonging to the major clones CC2 and CC6, while *cadA2* was not detected in another leading clone, CC1, suggesting proclivity of different clonal groups for specific cadmium resistance determinants [36,39]. The underlying reasons for differential prevalence of different *cadA* determinants in various serotypes and clones of *L. monocytogenes* are worthy of further investigation and may reflect differences in their ecology, including the microbial community that may include donors for the plasmids, and the accompanying selective pressures.

8. Chromosomal Cadmium Resistance Determinants

The proliferation of whole genome sequencing (WGS) data resulted in the discovery of several cadmium resistance determinants harbored chromosomally in *L. monocytogenes*. The first such determinant was *cadA3*, harbored on a mobile integrated conjugative element in strain EGDe [31]. While direct experimental evidence for its role in cadmium resistance is still lacking, the presence of this gene has been associated with tolerance to cadmium of >140 µg/mL [36]. Thus far, this determinant has been encountered infrequently, having been identified only in EGDe and a few additional strains [35,36,53]. A survey of 136 serotype 4b isolates from human sporadic listeriosis in the U.S. revealed 45 that were cadmium resistant, of which only one harbored *cadA3* [36].

Another chromosomal *cadA* family member (*cadA4*) was identified within the previously-discussed LGI2 in the chromosome of strain Scott A [35,36,48] (Figure 1). While *cadA1–3* are all associated with tolerance to cadmium of >140 µg/mL and can be detected through routine screening of the isolates on 70 µg/mL [29,52], *cadA4* only confers resistance to approximately 50 µg/mL of cadmium and thus can be detected by growth at 35 but not at 70 µg/mL [29,36,49,52,53]. Thus, resistance mediated by *cadA4* would have been undetected if isolates were screened at a level of 70 µg/mL as employed in studies prior to 2013 [28,29,52,53]. For this reason, the prevalence of cadmium resistance in *L. monocytogenes* may have been underestimated in earlier studies. Mechanisms underlying differences in resistance level between *cadA4* and *cadA1–A3* remain to be identified, but may reflect the divergent nature of the deduced *cadA4* product. Pairwise comparisons revealed that the *cadA4* amino acid sequence had approximately 36% identity to those encoded by *cadA1–A3*, while the latter exhibited approximately 70% identity to each other [36,49].

Due to its more recent characterization, *cadA4* has not been surveyed as extensively as *cadA1* and *cadA2*, and its distribution was mostly investigated in serotype 4b strains often associated with the arsenic resistance island LGI2 [30,36,39]. In a survey of 136 serotype 4b isolates from sporadic cases of human listeriosis in the United States discussed above, *cadA4* accounted for approximately 10% of the isolates and 29% of the cadmium-resistant isolates [36]. Interestingly, *cadA4* was always located on LGI2 downstream of arsenic resistance genes [39]. Even though *cadA4* has been primarily reported in serotype 4b, it has also been detected in a few strains of other serotypes (1/2a, 1/2b and 1/2c), including the persistent clone CC14 (serotype 1/2a), where *cadA4* was located on LGI2 downstream to arsenic resistance genes, as in serotype 4b [39,41]. To date, we lack reports of *cadA4*-harboring isolates of *Listeria* spp. other than *L. monocytogenes*.

As discussed above, a diversified LGI2 derivative (LGI2-1) was observed in a subset of arsenic-resistant serotype 4b CC1 isolates, and another derivative was also identified in a serotype 1/2c strain of CC9 [39]. The *cadA4* homolog in these divergent islands exhibited ~90% amino acid identity with the *cadA4* of Scott A (Figure 2). Together, LGI2-associated *cadA4* and its divergent homolog in LGI2-1 accounted for half of serotype 4b cadmium-resistant isolates in a previous study [30]. Further studies need to be conducted to investigate the prevalence and functional characteristics of this or additional *cadA4* homologs.

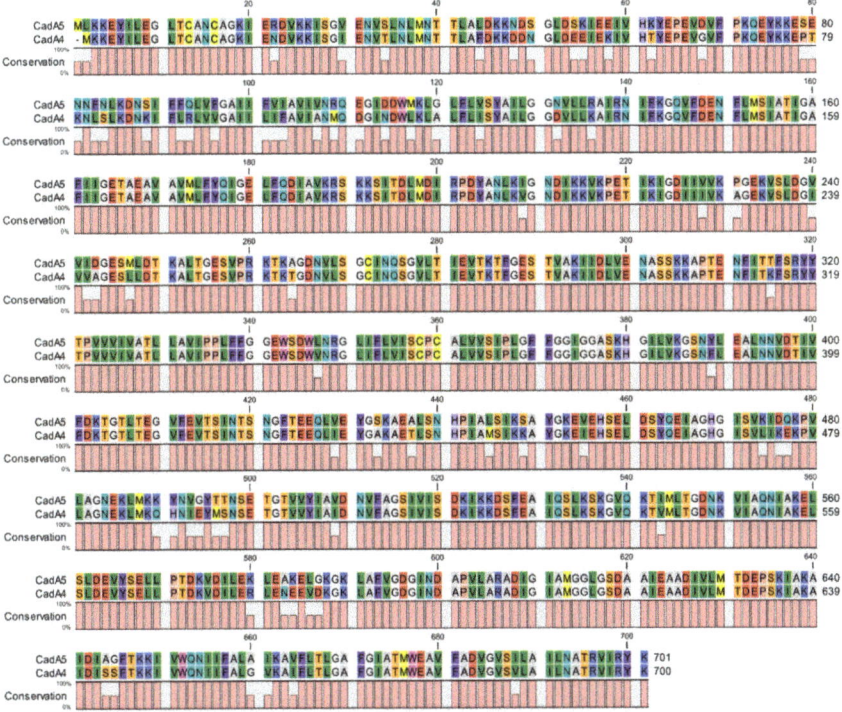

Figure 2. Amino acid alignment between *cadA4* in strain Scott A and its divergent counterpart (here labeled as *cadA5*) encountered in LGI2-1. Amino acid alignment was generated using CLC Genomics Workbench 11.0 [64].

9. Impacts of Heavy Metal Resistance Determinants on other Adaptations, Including Virulence

There is a growing body of evidence that suggests links between heavy metal resistance and the ability of *L. monocytogenes* to cause disease. For instance, an investigation of the prevalence of arsenic and cadmium resistance among serotype 4b isolates from human listeriosis patients in the United States found a high prevalence of resistance and a strong association with clones repeatedly implicated in outbreaks [30]. More direct evidence supporting the involvement of cadmium resistance with pathogenicity has been provided in several studies. The first experimental evidence for the possible involvement of cadmium resistance determinants in virulence was obtained from *in-vivo* transcriptional analysis of *L. monocytogenes* from the livers of mice infected with strain EGDe, which harbors *cadA3* [65]. The putative repressor encoded by *cadC3* was markedly upregulated in the liver of the infected mice, and deletion of this gene resulted in a decrease in virulence when the bacteria were administered intravenously [65]. Similar findings were also reported by Pombinho et al. in 2017, who again identified *cadC3* as being essential for virulence of *L. monocytogenes* [66]. They found that, in addition to repressing *cadA3*, *cadC3* also repressed *ispB*, which is involved in initiating an immune response, thus helping *L. monocytogenes* avoid detection by the host immune system [66]. Interestingly, a transposon insertion mutant of *cadA4* [49] showed increased virulence in the *Galleria mellonella* model, suggesting an inverse relationship between *cadA4* and virulence [49]. This finding is consistent with the previous study, which demonstrated that the putative repressor *cadC3* was required for full virulence [65]. Taken together, the results of these studies suggest an association between cadmium resistance determinants and the ability of *L. monocytogenes* to cause disease. It is important to note, however, that these studies have focused on *cadC3* and *cadA4*. We currently lack information on the potential virulence or pathogenicity roles of the predominant cadmium resistance determinants *cadA1* or *cadA2*, and their cognate *cadC* repressors (*cadC1* and *cadC2*, respectively).

The cadmium and arsenic resistance genes of *L. monocytogenes* discussed here are accompanied by transcriptional regulators of the ArsR family of metal-associated transcriptional regulators [67]. In other bacterial systems, these regulators have been found to regulate expression of single genes in some circumstances, while mediating a global transcriptional response in others [67]. Most often, they regulate expression of genes directly involved in metal detoxification, but they can also impact expression of genes with a variety of other functions including oxidative stress tolerance, acid adaptation, respiration and ribosome biogenesis [67].

Members of the ArsR family were involved in the regulation of virulence-associated genes in various species. For instance, the PhoPR two-component regulatory system which is responsible for the regulation of virulence and persistence genes in *Mycobacterium* spp. was shown to be under the control of an ArsR transcriptional factor [68]. In *L. monocytogenes*, the aforementioned cadmium resistance regulator *cadC* belongs to the ArsR family and, as previously stated, *cadC3* was involved in virulence via its impacts on *ispB* [49,65,67].

In the case of the arsenic resistance genomic island LGI2, it is tempting to speculate that this element may have roles in virulence and pathogenicity, based on the fact that, as discussed above, LGI2 has been detected exclusively in *L. monocytogenes* and primarily in serotype 4b, which makes significant contributions to human listeriosis [56]. However, direct experimental evidence is needed to assess LGI2's roles in virulence, e.g., by comparing virulence of isogenic strains with and without specific LGI2-associated genes. This was pursued with *cadA4* [49], but similar investigations with the arsenic resistance genes on LGI2 are lacking.

The potential contributions of heavy metal resistance to environmental persistence of *L. monocytogenes* remains to be elucidated. As discussed earlier, one study found that isolates repeatedly isolated from contaminated foods were more likely to be cadmium-resistant than those that were only sporadically encountered [56], and 74% or more of the serotype 1/2a and 1/2b isolates from turkey processing facilities were cadmium resistant, harboring *cadA1* and/or *cadA2* [52,62]. It is thus tempting to speculate that cadmium resistance via these determinants may enhance the capacity of the isolates to persist in the contaminated food or food processing environments, but the underlying mechanisms remain unknown.

In addition to their overt impacts on enhanced tolerance to heavy metals, heavy metal resistance genomic islands may indirectly influence environmental fitness or pathogenicity by promoting the horizontal transfer of accessory genes. In bacteria, metal resistance genes have been found to co-localize with antibiotic and other resistance genes on mobile genetic elements, such as plasmids, genomic islands, and transposons [69–72] (Figure 1). As discussed above, pLM80 and related plasmids harbor not only cadmium resistance genes but also genes mediating enhanced tolerance to QACs and toxic dyes [32,34,60]. Sequence data suggest that these elements have been introduced to *L. monocytogenes* from other species [48,60], and there is direct experimental evidence for the transfer of such elements between *L. welshimeri* or *L. innocua* and *L. monocytogenes* [61]. This creates the possibility that metal contamination and/or metal resistance genes could facilitate the acquisition and transfer of other resistance genes to *L. monocytogenes*, or from *L. monocytogenes* and other *Listeria* spp. to other bacterial agents of public health concern. It has also been shown that extremely low levels of metals can induce transcription of metal resistance genes and exert sufficient selective pressure to result in the retention of these elements [49,57,73]. These data suggest that in minute amounts such as might be encountered in the environment or in an animal host, heavy metals can potentially exert selective pressure, which in turn could direct the acquisition or transfer of mobile genetic elements that can impact the environmental or *in-vivo* fitness of *L. monocytogenes*.

10. Conclusions

Metals play a key role in the survival of *L. monocytogenes* both in the environment and in animal hosts [8]. Essential metals must be acquired, and toxic effects of excess metals must be mitigated. While the cellular functions of *L. monocytogenes* associated with essential metals have been extensively studied and reviewed, those involved with exclusively toxic metals such as cadmium and arsenic are poorly understood. The significance of these determinants is shown by their wide distribution within *L. monocytogenes*, as well as their association with food, food processing plants, clinical strains and clonal groups involved in outbreaks [36,53–55,62]. Evidence from other microorganisms suggests the involvement of metal resistance genes in a variety of functions beyond just metal detoxification [24,74]. Several studies discussed here would also suggest alternate and additional functions for these genes in *L. monocytogenes* [49,65]. Given their prevalence, potential involvement in selection and population dynamics, as well as their growing implication in important alternative cellular functions such as virulence, heavy metal resistance genes are an ideal candidate for further study.

Author Contributions: C.P. authored the initial draft of the manuscript and worked on later edits and revisions, S.L. contributed significant portions of original material to the manuscript and worked on later edits and revisions, S.K. contributed to initial concept and worked on later edits and revisions.

Funding: This research was funded by USDA NIFA grant 2012-67017-30218.

Conflicts of Interest: The authors declare no conflict of interest.

References

1. Painter, J.; Slutsker, L. Listeriosis in humans. In *Listeria, Listeriosis and Food Safety*; CRC Press: Boca Raton, FL, USA, 2007; Volume 30, pp. 85–109. ISBN 978-0-8247-5750-2.
2. Kathariou, S. *Listeria monocytogenes* virulence and pathogenicity, a food safety perspective. *J. Food Prot.* **2002**, *65*, 1811–1829. [CrossRef] [PubMed]
3. Charlier, C.; Perrodeau, É.; Leclercq, A.; Cazenave, B.; Pilmis, B.; Henry, B.; Lopes, A.; Maury, M.M.; Moura, A.; Goffinet, F.; et al. Clinical features and prognostic factors of listeriosis: The MONALISA national prospective cohort study. *Lancet Infect. Dis.* **2017**. [CrossRef]
4. Scallan, E.; Hoekstra, R.M.; Angulo, F.J.; Tauxe, R.V.; Widdowson, M.A.; Roy, S.L.; Jones, J.L.; Griffin, P.M. Foodborne illness acquired in the United States-major pathogens. *Emerg. Infect. Dis.* **2011**, *17*, 7–15. [CrossRef] [PubMed]

5. Gandhi, M.; Chikindas, M.L. *Listeria*: A foodborne pathogen that knows how to survive. *Int. J. Food Microbiol.* **2007**, *113*, 1–15. [CrossRef] [PubMed]
6. Gray, M.J.; Freitag, N.E.; Boor, K.J. How the bacterial pathogen *Listeria monocytogenes* mediates the switch from environmental Dr. Jekyll to pathogenic Mr. Hyde. *Infect. Immun.* **2006**, *74*, 2506–2512. [CrossRef] [PubMed]
7. Freitag, N.E.; Port, G.C.; Miner, M.D. *Listeria monocytogenes*—From saprophyte to intracellular pathogen. *Nat. Rev. Microbiol.* **2009**, *7*, 623–628. [CrossRef] [PubMed]
8. Jesse, H.E.; Roberts, I.S.; Cavet, J.S. Metal ion homeostasis in *Listeria monocytogenes* and importance in host-pathogen interactions. *Adv. Microb. Physiol.* **2014**, *65*, 83–123. [CrossRef]
9. Alloway, B.J. *Heavy Metals in Soils*; Springer: Dordrecht, The Netherlands, 2013; p. 614. [CrossRef]
10. Nunes, I.; Jacquiod, S.; Brejnrod, A.; Holm, P.E.; Brandt, K.K.; Priemé, A.; Sørensen, S.J. Coping with copper: Legacy effect of copper on potential activity of soil bacteria following a century of exposure. *FEMS Microbiol. Ecol.* **2016**, *92*. [CrossRef]
11. Lindh, S.; Razmara, P.; Bogart, S.; Pyle, G. Comparative tissue distribution and depuration characteristics of copper nanoparticles and soluble copper in rainbow trout (*Oncorhynchus mykiss*). *Environ. Toxicol. Chem.* **2018**. [CrossRef]
12. Jordanova, M.; Hristovski, S.; Musai, M.; Boškovska, V.; Rebok, K.; Dinevska-Ḱovkarovska, S.; Melovski, L. Accumulation of heavy metals in some organs in barbel and chub from Crn Drim River in the Republic of Macedonia. *Bull. Environ. Contam. Toxicol.* **2018**, *101*, 392–397. [CrossRef]
13. Li, P.; Du, B.; Chan, H.M.; Feng, X.; Li, B. Mercury bioaccumulation and its toxic effects in rats fed with methylmercury polluted rice. *Sci. Total Environ.* **2018**, *633*, 93–99. [CrossRef] [PubMed]
14. Zalewski, P.; Truong-Tran, A.; Lincoln, S.; Ward, D.; Shankar, A.; Coyle, P.; Jayaram, L.; Copley, A.; Grosser, D.; Murgia, C.; et al. Use of a zinc fluorophore to measure labile pools of zinc in body fluids and cell-conditioned media. *Biotechniques* **2006**, *40*, 509–520. [CrossRef] [PubMed]
15. Carrigan, P.E.; Hentz, J.G.; Gordon, G.; Morgan, J.L.; Raimondo, M.; Anbar, A.D.; Miller, L.J. Distinctive heavy metal composition of pancreatic juice in patients with pancreatic carcinoma. *Cancer Epidemiol. Biomark. Prev.* **2007**, *16*, 2656–2663. [CrossRef] [PubMed]
16. White, C.; Lee, J.; Kambe, T.; Fritsche, K.; Petris, M.J. A role for the ATP7A copper-transporting ATPase in macrophage bactericidal activity. *J. Biol. Chem.* **2009**, *284*, 33949–33956. [CrossRef]
17. Argüello, J.M.; Raimunda, D.; Padilla-Benavides, T. Mechanisms of copper homeostasis in bacteria. *Front. Cell. Infect. Microbiol.* **2013**, *3*, 73. [CrossRef]
18. McLaughlin, H.P.; Hill, C.; Gahan, C.G.M. The impact of iron on *Listeria monocytogenes*; inside and outside the host. *Curr. Opin. Biotechnol.* **2011**, *22*, 194–199. [CrossRef]
19. Lechowicz, J.; Krawczyk-Balska, A. An update on the transport and metabolism of iron in *Listeria monocytogenes*: The role of proteins involved in pathogenicity. *Biometals* **2015**, *28*, 587–603. [CrossRef]
20. Latorre, M.; Olivares, F.; Reyes-Jara, A.; López, G.; González, M. CutC is induced late during copper exposure and can modify intracellular copper content in *Enterococcus faecalis*. *Biochem. Biophys. Res. Commun.* **2011**, *406*, 633–637. [CrossRef]
21. Pi, H.; Patel, S.J.; Argüello, J.M.; Helmann, J.D. The *Listeria monocytogenes* Fur-regulated virulence protein FrvA is an Fe(II) efflux P_{1B4}-type ATPase. *Mol. Microbiol.* **2016**, *100*, 1066–1079. [CrossRef]
22. McLaughlin, H.P.; Xiao, Q.; Rea, R.B.; Pi, H.; Casey, P.G.; Darby, T.; Charbit, A.; Sleator, R.D.; Joyce, S.A.; Cowart, R.E.; et al. A putative P-type ATPase required for virulence and resistance to haem toxicity in *Listeria monocytogenes*. *PLoS ONE* **2012**, *7*. [CrossRef]
23. Santo, C.E.; Quaranta, D.; Grass, G. Antimicrobial metallic copper surfaces kill *Staphylococcus haemolyticus* via membrane damage. *Microbiologyopen* **2012**, *1*, 46–52. [CrossRef] [PubMed]
24. Singh, K.; Senadheera, D.B.; Lévesque, C.M.; Cvitkovitch, D.G. The *copYAZ* operon functions in copper efflux, biofilm formation, genetic transformation, and stress tolerance in *Streptococcus mutans*. *J. Bacteriol.* **2015**, *197*, 2545–2557. [CrossRef]
25. Yousuf, B.; Ahire, J.J.; Dicks, L.M.T. Understanding the antimicrobial activity behind thin- and thick-rolled copper plates. *Appl. Microbiol. Biotechnol.* **2016**, *100*, 5569–5580. [CrossRef] [PubMed]
26. Parsons, C.; Costolo, B.; Brown, P.; Kathariou, S. Penicillin-binding protein encoded by *pbp4* is involved in mediating copper stress in *Listeria monocytogenes*. *FEMS Microbiol. Lett.* **2017**, *364*. [CrossRef] [PubMed]

27. Buchanan, R.L.; Klawitter, L.A.; Bhaduri, S.; Stahl, H.G. Arsenite resistance in *Listeria monocytogenes*. *Food Microbiol.* **1991**, *8*, 161–166. [CrossRef]
28. Lebrun, M.; Loulergue, J.; Chaslus-Dancla, E.; Audurier, A. Plasmids in *Listeria monocytogenes* in relation to cadmium resistance. *Appl. Environ. Microbiol.* **1992**, *58*, 3183–3186.
29. McLauchlin, J.; Hampton, M.D.; Shah, S.; Threlfall, E.J.; Wieneke, A.A.; Curtis, G.D. Subtyping of *Listeria monocytogenes* on the basis of plasmid profiles and arsenic and cadmium susceptibility. *J. Appl. Microbiol.* **1997**, *83*, 381–388. [CrossRef] [PubMed]
30. Lee, S.; Ward, T.J.; Graves, L.M.; Tarr, C.L.; Siletzky, R.M.; Kathariou, S. Population structure of *Listeria monocytogenes* serotype 4b isolates from sporadic human listeriosis in the United States, 2003–2008. *Appl. Environ. Microbiol.* **2014**, *80*, AEM-00454. [CrossRef] [PubMed]
31. Glaser, P.; Frangeul, L.; Buchrieser, C.; Rusniok, C.; Amend, A.; Baquero, F.; Berche, P.; Bloecker, H.; Brandt, P.; Chakraborty, T.; et al. Comparative genomics of *Listeria* species. *Science* **2001**, *294*, 849–852. [CrossRef] [PubMed]
32. Elhanafi, D.; Dutta, V.; Kathariou, S. Genetic characterization of plasmid-associated benzalkonium chloride resistance determinants in a *Listeria monocytogenes* strain from the 1998–1999 outbreak. *Appl. Environ. Microbiol.* **2010**, *76*, 8231–8238. [CrossRef] [PubMed]
33. Nelson, K.E.; Fouts, D.E.; Mongodin, E.F.; Ravel, J.; DeBoy, R.T.; Kolonay, J.F.; Rasko, D.A.; Angiuoli, S.V.; Gill, S.R.; Paulsen, I.T.; et al. Whole genome comparisons of serotype 4b and 1/2a strains of the food-borne pathogen *Listeria monocytogenes* reveal new insights into the core genome components of this species. *Nucleic Acids Res.* **2004**, *32*, 2386–2395. [CrossRef] [PubMed]
34. Kuenne, C.; Voget, S.; Pischimarov, J.; Oehm, S.; Goesmann, A.; Daniel, R.; Hain, T.; Chakraborty, T. Comparative analysis of plasmids in the genus *Listeria*. *PLoS ONE* **2010**, *5*, e12511. [CrossRef] [PubMed]
35. Kuenne, C.; Billion, A.; Mraheil, M.A.; Strittmatter, A.; Daniel, R.; Goesmann, A.; Barbuddhe, S.; Hain, T.; Chakraborty, T. Reassessment of the *Listeria monocytogenes* pan-genome reveals dynamic integration hotspots and mobile genetic elements as major components of the accessory genome. *BMC Genom.* **2013**, *14*, 47. [CrossRef] [PubMed]
36. Lee, S.; Rakic-Martinez, M.; Graves, L.M.; Ward, T.J.; Siletzky, R.M.; Kathariou, S. Genetic determinants for cadmium and arsenic resistance among *Listeria monocytogenes* serotype 4b isolates from sporadic human listeriosis patients. *Appl. Environ. Microbiol.* **2013**, *79*, 2471–2476. [CrossRef] [PubMed]
37. Rosen, B.P. Families of arsenic transporters. *Trends Microbiol.* **1999**, *7*, 207–212. [CrossRef]
38. Lebrun, M.; Audurier, A.; Cossart, P. Plasmid-borne cadmium resistance genes in *Listeria monocytogenes* are similar to *cadA* and *cadC* of *Staphylococcus aureus* and are induced by cadmium. *J. Bacteriol.* **1994**, *176*, 3040–3048. [CrossRef]
39. Lee, S.; Ward, T.J.; Jima, D.D.; Parsons, C.; Kathariou, S. The arsenic resistance-associated *Listeria* genomic island LGI2 exhibits sequence and integration site diversity and a propensity for three *Listeria monocytogenes* clones with enhanced virulence. *Appl. Environ. Microbiol.* **2017**, *83*. [CrossRef]
40. Cheng, Y.; Siletzky, R.M.; Kathariou, S. Genomic divisions/lineages, epidemic clones, and population structure. In *Handbook of Listeria monocytogenes*; CRC Press: Boca Raton, FL, USA, 2008; pp. 337–358.
41. Pasquali, F.; Palma, F.; Guillier, L.; Lucchi, A.; De Cesare, A.; Manfreda, G. *Listeria monocytogenes* sequence types 121 and 14 repeatedly isolated within one year of sampling in a rabbit meat processing plant: Persistence and ecophysiology. *Front. Microbiol.* **2018**, *9*, 596. [CrossRef]
42. Kaur, S.; Kamli, M.R.; Ali, A. Role of arsenic and its resistance in nature. *Can. J. Microbiol.* **2011**, *57*, 769–774. [CrossRef]
43. Ordóñez, E.; Letek, M.; Valbuena, N.; Gil, J.A.; Mateos, L.M. Analysis of genes involved in arsenic resistance in *Corynebacterium glutamicum* ATCC 13032. *Appl. Environ. Microbiol.* **2005**, *71*, 6206–6215. [CrossRef]
44. Tisa, L.S.; Rosen, B.P. Molecular characterization of an anion pump. The ArsB protein is the membrane anchor for the ArsA protein. *J. Biol. Chem.* **1990**, *265*, 190–194. [PubMed]
45. López-Maury, L.; Florencio, F.J.; Reyes, J.C. Arsenic sensing and resistance system in the cyanobacterium *Synechocystis* sp. strain PCC 6803. *J. Bacteriol.* **2003**, *185*, 5363–5371. [CrossRef] [PubMed]
46. Gladysheva, T.B.; Oden, K.L.; Rosen, B.P. Properties of the arsenate reductase of plasmid R773. *Biochemistry* **1994**, *33*, 7288–7293. [CrossRef] [PubMed]
47. Wu, J.; Rosen, B.P. The *arsD* gene encodes a second trans-acting regulatory protein of the plasmid-encoded arsenical resistance operon. *Mol. Microbiol.* **1993**, *8*, 615–623. [CrossRef] [PubMed]

48. Briers, Y.; Klumpp, J.; Schuppler, M.; Loessner, M.J. Genome sequence of *Listeria monocytogenes* Scott A, a clinical isolate from a food-borne listeriosis outbreak. *J. Bacteriol.* **2011**, *193*, 4284–4285. [CrossRef] [PubMed]
49. Parsons, C.; Lee, S.; Jayeola, V.; Kathariou, S. Novel cadmium resistance determinant in *Listeria monocytogenes*. *Appl. Environ. Microbiol.* **2017**, *83*, e02580-16. [CrossRef]
50. Maury, M.M.; Tsai, Y.-H.; Charlier, C.; Touchon, M.; Chenal-Francisque, V.; Leclercq, A.; Criscuolo, A.; Gaultier, C.; Roussel, S.; Brisabois, A.; et al. Uncovering *Listeria monocytogenes* hypervirulence by harnessing its biodiversity. *Nat. Genet.* **2016**. [CrossRef]
51. Lee, S.; Chen, Y.; Gorski, L.; Ward, T.J.; Osborne, J.; Kathariou, S. *Listeria monocytogenes* source distribution analysis indicates regional heterogeneity and ecological niche preference among serotype 4b clones. *MBio* **2018**, *9*. [CrossRef]
52. Mullapudi, S.; Siletzky, R.M.; Kathariou, S. Heavy-metal and benzalkonium chloride resistance of *Listeria monocytogenes* isolates from the environment of turkey-processing plants. *Appl. Environ. Microbiol.* **2008**, *74*, 1464–1468. [CrossRef]
53. Ratani, S.S.; Siletzky, R.M.; Dutta, V.; Yildirim, S.; Osborne, J.A.; Lin, W.; Hitchins, A.D.; Ward, T.J.; Kathariou, S. Heavy metal and disinfectant resistance of *Listeria monocytogenes* from foods and food processing plants. *Appl. Environ. Microbiol.* **2012**, *78*, 6938–6945. [CrossRef]
54. Xu, D.; Li, Y.; Zahid, M.S.H.; Yamasaki, S.; Shi, L.; Li, J.; Yan, H. Benzalkonium chloride and heavy-metal tolerance in *Listeria monocytogenes* from retail foods. *Int. J. Food Microbiol.* **2014**, *190*, 24–30. [CrossRef] [PubMed]
55. Ferreira, V.; Barbosa, J.; Stasiewicz, M.; Vongkamjan, K.; Moreno Switt, A.; Hogg, T.; Gibbs, P.; Teixeira, P.; Wiedmann, M. Diverse geno-and phenotypes of persistent *Listeria monocytogenes* isolates from fermented meat sausage production facilities in Portugal. *Appl. Environ. Microbiol.* **2011**, *77*, 2701–2715. [CrossRef] [PubMed]
56. Harvey, J.; Gilmour, A. Characterization of recurrent and sporadic *Listeria monocytogenes* isolates from raw milk and nondairy foods by pulsed-field gel electrophoresis, monocin typing, plasmid profiling, and cadmium and antibiotic resistance determination. *Appl. Environ. Microbiol.* **2001**, *67*, 840–847. [CrossRef] [PubMed]
57. Lebrun, M.; Audurier, A.; Cossart, P. Plasmid-borne cadmium resistance genes in *Listeria monocytogenes* are present on Tn*5422*, a novel transposon closely related to Tn*917*. *J. Bacteriol.* **1994**, *176*, 3049–3061. [CrossRef] [PubMed]
58. Canchaya, C.; Giubellini, V.; Ventura, M.; De Los Reyes-Gavilán, C.G.; Margolles, A. Mosaic-Like sequences containing transposon, phage, and plasmid elements among *Listeria monocytogenes* plasmids. *Appl. Environ. Microbiol.* **2010**, *76*, 4851–4857. [CrossRef] [PubMed]
59. Dutta, V.; Elhanaf, D.; Kathariou, S. Conservation and distribution of the benzalkonium chloride resistance cassette *bcrABC* in *Listeria monocytogenes*. *Appl. Environ. Microbiol.* **2013**. [CrossRef] [PubMed]
60. Dutta, V.; Elhanafi, D.; Osborne, J.; Martinez, M.R.; Kathariou, S. Genetic characterization of plasmid-associated triphenylmethane reductase in *Listeria monocytogenes*. *Appl. Environ. Microbiol.* **2014**, *80*, 5379–5385. [CrossRef]
61. Katharios-Lanwermeyer, S.; Rakic-Martinez, M.; Elhanafi, D.; Ratani, S.; Tiedje, J.M.; Kathariou, S. Coselection of cadmium and benzalkonium chloride resistance in conjugative transfers from nonpathogenic *Listeria* spp. to other listeriae. *Appl. Environ. Microbiol.* **2012**, *78*, 7549–7556. [CrossRef]
62. Mullapudi, S.; Siletzky, R.M.; Kathariou, S. Diverse cadmium resistance determinants in *Listeria monocytogenes* isolates from the turkey processing plant environment. *Appl. Environ. Microbiol.* **2010**, *76*, 627–630. [CrossRef]
63. Lawrence, C. *Surface-Active Quaternary Ammonium Germicides*; Academic Press Inc. Publishers: New York, NY, USA, 1950.
64. Qiagen CLC Genomics Workbench. Available online: https://www.qiagenbioinformatics.com (accessed on 17 December 2018).
65. Camejo, A.; Buchrieser, C.; Couvé, E.; Carvalho, F.; Reis, O.; Ferreira, P.; Sousa, S.; Cossart, P.; Cabanes, D. In vivo transcriptional profiling of *Listeria monocytogenes* and mutagenesis identify new virulence factors involved in infection. *PLoS Pathog.* **2009**, *5*, e1000449. [CrossRef]

66. Pombinho, R.; Camejo, A.; Vieira, A.; Reis, O.; Carvalho, F.; Almeida, M.T.; Pinheiro, J.C.; Sousa, S.; Cabanes, D. *Listeria monocytogenes* CadC regulates cadmium efflux and fine-tunes lipoprotein localization to escape the host immune response and promote infection. *J. Infect. Dis.* **2017**, *215*, 1468–1479. [CrossRef] [PubMed]
67. Osman, D.; Cavet, J.S. Bacterial metal-sensing proteins exemplified by ArsR-SmtB family repressors. *Nat. Prod. Rep.* **2010**, *27*, 668–680. [CrossRef] [PubMed]
68. Gao, C.H.; Yang, M.; He, Z.G. An ArsR-like transcriptional factor recognizes a conserved sequence motif and positively regulates the expression of *phoP* in mycobacteria. *Biochem. Biophys. Res. Commun.* **2011**, *411*, 726–731. [CrossRef] [PubMed]
69. Farias, P.; Santo, C.E.; Branco, R.; Francisco, R.; Santos, S.; Hansen, L.; Sorensen, S.; Morais, P.V. Natural hot spots for gain of multiple resistances: Arsenic and antibiotic resistances in heterotrophic, aerobic bacteria from marine hydrothermal vent fields. *Appl. Environ. Microbiol.* **2015**, *81*, 2534–2543. [CrossRef] [PubMed]
70. Poole, K. At the Nexus of Antibiotics and Metals: The Impact of Cu and Zn on antibiotic activity and resistance. *Trends Microbiol.* **2017**, *25*, 820–832. [CrossRef] [PubMed]
71. Deng, W.; Quan, Y.; Yang, S.; Guo, L.; Zhang, X.; Liu, S.; Chen, S.; Zhou, K.; He, L.; Li, B.; et al. Antibiotic resistance in *Salmonella* from retail foods of animal origin and its association with disinfectant and heavy metal resistance. *Microb. Drug Resist.* **2018**, *24*, 782–791. [CrossRef] [PubMed]
72. Billman-Jacobe, H.; Liu, Y.; Haites, R.; Weaver, T.; Robinson, L.; Marenda, M.; Dyall-Smith, M. pSTM6-275, a conjugative IncHI2 plasmid of *Salmonella enterica* that confers antibiotic and heavy-metal resistance under changing physiological conditions. *Antimicrob. Agents Chemother.* **2018**, *62*. [CrossRef]
73. Gullberg, E.; Albrecht, L.M.; Karlsson, C.; Sandegren, L.; Andersson, D.I. Selection of a multidrug resistance plasmid by sublethal levels of antibiotics and heavy metals. *MBio* **2014**, *5*, e01918-14. [CrossRef]
74. Binepal, G.; Gill, K.; Crowley, P.; Cordova, M.; Brady, L.J.; Senadheera, D.B.; Cvitkovitch, D.G. Trk2 Potassium transport system in *Streptococcus mutans* and its role in potassium homeostasis, biofilm formation, and stress tolerance. *J. Bacteriol.* **2016**, *198*, 1087–1100. [CrossRef]

© 2018 by the authors. Licensee MDPI, Basel, Switzerland. This article is an open access article distributed under the terms and conditions of the Creative Commons Attribution (CC BY) license (http://creativecommons.org/licenses/by/4.0/).

Article

Distribution of the *pco* Gene Cluster and Associated Genetic Determinants among Swine *Escherichia coli* from a Controlled Feeding Trial

Gabhan Chalmers [1], Kelly M. Rozas [2], Raghavendra G. Amachawadi [3], Harvey Morgan Scott [2], Keri N. Norman [4], Tiruvoor G. Nagaraja [5], Mike D. Tokach [6] and Patrick Boerlin [1,*]

1. Department of Pathobiology, Ontario Veterinary College, University of Guelph, 50 Stone Rd. E., Guelph, ON N1G 2W1, Canada; gchalmer@uoguelph.ca
2. Department of Veterinary Pathobiology, College of Veterinary Medicine and Biomedical Sciences, Texas A&M University, College Station, TX 77843, USA; kmrozas@cvm.tamu.edu (K.M.R.); hmscott@cvm.tamu.edu (H.M.S.)
3. Department of Clinical Sciences, College of Veterinary Medicine, Kansas State University, Manhattan, KS 66506, USA; agraghav@vet.k-state.edu
4. Department of Veterinary Integrative Biosciences, College of Veterinary Medicine and Biomedical Sciences, Texas A&M University, College Station, TX 77843, USA; knorman@cvm.tamu.edu
5. Department of Diagnostic Medicine/Pathobiology, College of Veterinary Medicine, Kansas State University, Manhattan, KS 66506, USA; tnagaraj@vet.k-state.edu
6. Department of Animal Sciences & Industry, College of Agriculture, Kansas State University, Manhattan, KS 66506, USA; mtokach@k-state.edu
* Correspondence: pboerlin@uoguelph.ca; Tel.: +1-519-824-4120 (ext. 54647)

Received: 24 September 2018; Accepted: 15 October 2018; Published: 18 October 2018

Abstract: Copper is used as an alternative to antibiotics for growth promotion and disease prevention. However, bacteria developed tolerance mechanisms for elevated copper concentrations, including those encoded by the *pco* operon in Gram-negative bacteria. Using cohorts of weaned piglets, this study showed that the supplementation of feed with copper concentrations as used in the field did not result in a significant short-term increase in the proportion of *pco*-positive fecal *Escherichia coli*. The *pco* and *sil* (silver resistance) operons were found concurrently in all screened isolates, and whole-genome sequencing showed that they were distributed among a diversity of unrelated *E. coli* strains. The presence of *pco/sil* in *E. coli* was not associated with elevated copper minimal inhibitory concentrations (MICs) under a variety of conditions. As found in previous studies, the *pco/sil* operons were part of a Tn7-like structure found both on the chromosome or on plasmids in the *E. coli* strains investigated. Transfer of a *pco/sil* IncHI2 plasmid from *E. coli* to *Salmonella enterica* resulted in elevated copper MICs in the latter. *Escherichia coli* may represent a reservoir of *pco/sil* genes transferable to other organisms such as *S. enterica*, for which it may represent an advantage in the presence of copper. This, in turn, has the potential for co-selection of resistance to antibiotics.

Keywords: copper; resistance; swine; *Escherichia coli*

1. Introduction

As restrictions on the use of antimicrobial agents for the purpose of growth promotion and disease prevention in farm animals are increasing, alternatives to these agents are becoming more popular. Feed supplementation with copper is one of the most frequently used, particularly in the swine industry [1]. The copper concentrations used in swine feed for growth promotion are relatively high and usually in the range of 100 to 250 ppm [2].

Bacteria developed mechanisms to cope with high concentrations of copper. In Gram-positive bacteria, the most well-known mechanism is the *tcrB* gene [3,4], which provides a selective advantage to intestinal enterococci in swine and cattle [5–7]. It also seems to be involved in the co-selection of bacteria resistant to antimicrobial agents of importance for both veterinary and human medicine [6,8]. Several tolerance and homeostasis mechanisms were described in Gram-negative bacteria and in *Enterobacteriaceae* in particular (for a review, see, for instance, References [9–11]). Although most are chromosomally encoded and present in the majority of bacteria from the species in which they reside, one of them initially found in *Escherichia coli* was shown to be plasmid-borne and not present in every isolate of the species [12]. The *pco* gene cluster associated with this system was later characterized in more detail [13] and shown to consist of seven genes (*pcoA, B, C, D, R, S,* and *E* [14–16]). This cluster was found in a variety of *Enterobacteriaceae* species and, depending on bacterial species and strain, the associated copper tolerance phenotype was variable, both in terms of copper minimal inhibitory concentration and inducibility [13]. Since then, several studies showed that *pco* genes are not always plasmid-borne but can also regularly be found on the chromosome of *Enterobacteriaceae* species, including *Salmonella enterica* and *E. coli* [17–19]. This spread and mobility may be related to the location of the *pco* genes on a Tn7-like transposon [17]. This Tn7-like element frequently carries both the *pco* gene cluster and the *sil* gene cluster [17] associated with silver tolerance [20]. Investigations on silver and copper tolerance in *S. enterica* isolates from Portugal showed a clear association between the presence of *sil* genes and copper tolerance, while the presence of *pco* genes did not seem to show any evident correlation with this phenotype [18,21]. Similarly, recent experimental studies on the effect of feed supplementation with copper on fecal *E. coli* and on the fecal metagenome of swine did not demonstrate any clear or systematic selective effect for *pco* genes [2,22]. These results suggest that either the concentrations of copper used in feed (125 ppm) may have been too low to have such an effect, or the presence/absence of the *pco* genes did not affect the tolerance of *E. coli* and other bacteria to elevated copper concentrations under the conditions found in the gut of the animals. However, a negative association was observed between copper supplementation and resistance to antimicrobials, as well as resistance to extended-spectrum cephalosporins in particular [2]. Also, an association between *pco* genes and the *tet*(B) tetracycline resistance gene was detected in *E. coli*, while these two genes were negatively associated with the bla_{CMY} and *tet*(A) genes encoding for extended-spectrum cephalosporins and tetracycline resistance, respectively [2].

Based on these observations, the objectives of this study were (a) to replicate the previous experiments of Agga and collaborators [2] and reassess the associations between the *pco* genes and *tet*(A), *tet*(B), bla_{CMY}, and bla_{CTX-M} among *E. coli* from groups of swine subjected to diverse combinations of copper and tetracycline feed supplementation; (b) to use whole-genome sequencing to assess the genetic diversity and clonal relationships of *E. coli* isolates recovered from these experiments and carrying diverse combinations of these genes; (c) to compare the copper susceptibility and genome sequences of selected isolates with plasmid-borne and chromosomally encoded *pco* genes; and (d) to transfer *E. coli* plasmids carrying the *pco* and *sil* gene clusters into *S. enterica* by conjugation, and assess the associated copper susceptibility. These objectives related to the use of copper in feed and its effect on copper tolerance in *E. coli* were part of a broader study on alternatives to antibiotics [23]. The latter also included the use of zinc and oregano oil, but is not discussed here.

2. Materials and Methods

2.1. Experiment Design

The Kansas State University Institutional Animal Care and Use Committee approved the protocol for this experiment (AUP # 3135). The study was conducted at the university's Segregated Early Weaning Facility in Manhattan, KS. Each pen (1.22 × 1.22 m) had metal tri-bar flooring, one four-hole self-feeder, and a cup waterer to provide ad libitum access to feed and water. This experiment was also described in a publication by Feldpausch and collaborators [23].

A total of 350 piglets (21 days old) were assigned to one of 70 pens (five piglets per pen), which were then randomly assigned to each of the 10 in-feed treatments arranged in a 2 × 2 × 2 (+2) factorial design. In detail, the ten dietary treatments were (1) a basal swine diet fully meeting National Research Council (NRC) nutritional guidelines, including 16.5 ppm of supplemental copper and 165 ppm of supplemental zinc (control group); (2) a basal diet supplemented with 125 ppm of copper provided by copper sulfate; (3) a basal diet supplemented with zinc at 3000 ppm of zinc provided by zinc oxide; (4) a basal diet supplemented with oregano premix containing 5% oregano oil (Regano 500; Ralco-mix Products, Marshall, MN, USA); (5) a basal diet with both 125 ppm of copper and zinc at 3000 ppm; (6) a basal diet with both 125 ppm of copper and oregano premix; (7) a basal diet with both zinc at 3000 ppm and oregano premix; (8) a basal diet containing copper, zinc, and oregano premix; (9) a basal diet containing a preventive level of chlortetracycline (CTC) (22 mg/kg body weight (BW); High CTC; and (10) a basal diet containing a subtherapeutic level of CTC (4 mg/kg BW; Low CTC). These latter treatment groups (9 and 10) did not interact with other main treatment factors (Zn, Cu, and oregano oil) in the study design, so as to assess the impact of antimicrobial alternatives versus both true negative controls and the "existing standard controls" represented by antimicrobial use groups. The basal diet consisted of corn, soybean meal, vitamins, amino acids, and trace mineral supplements per NRC requirements.

The study lasted 49 days with an initial seven days of acclimation, and 28 days of feeding trial, followed by 14 days of washout phase. Three fresh fecal samples were collected from random pigs in each pen by gentle rectal massage at days 0 and 28 of the feeding trial. Fecal samples were transported to the laboratory for further processing. The fecal samples were thoroughly mixed with 50% glycerol (1:1) and stored at −80 °C. Laboratory personnel were blinded to the treatment groups.

2.2. Selection of Isolates and Detection of pco

A total of 420 samples, 210 from day 0 (for pre-treatment effect) and 210 from day 28 (maximum treatment effect), were subjected to bacteriological culture and quantified for *E. coli* using standard isolation techniques and spiral plating. Briefly, one gram of 50:50 glycerol and feces were diluted in 9 mL of phosphate-buffered saline (PBS). A 50-μL aliquot of the fecal suspension was spiral-plated onto each of MacConkey agar, MacConkey agar supplemented with 16 mg/L tetracycline, and MacConkey agar supplemented with 4 mg/L ceftriaxone using an Eddy Jet 2 spiral plater (Neu-tec Group Inc., Farmingdale, NY, USA). Crude quantification values were determined by the Flash & Go Automatic Colony Counter (Neu-tec Group Inc.). A single, randomly selected colony was used from a plain MacConkey plate and confirmed as *E. coli* by lactose fermentation and an indole test; the species identity was also later confirmed with Illumina-based DNA sequencing. Isolates were preserved at −80 °C in protectant CryoBeads™ for further characterization.

Antimicrobial susceptibility testing was conducted by broth microdilution using the Sensititre™ system (TREK, Thermo Scientific Microbiology, Oakwood Village, OH, USA) and Sensititre™ NARMS Gram-negative plates (CMV3AGNF) on 403 *E. coli* isolates. *Escherichia coli* ATCC 25922, *Escherichia coli* ATCC 35218, *Pseudomonas aeruginosa* ATCC 27853, *Staphylococcus aureus* ATCC 29213, and *Enterococcus faecalis* ATCC 29212 were used as quality control strains. Plates were incubated at 37 °C for 18 h and read on a Sensititre OptiRead™ (TREK). The results were interpreted according to Clinical and Laboratory Standards Institute (CLSI) guidelines [24]. Intermediate isolates were interpreted as susceptible for binary statistical analyses.

Detection of *pco*, tetracycline, and extended-spectrum cephalosporin resistance genes was performed by PCR with the primers described in Table 1. Thermocycling conditions were the same as those defined in the respective references. Amplicons were visualized by horizontal gel electrophoresis and ultraviolet (UV) imaging.

Table 1. PCR targets and primers used for detection of *pco*, *sil*, and antimicrobial resistance genes.

Target	Primer	Sequence	Amplicon	Reference
pco	pcoD-F	CAGGAACGGTGATTGTTGTA	700 bp	[2]
	pcoD-R	CCGTAAAATCAAAGGGCTTA		
sil	silA_Fw	GCAAGACCGGTAAAGCAGAG	936 bp	[21]
	silA_Rv	CCTGCCAGTACAGGAACCAT		
tet(A)	TetA-L	GGCGGTCTTCTTCATCATGC	502 bp	[25]
	TetA-R	CGGCAGGCAGAGCAAGTAGA		
tet(B)	TetBGK-F2	CGCCCAGTGCTGTTGTTGTC	173 bp	[25]
	TetBGK-R2	CGCGTTGAGAAGCTGAGGTG		
bla$_{CMY}$	CMYF	GACAGCCTCTTTCTCCACA	1000 bp	[25]
	CMYR	TGGACACGAAGGCTACGTA		
bla$_{CTX-M}$	CTX-M-F	ATGTGCAGYACCAGTAA	512 bp	[26]
	CTX-M-R	CCGCTGCCGGTYTTATC		

2.3. Copper Susceptibility

Susceptibility to copper was analyzed by broth microdilution for four randomly selected *pco/sil*-positive *E. coli*, and two *pco/sil*-negative isolates. Isolates were from day 0 (KSC9, 27, 64, and 207) and day 28 (KSC857 and 1031), from animals within the copper treatment group (KSC27, 857, and 1031) and from those without (KSC9, 64, and 207). A stock solution of 400 mM copper(II) sulfate (Sigma-Aldrich, St. Louis, MO, USA) was prepared in double-distilled water (ddH$_2$O), and filter-sterilized. A non-serial dilution range (0, 4, 8, 16, 20, 24, 36, 48, 64, and 100 mM) was prepared in Mueller–Hinton II broth, cation-adjusted (Becton Dickinson, Franklin Lakes, NJ, USA), and each dilution was adjusted to pH 7.2 using 5 M NaOH [11]. Bacterial suspensions of a 0.5 McFarland standard were diluted 1/100, and 50 µL of this suspension was inoculated in a 96-well plate with 50 µL of the copper dilutions, resulting in halving of the initial copper concentrations. Microplates were incubated at 37 °C for 16 h, under both aerobic and anaerobic conditions. Minimum inhibitory concentration (MIC) was defined as the first concentration without visible growth. Minimum bactericidal concentrations (MBCs) were determined by removing 10 µL from wells that showed no visible growth, and plating them on Mueller–Hinton II agar plates for incubation at 37 °C for 16 h. The ATCC 25922 *E. coli* strain (*pco/sil*-negative) was used as a negative control for susceptibility testing.

Minimum inhibitory concentrations were also determined using agar plate dilutions of copper, as described by Mourão and collaborators [21]. Briefly, copper dilutions of 0, 0.5, 1, 2, 4, 8, 12, 16, 20, 24, 28, 32, and 36 mM were prepared in Mueller–Hinton II agar, and the pH was adjusted as above. One microliter of an approximate 10^7 colony forming units (CFU)/mL culture was pipetted onto the surface of each plate. Growth at 37 °C in both aerobic and anaerobic conditions was assessed after 16 h.

2.4. Expression of pco *by Complementary DNA Synthesis and Real-Time PCR*

Three *E. coli* isolates were selected randomly (two *pco/sil*-positive and one negative control, none of which were isolated from copper-treated animals) for determining the expression of *pco* under aerobic conditions, with and without induction with low concentrations of copper, performed as previously described [15,27]. Briefly, isolates were plated overnight at 37 °C on Luria–Bertani (LB) agar (Becton Dickinson) plates. A single loop of bacteria was inoculated into 1 mL of LB broth, and vortexed; 200 µL of this suspension was inoculated into 20 mL of LB broth supplemented with 0, 1 mM, and 5 mM copper(II) sulfate and incubated at 37 °C for approximately 2.5 h, until optical densities of 0.5 were reached at 600 nm. Broth microdilution MICs were performed again as above under aerobic conditions to observe any effect of this induction on copper tolerance. In parallel, 10 mL of broth was centrifuged, and the resulting pellet was resuspended in 1 mL of RNAlater (QIAGEN Inc., Valencia, CA, USA). Total RNA was extracted using an RNAeasy Mini kit (QIAGEN), according to the manufacturer's instructions. An additional DNase step was performed to ensure all traces of DNA were removed, and was verified by a *pco* PCR using 1 µL as a template. RNA was quantified using a BioAnalyzer 2100 instrument (Agilent Technologies, Santa Clara, CA, USA), and 100 ng of each RNA preparation was used for complementary DNA (cDNA) synthesis using an Applied Biosystems High-Capacity cDNA Reverse Transcription kit (Thermo Fisher Scientific, Carlsbad, CA, USA).

PCR was performed to amplify gene fragments to be used for cloning into a plasmid vector, for use as a standard curve for real-time PCR. Amplicons of the *pcoA* and *pcoD* genes were produced using primers forward *pcoA* (pcoA_F), CGGGTATGCAAAGTCATCCT; reverse *pcoA* (pcoA_R), TTGATCAGCGTGATCCTGAG; and pcoD_F, AAGCGGTGTCAGACATGAAA; pcoD_R, GATGGGTCAGATCGCTCAGT, respectively. As controls, two housekeeping gene amplicons for *hcaT* (HcaT major facilitator superfamily transporter) and *rrsA* (16S ribosomal RNA) were amplified using primers hcaT_F, CTGATGCTGGTGATGATTGG; hcaT_R, CAATGCAGAATTTGCACCAC; and rrsA_F, CGGACGGGTGAGTAATGTCT; rrsA_R, GTTAGCCGGTGCTTCTTCTG, respectively. Each amplicon was cloned into a pCR 2.1-TOPO plasmid vector, using an Invitrogen TOPO-TA cloning kit (Thermo Fisher Scientific). Inserted sequences were confirmed by DNA sequencing, and plasmid DNA was prepared using a Plasmid Midi Kit (QIAGEN). Plasmid DNA was quantified using Quant-IT Picogreen dsDNA reagent (Thermo Fisher Scientific) and read using a DTX 880 Multimode detector (Beckman Coulter, Brea, CA, USA). Gene copy numbers were then predicted by the DNA concentration divided by the molecular weight of the plasmid.

Real-time PCR was used to quantify the expression of each *pcoA*, *pcoD*, *hcaT*, and *rrsA* gene using primers internal to the fragments described above. In triplicate, 1 µL of cDNA or plasmid standards were added to 19 µL of LightCycler 480 SYBR Green I Master (Roche Diagnostics, Indianapolis, IN, USA) containing 250 nM of each primer. Primers used for the quantification of *pco* expression were RT_pcoAF, TGGTTGATATGCAGGCGATG; RT_pcoAR, TCCGCGTACGTGAGAACCTT; and RT_pcoDF, GTCAGGCTCTGTGCCCTGTT; RT_pcoDR, CCCACTCATCGTCATCAGCA. Housekeeping gene primers used for *hcaT* and *rrsA* were those described by Zhou and collaborators [28].

2.5. Next-Generation Sequencing

A subset of 82 isolates was selected to represent suspected extended-spectrum β-lactamase (ESBL)-producing isolates and isolates with elevated ciprofloxacin MICs (based on Sensititre phenotypes and ciprofloxacin MICs of ≥ 0.05 mg/L; $n = 26$), isolates carrying the bla_{CMY} gene ($n = 26$), and a representative sample of isolates with resistance phenotypes determined by Sensititre ($n = 30$). In addition, all 34 *E. coli* carrying the *pcoD* gene were also included. Genomic DNA was prepared for MiSeq sequencing (Illumina, San Diego, CA, USA) for all of these 116 isolates using a QIAamp DNA extraction kit (QIAGEN), and libraries were prepared using a Nextera XT kit (Illumina). Achtman sequence types were determined with the SRST2 plugin for BaseSpace Labs (Illumina) using the MiSeq paired-end reads, where sufficient read quality was obtained. These reads were also used for core-genome multilocus sequence typing (cgMLST) (EnteroBase typing scheme) using the wgMLST application for BioNumerics v7.6 (Sint-Martens-Latem, Belgium). Single-nucleotide polymorphism (SNP) analysis of each *pco* gene cluster (*pcoEABCDRSE*) was also performed using the wgSNP analysis tool from BioNumerics.

DNA was also prepared for four of the isolates tested for susceptibility to copper and harboring *pco* (KSC9, KSC64, KSC207, and KSC1031) using a MasterPure DNA Purification Kit (Epicentre, Madison, WI, USA) for PacBio RS II sequencing (Pacific Biosciences, Menlo Park, CA, USA). Sequencing and assembly of these four isolate genomes were performed at the McGill University and Génome Québec Innovation Centre, Montreal, QC, Canada. PacBio sequencing assembly was completed on chromosome and plasmid assemblies of sheared large inserts (~20 kbp) using the de novo genome assembly pipeline Hierarchical Genome Assembly Process (HGAP); alignments were further polished using the Quiver consensus algorithm. Genomes and/or plasmids (pMRGN207 and pMRGN1031) containing the *pco* gene cluster were uploaded to GenBank under BioProject

PRJNA355857. Annotations were performed using the NCBI Prokaryotic Genome Annotation Pipeline (PGAP) version 4.0.

2.6. Conjugation of a pco/sil *Plasmid into Salmonella*

Based on the known plasmid-borne location in two of our sequenced *E. coli* isolates, we attempted to transfer these plasmids into *Salmonella* recipients to observe any change to copper susceptibility. Six *Salmonella* isolates were selected at random from a large collection of isolates maintained by the Canadian Integrated Program for Antimicrobial Resistance Surveillance (CIPARS) between 2008 and 2011. Isolates were screened for *pco* and *sil* by PCR, and three *pco/sil*-positive and three *pco/sil*-negative were used for copper susceptibility testing. *Escherichia coli* isolates KSC207 and KSC1031 were used as plasmid donor strains, with *pco/sil* plasmid incompatibility types HI2 and FII, respectively (as determined by DNA sequence analysis), while *pco/sil*-negative *Salmonella* isolates SA8197 (serovar Kentucky) and SA82540 (serovar Infantis) were used as recipients. Briefly, 50 µL of donor and 100 µL of recipient overnight broth cultures were mixed, plated on LB agar, and incubated overnight to facilitate conjugation. Growth was then resuspended in LB broth, and plated on Brilliant Green agar to inhibit the growth of donor *E. coli*, as both plasmids carried a tetracycline resistance determinant—*tet*(A) and *tet*(B), respectively. Additionally, 12 µg/mL tetracycline was included in the selective plates. Up to five pink-colored colonies were then sub-cultured for purity, and indole testing and PCR for *pco* and *sil* were used to confirm transfer. These transconjugant *Salmonella* were then used for copper susceptibility testing as before using anaerobic agar dilution, and performed in triplicate.

2.7. Statistics

Descriptive and inferential statistical methods were performed using Stata version 15 (StataCorp, College Station, TX, USA). Categorical data (i.e., resistance phenotypes and presence of genes) were tabulated and cross-tabulated to explore bivariable associations between treatments and the following outcomes: (1) presence of phenotypic resistance to (a) tetracyclines, (b) cefoxitin, for preliminary classification of ESBL (susceptible) versus AmpC (resistant) producing β-lactamases, (c) third-generation cephalosporins, and (d) multidrug resistance (MDR) count (integer count), out of 14 antimicrobials tested on a broth microdilution panel; and (2) presence of resistance genes for (a) tetracyclines (*tet*(A) and *tet*(B)), (b) extended-spectrum cephalosporins (bla_{CMY-2} and bla_{CTX-M}), and (c) metal resistance genes (*pco* and *sil*). Likelihood ratio χ^2 or Fisher's exact tests (when zero-cells were abundant) were used in bivariable analyses; statistical significance was determined at $p < 0.05$. Differences in growth ($\log_{10}$ CFU) of coliforms on plain versus antimicrobial (tetracycline or ceftriaxone) impregnated plates were examined by unpaired *t*-tests. Multivariable mixed logistic (binary outcomes) and linear models (count and $\log_{10}$ CFU outcomes) using a four-way factorial design (plus two additional indicator variables for low- and high-dose CTC) were built and assessed for each of the binary response (logistic) and quantitative ($\log_{10}$ CFU differences) endpoints. Full factorial models were subjected to reduction, firstly removing non-significant interaction terms and then main effects. Of note, in-feed copper (low versus high), sampling day (0 versus 28), and their interaction were always retained marginal means estimated with *p*-values representing the post hoc multiple comparisons adjusted using Bonferroni's correction.

3. Results

3.1. Resistance Determinants

Not all samples (*n* = 420) yielded lactose- and indole-positive isolates. A total of 403 *E. coli* isolates were included in this analysis. Thirty-four of the 403 isolates (8.4%) were positive for *pco* by PCR. Twelve isolates carried a bla_{CTX-M} gene (3.0%), and 60 carried bla_{CMY} (14.9%, all variant CMY-2). All but two isolates (99.5%) were positive for *tet*(A) or *tet*(B) (121 (30.0%) and 267 (66.3%), respectively); 13 isolates (3.2%) carried both of these tetracycline resistance genes.

All bla_{CTX-M} sequences encoded the CTX-M-27 variant, and all bla_{CTX-M}-positive isolates were pco-negative, but tet(B)-positive; 11 of these 12 isolates were ST744 and had an ampicillin/ceftriaxone/ciprofloxacin/nalidixic acid/tetracycline resistance phenotype.

3.2. Associations of Copper Treatment Groups with pco Prevalence

Multi-level model-adjusted estimates of the occurrence of the pco gene were initially unstable in the presence of the full factorial specification. A reduced model containing copper (forced into model), day (and its interaction), and low- versus high-dose CTC yielded a model significant at $p < 0.03$. Copper did not select for the pco gene ($p = 0.249$); that is, copper supplemented at NRC requirements yielded 0.07 (95% confidence intervals (CIs): 0.04–0.11) of isolates with the gene, versus 0.11 (95% CIs: 0.04–0.17) in the group supplemented with copper beyond nutrient needs. Likewise, copper did not appear to select for any of the additional microbiological endpoints ($p > 0.05$; data not shown). Low-dose CTC did select ($p = 0.001$) for increased pco with 0.28 (95% CIs: 0.11–0.46) of the isolates in the low-dose group harboring the gene versus 0.06 (95% CIs: 0.04–0.09) in pigs not receiving any CTC. The difference between the high-dose CTC and each of the other two levels was not significant (0.12; 95% CIs: 0.00–0.24). Of note, though unexplained by the field trial study design, the pco gene was significantly ($p = 0.012$) associated with a lower MDR count; however, most of this difference was due to a lack of the highest MDR counts (maximum with pco present = 7 versus 11 in pco-negative isolates). No associations were significant ($p > 0.05$) among the genes tested, with the notable exception of pco/sil for which there was complete agreement (34/34; Fisher's exact test, $p < 0.0001$).

3.3. Susceptibility to Copper and Expression of pco

Susceptibility to copper by broth microdilution ranged from 12 to 18 mM in all seven E. coli isolates tested (Table 2). The presence of the pco gene cluster did not appear to have an effect on susceptibility to copper under both aerobic and anaerobic conditions, nor did the use of broth or solid media. All MBC values were identical to their respective MIC results. The ATCC 25922 (pco/sil-negative) isolate had an MIC and MBC of 12 mM. For the Salmonella isolates, no systematic differences were observed when using broth dilution under anaerobic conditions (MICs of 12 to 18 mM), but the MICs of the pco/sil-positive and -negative isolates differed when tested on agar, with values of 24 mM and 4 mM, respectively (Table 2).

Table 2. Isolates used in this study for copper susceptibility testing. All reported minimum inhibitory concentration (MIC) values were determined under anaerobic conditions. MBC—minimum bactericidal concentration.

Isolate	Bacteria	Serovar	pco	sil	Location (Similar to)	Broth MIC [1]	Broth MBC [1]	Agar MIC [1]
KSC9	Escherichia coli		+	+	chromosome (IAI1)	18	18	16
KSC64	E. coli		+	+	chromosome (E24377A)	18	18	16
KSC207	E. coli		+	+	278 kbp plasmid (pR478)	18	18	16
KSC1031	E. coli		+	+	149 kbp plasmid (p1540)	12	12	12
KSC27	E. coli		−	−		12	12	16
KSC857	E. coli		−	−		12	12	16
ATCC 25922	E. coli		−	−		12	12	16
SA10689	Salmonella	Senftenberg	+	+	unknown	18	18	24
SA12224	Salmonella	Ouakam	+	+	unknown	18	18	24
SA13423	Salmonella	Ouakam	+	+	unknown	18	18	24
SA82699	Salmonella	Kentucky	−	−		18	18	4
SA81917	Salmonella	Kentucky	−	−		12	18	4
SA82540	Salmonella	Infantis	−	−		12	18	4
SA81917-TC	Salmonella	Kentucky	+	+	plasmid from E. coli KSC207	ND [2]	ND	24
SA82540-TC	Salmonella	Infantis	+	+	plasmid from E. coli KSC207	ND	ND	24

[1] concentration in mM; MICs are averages of three complete biological replicates. [2] ND (not done).

Induction with 1 mM and 5 mM under aerobic conditions appeared to have no effect on the MIC of each isolate tested (data not shown). RNA extraction and cDNA analysis also showed no significant change in pcoA or pcoD transcription, observed by real-time PCR. Using hcaT as the reference gene (expression of *rrsA* expression was considerably higher than all other genes, and was not used in the analysis), expressions of *pcoA* and *pcoD* were measured as the "target" in all samples using relative quantification. The average adjusted crossing point (CP) across all three induction concentrations was 31.9 (±1.7) for *pcoA*, and 32.1 (±1.9) for *pcoD*.

3.4. Next-Generation Sequencing and cgMLST

All 116 *E. coli* isolates selected for MiSeq sequencing were analyzed using the BioNumerics software (Figure 1). All 34 *pco*-positive isolates recovered during this study also carried the *sil* gene cluster (determined by reference mapping of MiSeq reads), and none of the 82 *pco*-negative isolates carried any *sil* gene. Core-genome MLST analysis of the MiSeq data showed a random distribution of *pco/sil* among sequence types and no association with a specific clonal lineage (Figure 1). Further *pco* SNP analysis showed the gene cluster to be highly conserved within a sequence type (ST), but had some variation between most STs (Figure 2).

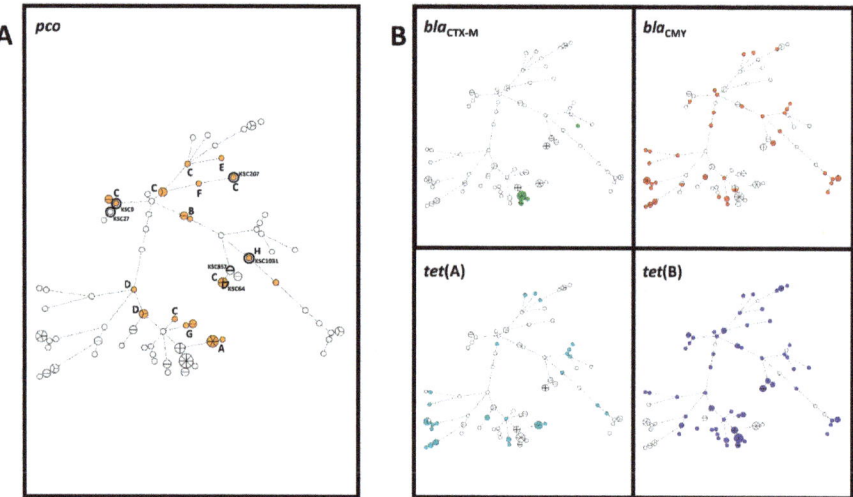

Figure 1. Minimum spanning tree of 116 *Escherichia coli* isolates, using core-genome multilocus sequence typing (MLST) analysis comprising 2513 genes (BioNumerics *E. coli/Shigella* EnteroBase scheme). A tree with the highest resampling support is shown, using 1000-resampling bootstrapping. (**A**) Isolates carrying the *pco* gene cluster are highlighted in orange; (**B**) isolates carrying the resistance genes bla_{CTX-M}, bla_{CMY}, *tet*(A), and *tet*(B) are indicated. Letters in 1A indicated single-nucleotide polymorphism (SNP) types found in Figure 2. Circles containing multiple sections indicate multiple isolates within a core-genome sequence type. Isolates used for minimum inhibitory concentration (MIC) testing are also highlighted and labeled in 1A.

Pacific Biosciences long-read sequencing and assembly showed the *pco* gene cluster to be both plasmid and chromosomally encoded (Figure 3). The gene cluster was accompanied by the *sil* gene cluster in all four *pco*-positive isolates, and was flanked by a Tn7-like transposable element in three of them. Using this *pco–sil–tns* sequence as a template, short-read Illumina sequences were successfully mapped onto all but one *pco*-positive isolate (KSC1031) for the complete structure, demonstrating the highly conserved nature of this transposable element across multiple STs and plasmid types found in this study.

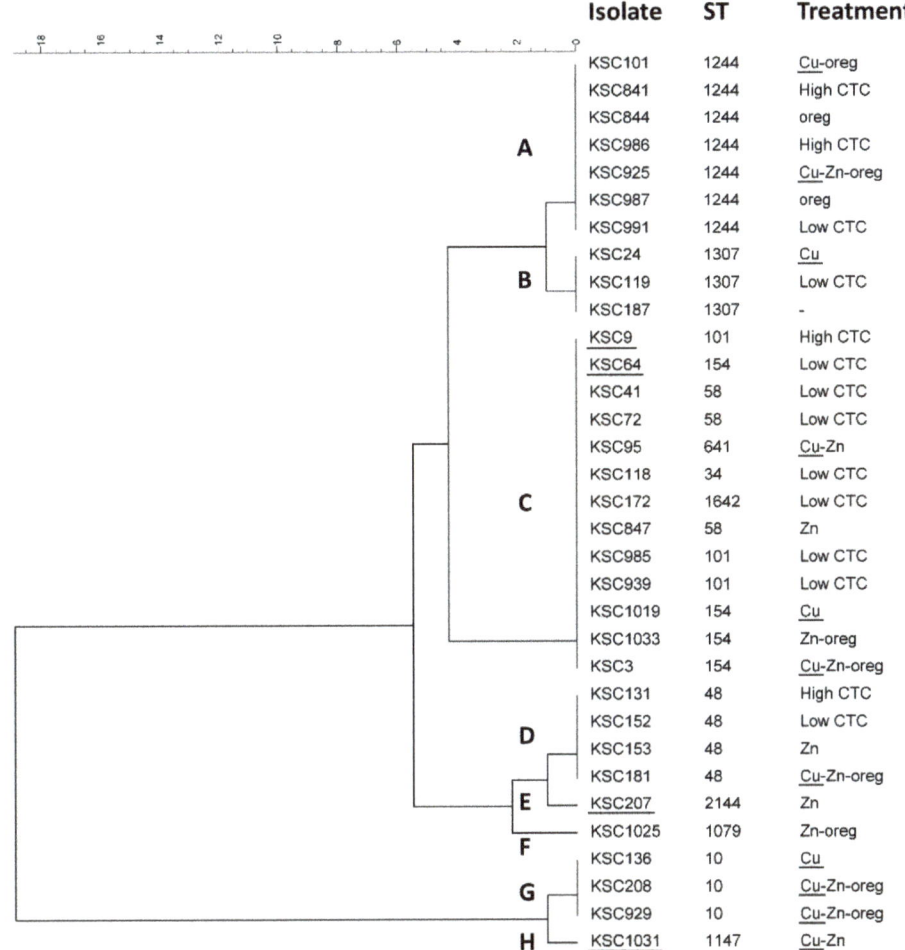

Figure 2. Phylogenetic analysis of the *pcoEABCDRSE* gene cluster (5487 bp) of all *pco*-positive isolates in this study, using a categorical (differences) similarity coefficient and unweighted pair group with arithmetic mean (UPGMA) cluster analysis. Treatment groups, including Cu (copper), Zn (zinc), oreg (oregano oil), and high/low-dose chlortetracycline (CTC) are shown. Letters indicating identical SNP groups are also shown in Figure 1A. Isolates used for MIC testing are underlined. ST: sequence type.

Both of the two chromosomal Tn7-like elements were found in approximately the same position, 300 kbp downstream of the preferential *glmS* insertion point for Tn7 [29], much farther than previously reported [30,31]. Neither plasmid harboring the *pco/sil* gene cluster contained any known virulence factors (using Virulence Finder v1.5 [32]). In one instance, the *pcoA* gene was interrupted by a transposase, while the *sil* gene cluster was always intact (Figure 3).

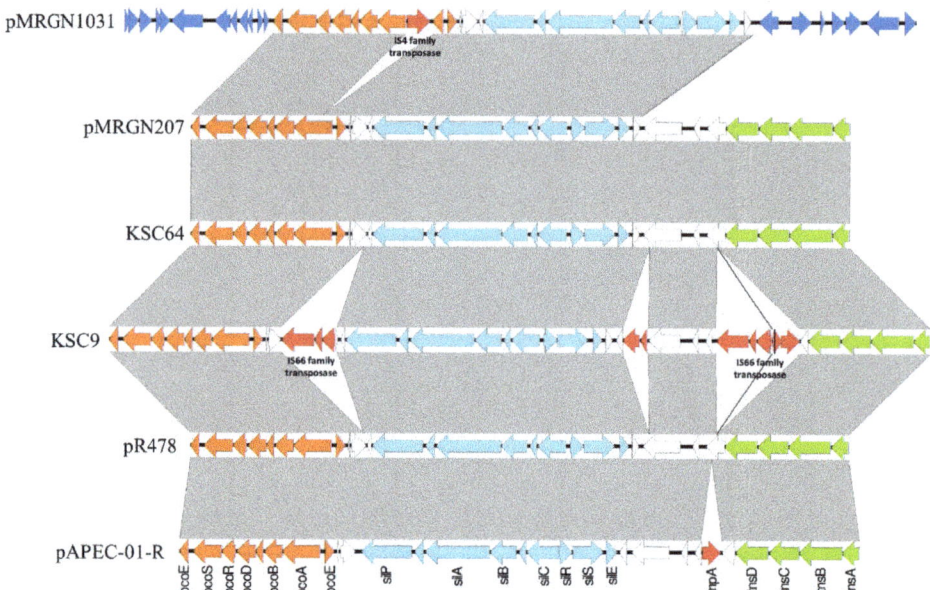

Figure 3. Genetic context of the *pco–sil–tns* area of four isolates from this study sequenced by PacBio. KSC1031 (plasmid pMRGN1031; GenBank accession number CP019561), KSC207 (plasmid pMRGN207; CP019559), KSC64 (CP018840), and KSC9 (CP018323) were compared to the pR478 (BXX664015) and pAPEC-O1-R (DQ517526) previously published sequences. Colors indicate the *pco* operon (orange), the *sil* operon (light blue), transposases (red), Tn7 genes (green), and others (dark blue).

3.5. Conjugation of a pco/sil Plasmid into Salmonella

The *pco/sil* plasmid (IncHI2) from *E. coli* isolate KSC207 was successfully transferred to both *Salmonella* recipient isolates, as confirmed by PCR. Using agar dilution under anaerobic conditions, a clear difference was observed between the transconjugants carrying the *sil/pco* gene clusters (MIC = 24), and the recipient isolates (MIC = 4; Table 2). The IncFII plasmid from *E. coli* isolate KSC1031 could not be successfully transferred to either *Salmonella* recipient.

4. Discussion

The plasmid-borne nature of the *pco* gene cluster [12,13,19] and the recent demonstration of its linkage with important antimicrobial resistance genes [11,17] warrant further investigations on the potential medical and public health implications of copper use in animal feed. A previous set of experiments under controlled conditions failed to demonstrate any significant selection of *pco*-positive *E. coli* isolates [2] or increase in *pco* copy numbers [22] in feces from pigs fed copper after weaning. However, *pco*-positive isolates were more frequently associated with the tetracycline resistance gene *tet*(B) than with its *tet*(A) counterpart [2]. This suggested some possible gene linkage between *pco* genes and *tet*(B) on mobile elements or clonal expansion of strains carrying both genes. The replication of these experiments described here showed the same lack of selection of *pco*-positive *E. coli* with copper concentrations in feed (125 mM) similar to those used in the field (100–250 mM; [2]). Although this does not exclude some selection in the long term or the selection of other genes, no significant effect could be detected during a single feeding period. Since none of the 82 *pco*-negative isolates sequenced carried the *sil* operon (31 of which were from samples of animals receiving copper supplementation), it is also unlikely that the copper treatment would have selected for this latter operon alone. A *pco–tet*(B) positive association, as well as a negative association between *pco* and bla_{CMY}, was observed previously by Agga et al. [2]; however, while the direction of association was similar here, the associations were not significant ($p = 0.344$ and 0.087, respectively). We did see much higher levels of *pco* among isolates from pigs subjected to low doses of CTC (a dosage regimen for growth promotion purposes not permitted in the US since 1 January 2017), which may suggest that indirect selection of *pco* and *pco*-positive strains could occur when using tetracyclines in swine. In the present study, the prevalence of *tet*(A) was highest in the group receiving high-dose CTC (45.0% versus 22.5%) whereas the prevalence of *tet*(B) was highest in the group receiving low-dose CTC (77.5% versus 62.5%). This finding may help explain the relationship of tetracycline uses with *pco*, though such hypotheses are largely based on the previous findings of Agga et al. [2].

Analysis of genomic similarities between *pco*-positive isolates through cgMLST demonstrates that the *pco* genes are distributed across a variety of clonal lineages and do not cluster in only a few clear discrete groups of closely related isolates. The most frequent *pco* single-nucleotide polymorphism (SNP group C in Figures 1A and 2) is also present in several STs and unrelated clonal lineages. Both observations illustrate the active horizontal transfer of the *pco* gene cluster in *E. coli* populations. However, the associations between most of the other *pco* SNP groups and STs (Figure 2) or clonal lineages (Figure 1) suggest that both a combination of short-term or local clonal spread and broader long-term horizontal gene transfer (HGT) play a role in the distribution of this gene cluster in *E. coli* from the swine population examined. Similar to the *pco–sil* clusters, the *tet*(A), *tet*(B), and bla_{CMY} genes which have been present in Enterobacteriaceae from farm animals in North America for several decades also appeared to be distributed randomly and did not cluster clearly together with *pco* genes in a discrete number of clonal lineages (Figure 1). Overall, these observations suggest that the positive and negative statistical associations observed between *pco* and *tet*(B) or *tet*(A) and bla_{CMY}, respectively, do not rely on the expansion and contraction of a very limited number of major clonal lineages. This differs from the $bla_{CTX-M-27}$ gene which was found mainly (11/12) in closely related isolates. CTX-M β-lactamases were reported in food animals much later in North America than in other continents [33] and may have emerged in swine in the US only recently. It may, therefore, be only in the early stages of its spread through HGT in bacteria from swine and still limited to a small number of clonal lineages.

The *pco* genes were located together with the *sil* cluster on a Tn7-like transposon structure [17,19,34] in all but one isolate in this study (KSC1031). The high transposition frequency of Tn7 and related elements [29] may be an important reason for the distribution of the *pco–sil* cluster in a wide diversity of strains illustrated in the present study. Tn7 transposons developed refined strategies to insert preferentially on mobile plasmids [29]. It may, therefore, not appear entirely surprising that the *pco* plasmid we were able to transfer by conjugation (pMRGN207) carried the full *pco/sil/*Tn7-like element, while pMRGN1031 missing the Tn7 part of the element was not transferable. Coincidentally, the plasmid we were able to transfer was an IncHI2 plasmid, an incompatibility group already shown by others to carry *pco* genes in different geographic locations and bacterial species [17,18,35,36]. Tn7 transposons also developed refined strategies to insert preferentially into the same selectively neutral *attTn7* chromosomal site located in proximity of the *glmS* gene [29]. However, the locations of the two chromosomal Tn7-like elements associated with the *pco–sil* cluster in the closed genome sequences generated with PacBio long reads show that this mobile element does not always insert in the same *attTn7* site or in the proximity of *glmS*. This may warrant further investigations on the transposition mechanisms of this Tn7-like transposable element. Together, these findings further stress the likely important role of IncHI2 plasmids and Tn7-like elements in the spread of the *pco–sil* gene clusters.

The overall structure of the region encompassing the *pco–sil* clusters was highly conserved and identical to pR478 [36] in two of the four isolates we investigated in detail (one plasmid-borne and the other chromosomal). This conserved region also included the *tns* gene cluster of Tn7 and the intervening region between the *tns* and *sil* genes. This structure was described by others on several plasmids [17,19]. Insertions were present in the *pco–sil* clusters for the two other isolates. In one of them (chromosomal), three insertions were present in this region, but all were within open reading frames encoding putative proteins of unknown function, and were not affecting the *pco* or the *sil* gene clusters. However, in pMRGN1031, an insertion was disrupting the *pcoA* gene. This latter insertion would be expected to inactivate the copper resistance if a phenotype were detectable [14].

In addition to the loss of the *tns* gene cluster and parts of the genes upstream of the *sil* cluster already mentioned above, SNP analysis also showed that the *pco* genes in pMRGN1031 are clearly divergent from the majority of those from the other isolates of this study. This strongly supports the hypothesis that the *pco–sil* gene clusters on this plasmid have a longer or different evolutionary history than those found on other plasmids, and that parts of it may possibly be decaying.

The surprising initial lack of difference in susceptibility to copper between *pco/sil*-positive and *pco/sil*-negative isolates that we obtained in broth under aerobic growth conditions triggered further investigations under a variety of other conditions. Previous publications showed that the copper resistance phenotype of *pco*-positive isolates is inducible and can be triggered by preliminary incubation in subinhibitory concentrations of copper [13,36]. Subjecting our isolates to subinhibitory concentrations of copper similar to those described in these studies did not result in any change in copper MIC, and our isolates did not show any significant change in RNA transcription of the *pcoA* and *pcoD* genes after induction. Copper susceptibility of *E. coli* and *S. enterica* was tested by others with a variety of methods, including broth [17] and agar dilutions [21,37], as well as under aerobic [17,21] and anaerobic conditions [21,37]. No differences in copper MICs were observed by these authors between *pco/sil*-positive and -negative isolates under aerobic conditions, neither for *E. coli*, nor for *S. enterica*. However, differences were consistently observed for *S. enterica* when agar dilutions were used under anaerobic conditions [18,21]. Therefore, we also tested our *E. coli* and a few *S. enterica* isolates by agar dilution under anaerobic conditions. As expected, an evident dichotomization of MICs was visible under these conditions for *S. enterica*, but this was not the case for *E. coli*. These data are in agreement with results from others showing that copper resistance associated with the *pco* gene cluster is host-dependent [13]. The increase in MIC observed in *S. enterica* after transfer of a *pco/sil* plasmid from *E. coli* clearly confirmed this hypothesis.

Overall, the results from this study strongly suggest that the *pco/sil* gene clusters may have only a minor effect on copper MICs in typical wild-type intestinal *E. coli*, and may not represent a major selective advantage in this bacterial species in the gut of swine fed high concentrations of copper. Some of our findings are based on a relatively limited number of isolates, and confirmation on larger numbers of isolates is needed. *Escherichia coli* may represent a reservoir of mobile copper resistance determinants of potential importance for *S. enterica*. As illustrated here with pMRGN207 and by other researchers [17,18], transferable *pco/sil* plasmids concomitantly carry antimicrobial resistance determinants. These antimicrobial resistance determinants may help maintain these mobile plasmids, and indirectly, the *pco–sil* cluster in *E. coli* populations. Antimicrobial resistance may, in turn, be maintained and selected in *S. enterica* harboring these plasmids by the supplementation of feed with copper. Further animal experiments are needed to clarify the latter points. The role of IncHI2 plasmids in this context and the exact mechanisms and dynamics of transposition of Tn7-like transposons associated with the *pco–sil* gene clusters certainly also warrant further investigations, as do the respective roles and contribution of the *pco* versus *sil* genes in the observed copper resistance in *S. enterica*.

Author Contributions: H.M.S. and P.B. conceived and designed the experiments; H.M.S., R.G.A., T.G.N. and M.D.T. designed and performed the animal study; K.M.R., K.N.N., R.G.A. and G.C. performed the experiments; K.N.N., H.M.S., P.B. and G.C. analyzed the data; G.C., H.M.S. and P.B. wrote the original draft of the manuscript. All authors read and approved the final manuscript.

Funding: This work was supported by the USDA National Institute of Food and Agriculture, AFRI Food Safety Challenge Grant project #2013-68003-21257.

Acknowledgments: The contents of this manuscript are solely the responsibility of the authors and do not necessarily represent the official views of the USDA or NIFA.

Conflicts of Interest: The authors declare no conflicts of interest. The funders had no role in the design of the study; in the collection, analyses, or interpretation of data; in the writing of the manuscript, or in the decision to publish the results.

References

1. National Research Council. *Nutrient Requirements of Swine*, Eleventh Revised Edition; The National Academies Press: Washington, DC, USA, 2012; ISBN 978-0-309-22423-9.
2. Agga, G.E.; Scott, H.M.; Amachawadi, R.G.; Nagaraja, T.G.; Vinasco, J.; Bai, J.; Norby, B.; Renter, D.G.; Dritz, S.S.; Nelssen, J.L.; et al. Effects of chlortetracycline and copper supplementation on antimicrobial resistance of fecal *Escherichia coli* from weaned pigs. *Prev. Vet. Med.* **2014**, *114*, 231–246. [CrossRef] [PubMed]
3. Hasman, H.; Aarestrup, F.M. *tcrB*, a gene conferring transferable copper resistance in *Enterococcus faecium*: occurrence, transferability, and linkage to macrolide and glycopeptide resistance. *Antimicrob. Agents Chemother.* **2002**, *46*, 1410–1416. [CrossRef] [PubMed]
4. Hasman, H. The *tcrB* gene is part of the tcrYAZB operon conferring copper resistance in *Enterococcus faecium* and *Enterococcus faecalis*. *Microbiology* **2005**, *151*, 3019–3025. [CrossRef] [PubMed]
5. Amachawadi, R.G.; Scott, H.M.; Alvarado, C.A.; Mainini, T.R.; Vinasco, J.; Drouillard, J.S.; Nagaraja, T.G. Occurrence of the transferable copper resistance gene *tcrB* among fecal enterococci of U.S. feedlot cattle fed copper-supplemented diets. *Appl. Environ. Microbiol.* **2013**, *79*, 4369–4375. [CrossRef] [PubMed]
6. Hasman, H.; Kempf, I.; Chidaine, B.; Cariolet, R.; Ersbøll, A.K.; Houe, H.; Hansen, H.C.B.; Aarestrup, F.M. Copper resistance in *Enterococcus faecium*, mediated by the *tcrB* gene, is selected by supplementation of pig feed with copper sulfate. *Appl. Environ. Microbiol.* **2006**, *72*, 5784–5789. [CrossRef] [PubMed]
7. Amachawadi, R.G.; Shelton, N.W.; Shi, X.; Vinasco, J.; Dritz, S.S.; Tokach, M.D.; Nelssen, J.L.; Scott, H.M.; Nagaraja, T.G. Selection of fecal Enterococci exhibiting *tcrB*-mediated copper resistance in pigs fed diets supplemented with copper. *Appl. Environ. Microbiol.* **2011**, *77*, 5597–5603. [CrossRef] [PubMed]
8. Amachawadi, R.G.; Scott, H.M.; Aperce, C.; Vinasco, J.; Drouillard, J.S.; Nagaraja, T.G. Effects of in-feed copper and tylosin supplementations on copper and antimicrobial resistance in faecal enterococci of feedlot cattle. *J. Appl. Microbiol.* **2015**, *118*, 1287–1297. [CrossRef] [PubMed]
9. Rensing, C.; Franke, S. Copper homeostasis in *Escherichia coli* and other *Enterobacteriaceae*. *EcoSal Plus* **2007**, *2*. [CrossRef] [PubMed]
10. Hao, X.; Lüthje, F.L.; Qin, Y.; McDevitt, S.F.; Lutay, N.; Hobman, J.L.; Asiani, K.; Soncini, F.C.; German, N.; Zhang, S.; et al. Survival in amoeba—A major selection pressure on the presence of bacterial copper and zinc resistance determinants? Identification of a "copper pathogenicity island". *Appl. Microbiol. Biotechnol.* **2015**, *99*, 5817–5824. [CrossRef] [PubMed]
11. Rensing, C.; Moodley, A.; Cavaco, L.M.; McDevitt, S.F. Resistance to metals used in agricultural production. In *Antimicrobial Resistance in Bacteria from Livestock and Companion Animals*; Schwarz, S., Cavaco, L., Shen, J., Eds.; ASM Press: Washington, DC, USA, 2018; pp. 83–107.
12. Tetaz, T.J.; Luke, R.K. Plasmid-controlled resistance to copper in *Escherichia coli*. *J. Bacteriol.* **1983**, *154*, 1263–1268. [PubMed]
13. Williams, J.R.; Morgan, A.G.; Rouch, D.A.; Brown, N.L.; Lee, B.T. Copper-resistant enteric bacteria from United Kingdom and Australian piggeries. *Appl. Environ. Microbiol.* **1993**, *59*, 2531–2537. [PubMed]
14. Brown, N.L.; Barrett, S.R.; Camakaris, J.; Lee, B.T.O.; Rouch, D.A. Molecular genetics and transport analysis of the copper-resistance determinant (*pco*) from *Escherichia coli* plasmid pRJ1004. *Mol. Microbiol.* **1995**, *17*, 1153–1166. [CrossRef] [PubMed]
15. Rouch, D.A.; Brown, N.L. Copper-inducible transcriptional regulation at two promoters in the *Escherichia coli* copper resistance determinant *pco*. *Microbiology* **1997**, *143*, 1191–1202. [CrossRef] [PubMed]
16. Lee, S.M.; Grass, G.; Rensing, C.; Barrett, S.R.; Yates, C.J.D.; Stoyanov, J.V.; Brown, N.L. The Pco proteins are involved in periplasmic copper handling in *Escherichia coli*. *Biochem. Biophys. Res. Commun.* **2002**, *295*, 616–620. [CrossRef]
17. Fang, L.; Li, X.; Li, L.; Li, S.; Liao, X.; Sun, J.; Liu, Y. Co-spread of metal and antibiotic resistance within ST3-IncHI2 plasmids from *E. coli* isolates of food-producing animals. *Sci. Rep.* **2016**, *6*, 25312. [CrossRef] [PubMed]
18. Mourão, J.; Marçal, S.; Ramos, P.; Campos, J.; Machado, J.; Peixe, L.; Novais, C.; Antunes, P. Tolerance to multiple metal stressors in emerging non-typhoidal MDR *Salmonella* serotypes: A relevant role for copper in anaerobic conditions. *J. Antimicrob. Chemother.* **2016**, *71*, 2147–2157. [CrossRef] [PubMed]

19. Staehlin, B.M.; Gibbons, J.G.; Rokas, A.; O'Halloran, T.V.; Slot, J.C. Evolution of a heavy metal homeostasis/resistance island reflects increasing copper stress in Enterobacteria. *Genome Biol. Evol.* **2016**, *8*, 811–826. [CrossRef] [PubMed]
20. Gupta, A.; Matsui, K.; Lo, J.-F.; Silver, S. Molecular basis for resistance to silver cations in *Salmonella*. *Nat. Med.* **1999**, *5*, 183–188. [CrossRef] [PubMed]
21. Mourão, J.; Novais, C.; Machado, J.; Peixe, L.; Antunes, P. Metal tolerance in emerging clinically relevant multidrug-resistant *Salmonella enterica* serotype 4,[5],12:i:− clones circulating in Europe. *Int. J. Antimicrob. Agents* **2015**, *45*, 610–616. [CrossRef] [PubMed]
22. Agga, G.E.; Scott, H.M.; Vinasco, J.; Nagaraja, T.G.; Amachawadi, R.G.; Bai, J.; Norby, B.; Renter, D.G.; Dritz, S.S.; Nelssen, J.L.; et al. Effects of chlortetracycline and copper supplementation on the prevalence, distribution, and quantity of antimicrobial resistance genes in the fecal metagenome of weaned pigs. *Prev. Vet. Med.* **2015**, *119*, 179–189. [CrossRef] [PubMed]
23. Feldpausch, J.A.; Amachawadi, R.G.; Tokach, M.D.; Scott, H.M.; Dritz, S.S.; Goodband, R.D.; Woodworth, J.C.; DeRouchey, J.M. Effects of dietary chlortetracycline, Origanum essential oil, and pharmacological Cu and Zn on growth performance of nursery pigs. *Trans. Anim. Sci.* **2018**, *2*, 62–73. [CrossRef]
24. Clinical and Laboratory Standards Institute. *CLSI Performance Standards for Antimicrobial Disk and Dilution Susceptibility Tests for Bacteria Isolated from Animals*; Second Informational Supplement; VET01-S2; Clinical and Laboratory Standards Institute: Wayne, PA, USA, 2013.
25. Kozak, G.K.; Boerlin, P.; Janecko, N.; Reid-Smith, R.J.; Jardine, C. Antimicrobial resistance in *Escherichia coli* isolates from swine and wild small mammals in the proximity of swine farms and in natural environments in Ontario, Canada. *Appl. Environ. Microbiol.* **2009**, *75*, 559–566. [CrossRef] [PubMed]
26. Cottell, J.L.; Kanwar, N.; Castillo-Courtade, L.; Chalmers, G.; Scott, H.M.; Norby, B.; Loneragan, G.H.; Boerlin, P. $bla_{CTX-M-32}$ on an IncN plasmid in *Escherichia coli* from beef cattle in the United States. *Antimicrob. Agents Chemother.* **2013**, *57*, 1096–1097. [CrossRef] [PubMed]
27. Rouch, D.; Camakaris, J.; Lee, B.T.; Luke, R.K. Inducible plasmid-mediated copper resistance in *Escherichia coli*. *J. Gen. Microbiol.* **1985**, *131*, 939–943. [CrossRef] [PubMed]
28. Zhou, K.; Zhou, L.; Lim, Q.E.; Zou, R.; Stephanopoulos, G.; Too, H.-P. Novel reference genes for quantifying transcriptional responses of *Escherichia coli* to protein overexpression by quantitative PCR. *BMC Mol. Biol.* **2011**, *12*, 18. [CrossRef] [PubMed]
29. Peters, J.E.; Craig, N.L. Tn7: Smarter than we thought. *Nat. Rev. Mol. Cell Biol.* **2001**, *2*, 806–814. [CrossRef] [PubMed]
30. Mitra, R.; McKenzie, G.J.; Yi, L.; Lee, C.A.; Craig, N.L. Characterization of the TnsD-*attTn7* complex that promotes site-specific insertion of *Tn7*. *Mob. DNA* **2010**, *1*, 18. [CrossRef] [PubMed]
31. Chakrabarti, A.; Desai, P.; Wickstrom, E. Transposon Tn7 protein TnsD binding to *Escherichia coli attTn7* DNA and its eukaryotic orthologs. *Biochemistry* **2004**, *43*, 2941–2946. [CrossRef] [PubMed]
32. Joensen, K.G.; Scheutz, F.; Lund, O.; Hasman, H.; Kaas, R.S.; Nielsen, E.M.; Aarestrup, F.M. Real-time whole-genome sequencing for routine typing, surveillance, and outbreak detection of verotoxigenic *Escherichia coli*. *J. Clin. Microbiol.* **2014**, *52*, 1501–1510. [CrossRef] [PubMed]
33. Seiffert, S.N.; Hilty, M.; Kronenberg, A.; Droz, S.; Perreten, V.; Endimiani, A. Extended-spectrum cephalosporin-resistant *Escherichia coli* in community, specialized outpatient clinic and hospital settings in Switzerland. *J. Antimicrob. Chemother.* **2013**, *68*, 2249–2254. [CrossRef] [PubMed]
34. Randall, C.P.; Gupta, A.; Jackson, N.; Busse, D.; O'Neill, A.J. Silver resistance in Gram-negative bacteria: A dissection of endogenous and exogenous mechanisms. *J. Antimicrob. Chemother.* **2015**, *70*, 1037–1046. [CrossRef] [PubMed]
35. Falgenhauer, L.; Ghosh, H.; Guerra, B.; Yao, Y.; Fritzenwanker, M.; Fischer, J.; Helmuth, R.; Imirzalioglu, C.; Chakraborty, T. Comparative genome analysis of IncHI2 VIM-1 carbapenemase-encoding plasmids of *Escherichia coli* and *Salmonella enterica* isolated from a livestock farm in Germany. *Vet. Microbiol.* **2017**, *200*, 114–117. [CrossRef] [PubMed]

36. Gilmour, M.W.; Thomson, N.R.; Sanders, M.; Parkhill, J.; Taylor, D.E. The complete nucleotide sequence of the resistance plasmid R478: Defining the backbone components of incompatibility group H conjugative plasmids through comparative genomics. *Plasmid* **2004**, *52*, 182–202. [CrossRef] [PubMed]
37. Medardus, J.J.; Molla, B.Z.; Nicol, M.; Morrow, W.M.; Rajala-Schultz, P.J.; Kazwala, R.; Gebreyes, W.A. In-feed use of heavy metal micronutrients in U.S. swine production systems and its role in persistence of multidrug-resistant salmonellae. *Appl. Environ. Microbiol.* **2014**, *80*, 2317–2325. [CrossRef] [PubMed]

© 2018 by the authors. Licensee MDPI, Basel, Switzerland. This article is an open access article distributed under the terms and conditions of the Creative Commons Attribution (CC BY) license (http://creativecommons.org/licenses/by/4.0/).

Review

Harnessing Rhizobia to Improve Heavy-Metal Phytoremediation by Legumes

Camilla Fagorzi [1], Alice Checcucci [1,*], George C. diCenzo [1], Klaudia Debiec-Andrzejewska [2], Lukasz Dziewit [3], Francesco Pini [4] and Alessio Mengoni [1,*]

[1] Department of Biology, University of Florence, Via Madonna del Piano 6, 50019 Sesto Fiorentino, Italy; camilla.fagorzi@unifi.it (C.F.); georgecolin.dicenzo@unifi.it (G.C.D.)
[2] Laboratory of Environmental Pollution Analysis, Faculty of Biology, University of Warsaw, Miecznikowa 1, 02-096 Warsaw, Poland; k.debiec@biol.uw.edu.pl
[3] Department of Bacterial Genetics, Institute of Microbiology, Faculty of Biology, University of Warsaw, Miecznikowa 1, 02-096 Warsaw, Poland; ldziewit@biol.uw.edu.pl
[4] Department of Agri-food Production and Environmental Science, University of Florence, 50144 Florence, Italy; francesco.pini@unifi.it
* Correspondence: alice.checcucci@unifi.it (A.C.); alessio.mengoni@unifi.it (A.M.); Tel.: +39-055-457-4738 (A.M.)

Received: 28 September 2018; Accepted: 6 November 2018; Published: 8 November 2018

Abstract: Rhizobia are bacteria that can form symbiotic associations with plants of the Fabaceae family, during which they reduce atmospheric di-nitrogen to ammonia. The symbiosis between rhizobia and leguminous plants is a fundamental contributor to nitrogen cycling in natural and agricultural ecosystems. Rhizobial microsymbionts are a major reason why legumes can colonize marginal lands and nitrogen-deficient soils. Several leguminous species have been found in metal-contaminated areas, and they often harbor metal-tolerant rhizobia. In recent years, there have been numerous efforts and discoveries related to the genetic determinants of metal resistance by rhizobia, and on the effectiveness of such rhizobia to increase the metal tolerance of host plants. Here, we review the main findings on the metal resistance of rhizobia: the physiological role, evolution, and genetic determinants, and the potential to use native and genetically-manipulated rhizobia as inoculants for legumes in phytoremediation practices.

Keywords: soil bioremediation; heavy-metals; serpentine soils; serpentine vegetation; genome manipulation; *cis*-hybrid strains

1. Introduction

Plants are colonized by an extraordinarily high number of (micro)organisms, which may reach numbers much larger than those of plant cells [1]. This is particularly evident in the rhizosphere, the thin layer of soil surrounding and influenced by plant roots, where a staggering diversity of microorganisms is present. The collective communities of plant-associated microorganisms are referred to as the plant microbiota, and include the microbial communities of the rhizosphere, as well as those of the external and internal (the endosphere) plant tissues (for examples see [1–4]). The rhizobiome refers specifically to the microbial community of the rhizosphere, and microbes from this community have been deeply studied for their beneficial effects on plant growth and health. These mainly include mycorrhizal fungi (AMF) and plant-growth promoting rhizobacteria (PGPR), with the latter including the nitrogen fixing legume endosymbiotic bacteria known as rhizobia [5]. Rhizobia are a paraphyletic group of nitrogen fixing bacteria belonging to the Alpha- and Betaproteobacteria classes. Rhizobia can penetrate plant tissues and establish an intracellular population within specialized tissue (known as a nodule) on the root (or stem in a few cases) of leguminous plants. Once inside the cells, the rhizobia differentiate into forms known as bacteroids, which are able to perform nitrogen fixation (the formation of ammonia from di-nitrogen gas) [6]. This process, termed "symbiotic nitrogen fixation" (SNF), provides the plant with nitrogen to sustain its growth in nitrogen-deficient soils, and has been suggested as one of the factors contributing to the evolutionary success of the Fabaceae plant family [6]. Plant growth and crop yield in agricultural systems emerge as the net results of the interactions between the specific plant cultivar and its associated microbiome [7].

Heavy metals are naturally present in soils; however, their increase over certain thresholds has become a worldwide issue [8]. The major cause of heavy-metal contamination in soil is anthropogenic activities (i.e., atmospheric pollution, industrial and urban waste, mining, and some agricultural practices), while natural contamination is mainly due to weathering of metal-enriched rocks [9]. Plant-associated microbiomes play important roles in phytoremediation, allowing plants to thrive on contaminated soils, alleviating the stress associated with toxic levels of heavy-metals and metalloids (such as As), and increasing phytoextraction and phytostabilization [10–14]. Phytoextraction refers to the plants' ability to import soil contaminants through their roots, and to accumulate these compounds in the aboveground tissues [15]. Phytostabilization involves the immobilization of pollutants in the soil as a result of either their absorption and accumulation in the roots, their adsorption on the root surface, or their transformation within the rhizosphere into sparingly-soluble compounds [16]. In plants such as legumes, which are generally non-hyperaccumulating species, phytostabilization is likely the more relevant process when considering the remediation of contaminated soils [15–17]. Plant-associated bacteria may promote the chemical transformation, the chelation, or precipitation and sorption of heavy-metals [18] (Figure 1). For instance, some endophytic bacteria may reduce heavy-metal toxicity [19,20]. Improved growth and increased chlorophyll content were detected in several crop plants inoculated with siderophore-producing bacteria [19]. Additionally, enhanced plant biomass production and remediation has been observed in several hyperaccumulating plants following inoculation with rhizosphere or endophytic bacteria with plant growth promoting (PGP) capabilities [21], such as 1-aminocyclopropane-1-carboxylate (ACC) deaminase production (for detailed reviews, please see [11,12]).

The association between leguminous plants and symbiotic rhizobia has stirred the attention of researchers involved in the restoration of heavy-metal-contaminated sites [22]. The possibility to cultivate legumes on marginal and nutrient-poor soils thanks to the intimate association with PGPR, particularly with nitrogen-fixing rhizobia, has been seen as an opportunity to increase phytoremediation efficiencies while simultaneously reducing its costs [23]. Heavy-metals play central roles in symbiotic nitrogen fixation (see [24] for a review of on the role of metals in the symbiosis). Notably, the nitrogenase enzyme is dependent on a cofactor containing molybdenum and iron (FeMo-co), vanadium and iron (VFe-co), or two iron molecules (FeFe-co). There is also evidence for the role of nickel in the symbiosis. For instance, plants inoculated with a deletion mutant of the rhizobium *Sinorhizobium meliloti* lacking the *nreB*-encoded Ni^{2+} efflux system displayed increased growth under controlled conditions [25]. Additionally, a treatment with low doses of Ni^{2+} as the amendment was shown to stimulate nitrogen fixation and plant growth in soybean, and to increase hydrogenase activity in *Rhizobium leguminosarum* bv. *viciae* [26,27]. However, an excess of heavy-metals negatively impacts the symbiosis, reducing the number of symbiotic nodules, the rate of nodulation, and the rate of nitrogen fixation [28,29]. Consequently, in order to promote legume-based phytoremediation through the improvement of the host plant-symbiont partnership, there is a need to discover metal-resistant rhizobia and/or to manipulate existing rhizobial inoculants to increase their level of metal resistance.

In this review, we summarize the main findings on metal resistance in rhizobia: the physiological role, evolution, and genetic determinants of metal resistance, and the perspective to use native and genetically-manipulated rhizobia as inoculants for legumes in phytoremediation practices.

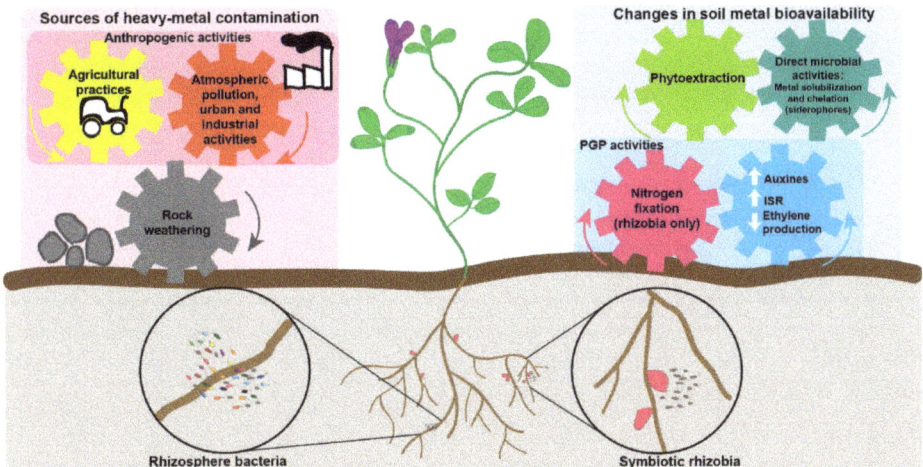

Figure 1. The multiple roles of bacteria in helping plants cope with heavy metals. Plant-associated bacteria may have various roles in both phytostabilization and plant growth. They may influence metal solubility by directly producing molecules for metal chelation (e.g., siderophores), or by influencing plant root growth, resulting in increased production of root exudates. Moreover, both rhizospheric and endophytic bacteria can positively affect plant growth by producing phytohormone molecules (e.g., auxins), alleviating plant stress (e.g., plant ethylene production), or through nitrogen fixation. This latter activity is especially relevant when leguminous plants and their rhizobial microsymbionts are considered. PGP: Plant growth promotion.

2. Legumes in Heavy-Metal Contaminated Areas

The family Leguminosae (Fabaceae) is one of the most diverse among land plants and includes over 700 genera and 20,000 species [30]. Legumes have been proposed as relevant species for phytoremediation, largely due to their ability to colonize marginal lands and nutrient-poor soils [28,31]. In particular, legumes are relevant for phytostabilization, as only a few species have been found to be metal hyperaccumulators (e.g., some species of the genus *Astragalus* isolated in the Western United States are selenium hyperaccumulators) for phytoextraction [23,28,32]. Normally, the symbiosis with rhizobia is inhibited by high levels of heavy-metals in the soil, and genetic engineering techniques have been suggested to improve symbiotic nitrogen fixation under such harsh environmental conditions [33]. However, although such biotechnological proposals are interesting in terms of molecular dissection of the system and theoretical application, currently, there are a number of limitations to the use of genetically-modified microorganisms, including their free release in nature. Analyses on legumes from heavy-metal-contaminated soils have led to the discovery of naturally-resistant rhizobia, which could be used as inoculants in these extreme environments. However, a deeper investigation of leguminous plants growing in metal-enriched sites is required to improve legume-based phytoremediation.

2.1. The Serpentine Vegetation: A Source of Legumes Evolved on Heavy-Metal Rich Soils

Serpentine rocks are an array of ultramafic rock types composed of a hydrous magnesium iron phyllosilicate mineral that originates from metamorphic alterations of peridotite and pyroxene with water. The soils derived from these rocks are characterized by: (i) high levels of nickel, cobalt, and chromium, (ii) low levels of N, P, K, and Ca, and (iii) a high Mg/Ca ratio [34]. This chemical composition strongly limits the growth of most plant species [35], as well as many microorganisms [3]. The presence of serpentine outcrops is scattered across the planet. Along a geological timescale, serpentine outcrops have prompted the evolution of peculiar plant adaptation mechanisms (such as metal hyperaccumulation [36]), which then gave rise to plant differentiation and speciation in a classical "ecological islands" model [37,38]. Serpentine vegetation in temperate ecosystems includes several leguminous species from various genera, including *Lotus, Lupinus, Trifolium, Vicia, Melilotus, Medicago, Lathyrus, Ononis, Dorychnium, Chamaecytisus, Astragalus, Anthyllis, Cytisus,* and *Acmispon* [39,40]. Serpentine endemic legumes have also been reported, such as *Errazurizia benthamii* [41] in North America, and *Serianthes calycina* [42] in New Caledonia. The microbiomes associated with serpentine plants contain a fraction of microorganisms that appear to have specifically evolved functions to cope with toxic levels of metals present in the soil and in the plant itself [3]. Moreover, some of these microorganisms have been shown to be effective in promoting host plant growth in serpentine soil and, for metal hyperaccumulating plants, to increase metal translocation to the aerial part [43]. Consequently, rhizobia from serpentine endemic legumes (such as Ni-resistant bradyrhizobia from *S. calycina* [42]) may already be adapted to optimizing the fitness of their host in serpentine environments through a long-term natural selection process [44]. Serpentine endemic legumes may therefore represent an ideal source of rhizobia that are naturally highly-competent symbiotic partners in heavy-metal contaminated soils.

2.2. The Search for Heavy-Metal Tolerant Rhizobia and Their Use as Inoculants

Legumes growing in contaminated areas such as mine deposits and serpentine soils have been a source of symbiotic rhizobial strains displaying resistance to heavy-metals, including Zn, Pb, and Cu [45–48]. Table 1 summarizes the main studies on the (positive) effects of rhizobial inoculation on the heavy-metal tolerance of host plants.

Anthyllis vulneraria is one of the most relevant legumes for isolating rhizobia that promote metal-tolerance by the host plant. *A. vulneraria* is a perennial herb from boreo-temperate climate areas in Europe, and it can be found colonizing rocky outcrops and establishing populations on heavy-metal (mainly Zn)-contaminated sites. *Anthyllis* is characterized by determinate nodules, where the meristematic activity disappears shortly after nodule formation, resulting in nodules of spherical shape. *Anthyllis* nodules contain a multilayer cortex: a glycoproteic parenchyma for diffusion, an endodermis, and the outer cortex, which mainly serves as a barrier against pathogens [49]. Nodule bacterial population of leguminous plants grown in Morocco metal-polluted soil displayed a great biodiversity, suggesting that, in these conditions, metal resistant non-rhizobia may efficiently colonize the nodules as endophytes [50]. This highlights the importance of heavy-metal resistance in rhizobia for the establishment of an effective symbiotic interaction in contaminated soils. *A. vulneraria* has been found to be associated with rhizobial symbionts from the genera *Mesorhizobium, Rhizobium,* and *Aminobacter*. These include novel rhizobial species, such as *Mesorhizobium metallidurans, Rhizobium metallidurans,* and *Aminobacter anthyllidis* [45,47–51]. Interestingly, these novel rhizobial species have so far been identified only in Pb-contaminated environments and not in unpolluted soils [47]. The symbiosis between *A. vulneraria* and its possibly exclusive metal-resistant bacterial species may provide the basis for the establishment of phytoremediation practices. This could involve the use of *A. vulneraria* metal-resistant germplasms, together with its specific natural rhizobial symbionts. Alternatively, the heavy-metal-resistant rhizobia isolated from *A. vulneraria* could be modified, either through laboratory-based experimental evolution studies [52] or direct genetic manipulation, to be capable of entering into an effective symbiosis with other host legumes.

Legumes of the genus *Medicago* have also been deeply investigated for their application in phytoremediation (see Table 1 and references therein). This is mainly because species from this genus are important forage crops for which cultivation techniques and genetics are well established, providing important advantages for future cost-effective applications [53]. Genetically-modified [54,55] and natural [56,57] inocula of *Sinorhizobium* (syn. *Ensifer*) *meliloti* and *Sinorhizobium medicae* [54] have been examined for their abilities to improve plant growth and metal accumulation in the presence of toxic levels of heavy metals such as Cu, Cd, and Zn. However, genetic manipulation is not absolutely required, as interesting results have also been obtained using indigenous *S. meliloti* and *S. medicae* strains directly isolated from contaminated soils [56,57]. For example, inoculation of *Medicago sativa* plants, grown under field conditions, with wild *S. meliloti* and *S. medicae* strains resulted in active nodulation and the promotion of metal bioaccumulation within the root nodules [56,57]. These results suggest that the exploitation of natural rhizobia could be a valuable tool for promoting land restoration and phytostabilization by legumes.

Legume-based phytoremediation may also be improved through inoculation with a consortium of metal-resistant rhizobia and other PGP bacteria. In metal polluted soil, inoculation of *Lupinus luteus* with *Bradyrhizobium* sp. 750 in consortium with *Pseudomonas* sp. Az13 and *Ochrobactrum cytisi* Azn6.2 increased plant biomass by greater than 100% with respect to uninoculated plants [10]. In contrast, inoculation with only *Bradyrhizobium* sp. 750 increased plant biomass by only 30%. Similarly, co-inoculation of *M. lupina* with *S. meliloti* CCNWSX0020 and *Pseudomonas putida* UW4 resulted in larger plants and greater total Cu accumulation than inoculation with just *S. meliloti* CCNWSX0020 [55]. Inoculation of *Vicia faba*, *Lens culinaris*, and *Sulla coronaria* with consortia of rhizobia and non-rhizobia was also effective at improving plant growth and pod yield when grown in metal-contaminated soil [58]. Moreover, the inoculated *S. coronaria* accumulated significantly more cadmium than uninoculated plants [58]. These results highlight the potential for root-associated microbial communities to influence the success of phytoremediation by rhizobium-inoculated legumes.

It may be concluded that there is great biotechnological potential in increasing the phytoremediation capabilities of legumes by their associated rhizobia. This may be mediated through at least two mechanisms: (i) reducing the toxic effects of the metals, and (ii) promoting the growth of the plant through PGP activities.

Table 1. Studies of phytoremediation mediated by rhizobium-inoculated legumes. NA, not analyzed.

Legume Species	Heavy-Metals in the Soil	Rhizobium Inoculant	Co-Inoculation with Other PGPR?	Evidence for Stimulation of Rhizosphere Microbiota	Type of Study	Effect	Reference
Glycine max	As	*Bradyrhizobium* sp. Per 3.61	No	NA	Lab scale (pot)	Reduce translocation factor	[59]
Lupinus luteus	Cu, Cd, Pb	*Bradyrhizobium* sp. 750	Yes	NA	*In situ*	Increased metal accumulation in root	[10]
Medicago lupulina	Cu	*Sinorhizobium meliloti* CCNWSX0020	No	NA	*In vitro* (pot)	Increased plant growth and copper tolerance	[55]
Medicago sativa	Cu	*Sinorhizobium meliloti* CCNWSX0020	No	NA	*In vitro*	Increased tolerance of seedlings	[60]
Medicago sativa	Cd	*Sinorhizobium meliloti* (from contaminated soil [61])	No	NA	Lab scale (pot)	Increased Cd phytoextraction	[56]
Medicago sativa	Zn	*Sinorhizobium meliloti* (from contaminated soil [61])	No	NA	Lab scale (pot with sterile sand)	Increased Zn accumulation in root	[57]
Medicago truncatula	Cu	*Sinorhizobium medicae* MA11 (genetically modified with *copAB* genes)	No	NA	*In vitro*	Increased metal accumulation in root	[54]
Robinia pseudoacacia	Cd, Zn, Pb	*Mesorhizobium loti* HZ76	No	Yes	Lab scale (pot)	Increased growth of the plant	[62]
Sulla coronaria	Cu, Zn, Pb	*Rhizobium sullae*	Yes	NA	*In situ*	Increased soil Zn stabilization	[58]
Vicia faba	Cu, Zn, Pb	*Rhizobium* sp. CCNWSX0481	Yes	NA	*In situ*	Increased soil Cu stabilization	[58]

3. Genetics and Genomics of Heavy-Metal Resistance in Symbiotic Rhizobia

A deep understanding of the genetics and molecular mechanisms of metal resistance remains one of the main goals in environmental biotechnology, with the final aim of promoting the bioremediation (including phytoremediation) of contaminated soils. Table 2 reports the main studies evaluating the genetic determinants of heavy metal resistance in rhizobia. Such studies have most commonly identified the presence of efflux systems that increase metal tolerance by reducing the intracellular concentrations of the metal(s). However, studies employing genome-scale methods, such as transcriptome analyses and transposon mutagenesis, have demonstrated that the cellular response to metal stress involves an intricate genetic network.

Mechanisms mediating resistance to Co and Ni have been identified in many metal resistant rhizobia through the identification of orthologs of metal resistance genes characterized in *Cupriavidus metallidurans* CH34 [63,64]. A gene encoding a DmeF ortholog has been identified in *R. leguminosarum* bv. *viciae* strain UPM791 [65]. DmeF proteins belong to the cation diffusion facilitator (CDF) protein family, which form metal/proton antiport systems to translocate heavy metals across the bacterial membrane [66]. Mutation of the *dmeRF* operon in *R. leguminosarum* resulted in increased sensitivity to Co and Ni, but not to Zn or Cu [65]. The mutant also appeared to be somewhat less effective in symbiosis with pea plants, but not lentil plants, when grown with high concentrations of Co or Ni [65]. Further experiments demonstrated that *dmeR* encodes a Ni- and Co-responsive transcriptional regulator that represses expression of the efflux system in the absence of these metals [65]. Despite being considered a metal-sensitive strain, the *S. meliloti* strain 1021 encodes various metal homeostasis mechanisms, including the DmeRF system, several P-ATPases that are highly common in bacteria, and an ortholog of the *C. metallidurans* NreB protein [25,65,67]. Mutation of *nreB*, encoding a Ni^{2+} efflux protein, resulted in increased sensitivity to Ni, Cu, and low pH, but increased tolerance to urea osmotic stress [25]. The P_{1B-5}-ATPase of *S. meliloti*, termed Nia (<u>n</u>ickel <u>i</u>ron <u>A</u>TPase), is positively induced by the presence of Ni^{2+} and Fe^{2+}, and its expression is higher within nodules relative to free-living cells, which may prevent toxic levels of iron accumulation in the symbiosomes. The wild type protein and recombinants with a deletion of the C-terminal Hr domain have been used to understand the metal specificity of the P_{1B-5}-ATPase family [67].

Genome-wide analyses have been used to investigate the genetics of the resistance mechanisms in *S. meliloti* strain CCNWSX0020, which is resistant to high levels of various heavy-metals (Cu, Zn, Cd and Pb). Gene mutation and transcriptome analyses have suggested the involvement of dozens of genes in the metal-resistance phenotypes of CCNWSX0020, including housekeeping genes [68–70]. Of particular note are the following three operons: the multicopper oxidase (MCO), CopG, and YadYZ operons. The MCO operon is highly expressed following exposure to Cu, and it encodes an outer membrane protein (Omp), the multicopper oxidase CueO, a blue copper azurin-like protein, and a copper chaperone involved in Cu homeostasis [70]. It was proposed that the CueO protein (showing 40% similarity with the CueO protein of *E. coli*) catalyzes Cu(I) oxidation in the periplasmic space, followed by the export of the excessive Cu(II) across the outer membrane [70,71]. The CopG operon consists of four genes: CopG, a CusA-like protein, a FixH-like protein, and a hypothetical protein. Mutation of any of the latter three genes resulted in elevated sensitivity to Zn, Pb, Cd, and Cu, although the mechanism of resistance of this operon remains unknown [70]. The CusA-like protein appears to be a highly-truncated ortholog of the CusA protein of the CusCBA Cu(I) efflux system of *E. coli* [72,73], and may act as a metal binding protein [70]. The FixH-like protein displays similarity to the FixH protein of the FixHGI membrane-bound system, a likely cation transporter that has been shown to be essential for symbiotic nitrogen fixation [74,75]. A FixH-like homolog is also encoded by the pSinB plasmid of *Ensifer* sp. M14 (formerly *Sinorhizobium* sp. M14), where it was also experimentally shown to be involved in metal resistance [76]. Deletion of the *yedYZ* operon resulted in increased sensitivity to Zn, Pb, Cd, and Cu [70]. This was the first report suggesting that YedYZ may be involved in heavy-metal tolerance. In *E. coli*, YedYZ forms a sulfite oxidoreductase [77], and expression of a homologous protein in *S. meliloti* 1021 is induced by taurine and thiosulfate [78]. Thus, the heavy-metal resistance phenotype may be mediated through disrupting sulfite metabolism, which may influence antioxidant defenses against reactive oxygen species (ROS) generated by heavy metals [70].

Many scientists have used population genetics approaches to identify loci associated with heavy-metal resistance. This was achieved by performing genome-wide association studies on a population's pan-genome, considering allelic variations in the core genome (the set of genes shared by the members of the population), and gene presence/absence in the dispensable genome fraction (the set of genes present in only a fraction of the population). Genomic variants statistically associated with nickel adaptation were identified in a *Mesorhizobium* population using this approach [79]. A population of 47 *Mesorhizobium* strains, isolated from root nodules and soils with different levels of nickel contamination, was studied. Most of the variants associated with metal adaptation were found in the dispensable genome fraction. This work highlights that adaptation to heavy metal stress is likely driven predominately by horizontal gene transfer, and is not due to mutations of pre-existing genes.

Multiple studies have demonstrated that the genetic determinants of metal-resistance in rhizobia are relevant for phytoremediation purposes. Mutation of *ceuO*, *yedYZ*, and the *fixH*-like gene negatively impacted the *M. lupulina* nodulation kinetics of *S. meliloti* CCNWSX0020 in the presence of Cu and/or Zn [70], while deletion of the *cusA*-like gene had a negative effect, even in the absence of heavy metals. It was separately observed that *M. lupulina* plants inoculated with *S. meliloti* CCNWSX0020 strains with independent mutations in five Cu resistance loci were smaller than plants inoculated with the wild type, when grown in the presence of Cu [80]. Notably, *M. lupulina* plants inoculated with any of the *S. meliloti* CCNWSX0020 mutants mentioned above accumulated lower amounts of Cu and/or Ni [78]. Similarly, *Robinia pseudoacacia* plants inoculated with a *Mesorhizobium amorphae* 186 *copA* mutant accumulated 10–15% less Cu than plants inoculated with the wild type [81]; however, no effect on plant growth was observed.

Table 2. Genes for heavy-metal (and metalloid) tolerance in symbiotic rhizobia. A summary of the main genes whose function in tolerance was confirmed experimentally is reported.

Strain	Host Plant	Isolation Site	Method of Identification	Gene(s)	Metal(s) Tolerance	Reference
Bradhyrhizobium spp.	*Serianthes calycina*	Serpentine (New Caledonia)	PCR amplification, site-directed mutagenesis	*cnr*/*nre* systems	Co, Ni	[42]
Mesorhizobium spp.	*Acmispon wrangelianus*	Serpentine (California)	Association mapping	Various	Ni	[79]
Mesorhizobium metallidurans	*Antyllis vulneraria*	Zinc mine (France)	Cosmid library	*cadA* (PIB-2-type ATPase)	Zn, Cd	[82]
Sinorhizobium meliloti 1021	*Medicago sativa*	Laboratory strain	Site-directed gene deletion	*nreB* (SMa1641)	Ni	[25]
Sinorhizobium meliloti 1021	*Medicago sativa*	Laboratory strain	Tn5 insertion, biochemical characterization	SMa1163 (P1B-5-ATPase)	Ni, Fe	[67]
Sinorhizobium meliloti CCNWSX0020	*Medicago lupulina*	Mine tailings (China)	Site-directed gene deletion and transcriptomics	P1B-type ATPases and others	Cu, Zn	[69,70]
Rhizobium leguminosarum bv. *viciae* UPM1137	*Pisum sativum*	Serpentine (Italy)	Transposon mutagenesis	14 loci (gene annotation corresponds to Rlv 3841 genome): RL2862, RL2436, RL2322, pRL110066, RL1351, RL4539, pRL90287, RL4188, RL2793, RL2100, RL0615, RL1589, pRL110071, RL1553	Ni, Co	[83]

4. Genomic Manipulation Strategies for Improving Legume Phytoremediation

Various attempts have been made to increase plant growth in the presence of toxic metal concentrations through genetic modification of their rhizobial microsymbionts. One approach is to introduce new genes conferring heavy-metal resistance into the rhizobium. For example, inoculation of a genetically-modified *M. truncatula* line (which expressed a metallothionein gene from *Arabidopsis thaliana* in its roots) with wild type *S. medicae* resulted in elevated Cu tolerance [84]. Copper tolerance was further increased using a *S. medicae* strain expressing the *P. fluorescence copAB* Cu resistance genes [84]. Inoculation with the latter strain also resulted in elevated Cu accumulation in the plant roots [84]. Similarly, the introduction of an algal As(III) methyltransferase gene (*arsM*) into the chromosome of *R. leguminosarum* bv. *trifolii* produced a strain that was able to methylate and volatilize inorganic arsenic in symbiosis with red clover (with no negative impact on nitrogen fixation) [85]. A second approach is the insertion of genes in rhizobia to modulate phytohormone production, thereby reducing plant stress perception. For example, an ACC deaminase overproducing *S. meliloti* strain increased Cu tolerance and promoted plant growth of the host plant *M. lupulina* [86]. This result was probably due to reduced production of ethylene by the host plant, in turn decreasing stress perception. However, it should be kept in mind that a relatively high number of genes may contribute to the heavy-metal stress response [87–89]. Consequently, a multigenic, genome-wide approach should be considered when attempting to genetically modify competitive rhizobial symbionts to have increased heavy-metal tolerance. One possibility along these lines is the introduction of entire, large resistance plasmids from a non-symbiotic (but highly resistant) strain to a phylogenetically-related, symbiotic metal-sensitive strain. A candidate plasmid for such studies is the pSinA plasmid of the non-symbiotic *Ensifer* sp. M14, which was isolated from an As-contaminated gold mine [76,90,91]. The pSinA plasmid is a self-transmissible replicon with a broad host range. It harbors a genomic island with genes for arsenite oxidation (*aio* genes) and arsenite resistance (*ars* genes), and its transfer to other species results in increased arsenic resistance [90]. Transfer of the pSinA plasmid to closely-related rhizobia, such as *S. meliloti*, may result in the construction of As-tolerant legume symbionts for use in arsenic remediation. Subsequent acquisition of pSinA by other members of the rhizospheric microbiota may further stimulate phytoremediation of arsenic contaminated soils through reducing the arsenic toxicity (oxidizing arsenites to arsenates) and biofortification (increase of the arsenic resistance level) of the autochthonic or augmented microflora.

Similarly, elite and metal-resistant rhizobia may be obtained through combining within one strain genomic elements from the species pangenome. The genomes of most rhizobia are extremely diverse, and many rhizobia have a divided genome structure consisting of at least two large DNA replicons [92]. Although there can be numerous inter-replicon functional, regulatory, and genetic interactions [93–95], in some ways, each replicon in a divided genome could be considered as an independent evolutionary and functional element [94,96–98]. Recently, it was shown that the genome and metabolism of *S. meliloti* is robust to the replacement of the symbiotic megaplasmid with the symbiotic megaplasmid of a different wild-type isolate [99]. Therefore, it may be possible to construct "hybrid" strains (Figure 2) with a collection of replicons derived from various wild-type isolates, potentially allowing for the development of elite strains with improved multifactorial phenotypes (e.g., resistance to heavy-metals, high symbiotic efficiency, and competition toward the indigenous soil microbiota).

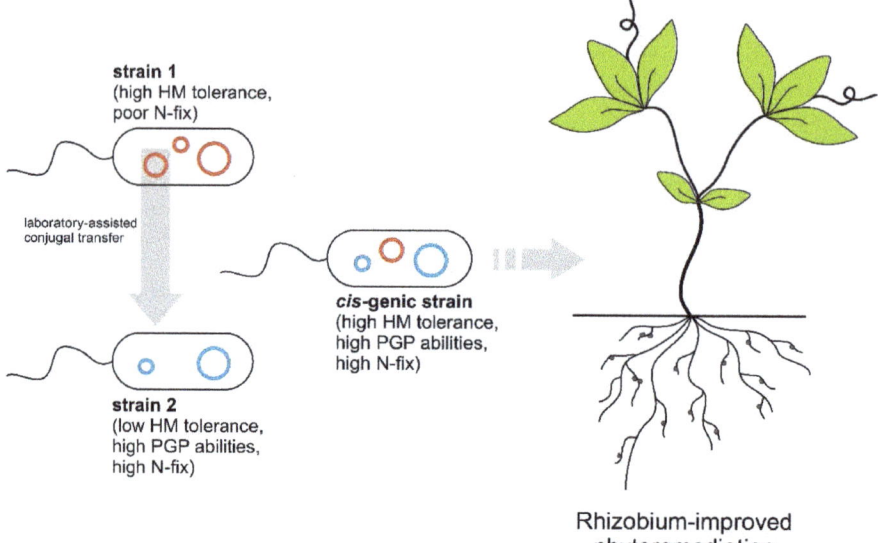

Figure 2. A synthetic biology-based proposal to increase rhizobial-mediated heavy-metal tolerance. Surveys of rhizobial phenotypic and genetic diversity in heavy-metal (HM) rich areas facilitates the discovery of strains (strain 1) with high levels of heavy-metal resistance. However, such strains may not be competitive or good nitrogen-fixers in the crops to be used for phytoremediation. The simultaneous transfer of a large collection of genomic determinants that contribute to HM tolerance, good PGP, and/or nitrogen fixation (N-fix) abilities between two or more strains (strain 2) could create hybrid strains (cis-genic strain) with improved features for application in the field for phytoremediation.

5. Conclusions

In recent years, the number of studies related to the potential exploitation of rhizobium–legume symbioses for phytoremediation practices have increased enormously as a result of environmental emergencies. In this brief review, we have presented state-of-the-art studies on heavy-metal tolerant rhizobia, and on their applications in phytoremediation as legume symbionts. A large number of investigations have indicated that rhizobia, and especially heavy-metal resistant rhizobia, can increase legume heavy-metal tolerance and promote improved legume growth in metal-rich soils, thereby resulting in greater removal of heavy-metals from the soil. Heavy-metal resistant rhizobia have been isolated from the nodules of legumes grown in soils that are rich in heavy-metals as a result of geological (e.g., serpentine outcrops) or anthropic causes (e.g., mine deposits). Genetic and genomic studies of heavy-metal resistant rhizobia have shown that although relatively few genes act as the main player in tolerance, a much larger set of genes may be involved in maximizing fitness in heavy metal rich growth conditions. Some of these genes, such as the systems for Ni^{2+} efflux in *S. meliloti*, may also contribute to a linkage between metal homeostasis and nitrogen-fixation efficiency. As such, systems-biology approaches are required to develop an overall picture of heavy-metal resistance and the ways that we can increase and exploit it in biotechnology. It will also be important to keep in mind that the engineering of rhizobia should consider several additional aspects, including the rhizobial genotype, the host plant genotype, and the interactions between the rhizobium with the soil and root microbiota [100].

Going forward, we suggest that large-scale genome-manipulation approaches may be considered in developing rhizobial strains with elite phenotypes (e.g., high heavy-metal resistance, high nitrogen-fixation ability, high competitiveness, etc.) for use in phytoremediation applications. As a pre-requisite to such studies, it will be necessary to increase efforts at creating culture collections of rhizobial strains from contaminated areas, since the strains isolated from these environments is quite limited in number and in terms of host plant (see also [23]). Such efforts would benefit from exploring areas that have evolved peculiar flora, such as serpentine outcrops, maximizing the chance to find well-adapted strains. Whole genome sequencing, genome-scale mutagenesis (such as Tn-seq or INseq [101]), and metabolic modeling of these strains could then be employed to fully characterize the genomic basis for tolerance against the contaminants.

Author Contributions: Conceptualization, A.M., C.F., A.C. and G.C.D.; Writing-Original Draft Preparation, A.M., C.F., A.C. and G.C.D.; Writing-Review & Editing, A.M., C.F., A.C., G.C.D., L.D. and K.D.A.; Funding Acquisition, A.M.

Funding: This research was partially funded by Ente Cassa di Risparmio di Firenze, grant "2017-0719" and by the University of Florence, project "Dinamiche dell'evoluzione dei genomi batterici: l'evoluzione del genoma multipartito e la suddivisione in moduli funzionali", call "PROGETTI STRATEGICI DI ATENEO ANNO 2014" to A.M. A.C. was supported by a grant from Fondazione Adriano Buzzati-Traverso. G.C.D. was supported by the Natural Sciences and Engineering Research Council of Canada (NSERC) through a PDF fellowship. K.D. was supported by the European Molecular Biology Organization in the frame of the EMBO Short-Term Fellowship program [grant number 7376] and the National Science Center (Poland) in the frame of the Preludium grant No. 2016/23/N/NZ9/01655.

Conflicts of Interest: The authors declare no conflict of interest.

References

1. Mendes, R.; Garbeva, P.; Raaijmakers, J.M. The rhizosphere microbiome: Significance of plant beneficial, plant pathogenic, and human pathogenic microorganisms. *FEMS Microbiol. Rev.* **2013**, *37*, 634–663. [CrossRef] [PubMed]
2. Bai, Y.; Müller, D.B.; Srinivas, G.; Garrido-Oter, R.; Potthoff, E.; Rott, M.; Dombrowski, N.; Münch, P.C.; Spaepen, S.; Remus-Emsermann, M.; et al. Functional overlap of the *Arabidopsis* leaf and root microbiota. *Nature* **2015**, *528*, 364. [CrossRef] [PubMed]
3. Mengoni, A.; Schat, H.; Vangronsveld, J. Plants as extreme environments? Ni-resistant bacteria and Ni-hyperaccumulators of serpentine flora. *Plant Soil* **2010**, *331*, 5–16. [CrossRef]
4. Pini, F.; Frascella, A.; Santopolo, L.; Bazzicalupo, M.; Biondi, E.G.; Scotti, C.; Mengoni, A. Exploring the plant-associated bacterial communities in *Medicago sativa* L. *BMC Microbiol.* **2012**, *12*, 78. [CrossRef] [PubMed]
5. Nadeem, S.M.; Ahmad, M.; Zahir, Z.A.; Javaid, A.; Ashraf, M. The role of mycorrhizae and plant growth promoting rhizobacteria (PGPR) in improving crop productivity under stressful environments. *Biotechnol. Adv.* **2014**, *32*, 429–448. [CrossRef] [PubMed]
6. Sprent, J.I. *Legume Nodulation: A Global Perspective*; John Wiley & Sons: Hoboken, NJ, USA, 2009; ISBN 1444316397.
7. Theis, K.R.; Dheilly, N.M.; Klassen, J.L.; Brucker, R.M.; Baines, J.F.; Bosch, T.C.G.; Cryan, J.F.; Gilbert, S.F.; Goodnight, C.J.; Lloyd, E.A.; et al. Getting the Hologenome Concept Right: An Eco-Evolutionary Framework for Hosts and Their Microbiomes. *Msystems* **2016**, *1*, e00028-16. [CrossRef] [PubMed]
8. Chibuike, G.U.; Obiora, S.C. Heavy metal polluted soils: Effect on plants and bioremediation methods. *Appl. Environ. Soil Sci.* **2014**. [CrossRef]
9. Lebrazi, S.; Fikri-Benbrahim, K. Rhizobium-Legume Symbioses: Heavy metal effects and principal approaches for bioremediation of contaminated soil. In *Legumes for Soil Health and Sustainable Management*; Springer: Heidelberg, Germany, 2018; pp. 205–233. ISBN 978-981-13-0252-7.
10. Dary, M.; Chamber-Pérez, M.A.; Palomares, A.J.; Pajuelo, E. "In situ" phytostabilisation of heavy metal polluted soils using *Lupinus luteus* inoculated with metal resistant plant-growth promoting rhizobacteria. *J. Hazard. Mater.* **2010**, *177*, 323–330. [CrossRef] [PubMed]
11. Kong, Z.; Glick, B.R. The Role of Plant Growth-Promoting Bacteria in Metal Phytoremediation. In *Advanced in Microbial Physiology*, 1st ed.; Elsevier Ltd.: New York, NY, USA, 2017; Volume 71, pp. 97–132. ISBN 0065-2911.
12. Sessitsch, A.; Kuffner, M.; Kidd, P.; Vangronsveld, J.; Wenzel, W.W.; Fallmann, K.; Puschenreiter, M. The role of plant-associated bacteria in the mobilization and phytoextraction of trace elements in contaminated soils. *Soil Biol. Biochem.* **2013**, *60*, 182–194. [CrossRef] [PubMed]

13. Weyens, N.; Lelie, D. Van Der; Taghavi, S.; Newman, L. Exploiting plant—Microbe partnerships to improve biomass production and remediation. *Trends Biotechnol.* **2009**, 1–8. [CrossRef] [PubMed]
14. Kidd, P.S.; Alvarez-Lopez, V.; Becerra-Castro, C.; Cabello-Conejo, M.; Prieto-Fernandez, A. Potential role of plant-associated bacteria in plant metal uptake and implications in phytotechnologies. In *Advances in Botanical Research*; Academic Press: London UK; New York, NY, USA, 2017; pp. 87–126.
15. Mahar, A.; Wang, P.; Ali, A.; Awasthi, M.K.; Lahori, A.H.; Wang, Q.; Li, R.; Zhang, Z. Challenges and opportunities in the phytoremediation of heavy metals contaminated soils: A review. *Ecotoxicol. Environ. Saf.* **2016**, *126*, 111–121. [CrossRef] [PubMed]
16. Bolan, N.S.; Park, J.H.; Robinson, B.; Naidu, R.; Huh, K.Y. Phytostabilization: A green approach to contaminant containment. *Adv. Agron.* **2011**, *112*, 145–204. [CrossRef]
17. Mahieu, S.; Frérot, H.; Vidal, C.; Galiana, A.; Heulin, K.; Maure, L.; Brunel, B.; Lefèbvre, C.; Escarré, J.; Cleyet-Marel, J.-C. *Anthyllis vulneraria/Mesorhizobium metallidurans*, an efficient symbiotic nitrogen fixing association able to grow in mine tailings highly contaminated by Zn, Pb and Cd. *Plant Soil* **2011**, *342*, 405–417. [CrossRef]
18. Gadd, G.M. Accumulation and transformation of metals by microorganisms. In *Biotechnology: Special Processes*; John Wiley & Sons: Hoboken, NJ, USA, 2008; Volume 10, pp. 226–264. ISBN 978-3-52-762093-7.
19. Mastretta, C.; Taghavi, S.; Van Der Lelie, D.; Mengoni, A.; Galardi, F.; Gonnelli, C.; Barac, T.; Boulet, J.; Weyens, N.; Vangronsveld, J. Endophytic bacteria from seeds of *Nicotiana tabacum* can reduce cadmium phytotoxicity. *Int. J. Phytoremediat.* **2009**, *11*, 251–267. [CrossRef]
20. Etesami, H. Bacterial mediated alleviation of heavy metal stress and decreased accumulation of metals in plant tissues: Mechanisms and future prospects. *Ecotoxicol. Environ. Saf.* **2018**, *147*, 175–191. [CrossRef] [PubMed]
21. Novo, L.A.B.; Castro, P.M.L.; Alvarenga, P.; da Silva, E.F. Plant Growth–Promoting Rhizobacteria-Assisted phytoremediation of mine soils. In *Bio-Geotechnologies for Mine Site Rehabilitation*; Elsevier: New York, NY, USA, 2018; pp. 281–295. ISBN 978-0-12-812987-6.
22. Teng, Y.; Wang, X.; Li, L.; Li, Z.; Luo, Y. Rhizobia and their bio-partners as novel drivers for functional remediation in contaminated soils. *Front. Plant Sci.* **2015**, *6*, 32. [CrossRef] [PubMed]
23. Checcucci, A.; Bazzicalupo, M.; Mengoni, A. Exploiting nitrogen-fixing rhizobial symbionts genetic resources for improving phytoremediation of contaminated soils. In *Enhancing Cleanup of Environmental Pollutants*; Springer: Heidelberg, Germany, 2017; Volume 1, pp. 275–288. ISBN 978-3-31-955426-6.
24. González-Guerrero, M.; Matthiadis, A.; Saez, Á.; Long, T. Fixating on metals: New insights into the role of metals in nodulation and symbiotic nitrogen fixation. *Front. Plant Sci.* **2014**, *5*, 45. [CrossRef] [PubMed]
25. Pini, F.; Spini, G.; Galardini, M.; Bazzicalupo, M.; Benedetti, A.; Chiancianesi, M.; Florio, A.; Lagomarsino, A.; Migliore, M.; Mocali, S.; et al. Molecular phylogeny of the nickel-resistance gene *nreB* and functional role in the nickel sensitive symbiotic nitrogen fixing bacterium *Sinorhizobium meliloti*. *Plant Soil* **2013**, *377*, 189–201. [CrossRef]
26. Lavres, J.; Castro Franco, G.; de Sousa Câmara, G.M. Soybean seed treatment with nickel improves biological nitrogen fixation and urease activity. *Front. Environ. Sci.* **2016**, *4*, 37. [CrossRef]
27. Ureta, A.-C.; Imperial, J.; Ruiz-Argüeso, T.; Palacios, J.M. *Rhizobium leguminosarum* biovar viciae symbiotic hydrogenase activity and processing are limited by the level of nickel in agricultural soils. *Appl. Environ. Microbiol.* **2005**, *71*, 7603–7606. [CrossRef] [PubMed]
28. Hao, X.; Taghavi, S.; Xie, P.; Orbach, M.J.; Alwathnani, H.A.; Rensing, C.; Wei, G. Phytoremediation of heavy and transition metals aided by legume-rhizobia symbiosis. *Int. J. Phytoremediat.* **2014**, *16*, 179–202. [CrossRef] [PubMed]
29. Ahmad, E.; Zaidi, A.; Khan, M.S.; Oves, M. Heavy metal toxicity to symbiotic nitrogen-fixing microorganism and host legumes. In *Toxicity of Heavy Metals to Legumes and Bioremediation*; Springer: Heidelberg, Germany, 2012; pp. 29–44. ISBN 3709107296.
30. Doyle, J.J.; Luckow, M.A. The rest of the iceberg. Legume diversity and evolution in a phylogenetic context. *Plant Physiol.* **2003**, *131*, 900–910. [CrossRef] [PubMed]
31. Bradshaw, A.D.; Chadwick, M.J. *The Restoration of Land: The Ecology and Reclamation of Derelict and Degraded Land*; University of California Press: Berkeley, CA, USA, 1980; ISBN 0520039610.
32. Reeves, R.D.; van der Ent, A.; Baker, A.J.M. Global distribution and ecology of hyperaccumulator plants. In *Agromining: Farming for Metals*; Springer: Heidelberg, Germany, 2018; pp. 75–92, ISBN 978-3-319-61898-2.

33. Pajuelo, E.; Rodríguez-Llorente, I.D.; Lafuente, A.; Caviedes, M.Á. Legume–rhizobium symbioses as a tool for bioremediation of heavy metal polluted soils. In *Biomanagement of Metal-Contaminated Soils*; Springer: Heidelberg, Germany, 2011; pp. 95–123. ISBN 978-94-007-1914-9.
34. Brooks, R.R. *Serpentine and Its Vegetation: A Multidisciplinary Approach*; Croom Helm: London, UK, 1987; ISBN 0709950632.
35. Brady, K.U.; Kruckeberg, A.R.; Bradshaw, H.D., Jr. Evolutionary ecology of plant adaptation to serpentine soils. *Annu. Rev. Ecol. Evol. Syst.* **2005**, *36*, 243–266. [CrossRef]
36. Words, K. Metal Hyperaccumulation in plants. *Annu. Rev. Plant Biol.* **2010**, *61*, 517–534. [CrossRef]
37. Harrison, S.; Rajakaruna, N. *Serpentine: The Evolution and Ecology of a Model System*; University of California Press: Berkeley, CA, USA, 2011; ISBN 0520268350.
38. Mengoni, A.; Mocali, S.; Surico, G.; Tegli, S.; Fani, R. Fluctuation of endophytic bacteria and phytoplasmosis in elm trees. *Microbiol. Res.* **2003**, *158*, 363–369. [CrossRef] [PubMed]
39. Alexander, E.B.; Coleman, R.G.; Harrison, S.P.; Keeler-Wolfe, T. *Serpentine Geoecology of Western North America: Geology, Soils, and Vegetation*; OUP: Oxford, UK, 2007; ISBN 019516508X.
40. Pustahija, F.; Brown, S.C.; Bogunić, F.; Bašić, N.; Muratović, E.; Ollier, S.; Hidalgo, O.; Bourge, M.; Stevanović, V.; Siljak-Yakovlev, S. Small genomes dominate in plants growing on serpentine soils in West Balkans, an exhaustive study of 8 habitats covering 308 taxa. *Plant Soil* **2013**, *373*, 427–453. [CrossRef]
41. Selvi, F. Diversity, geographic variation and conservation of the serpentine flora of Tuscany (Italy). *Biodivers. Conserv.* **2007**, *16*, 1423–1439. [CrossRef]
42. Chaintreuil, C.; Rigault, F.; Moulin, L.; Jaffré, T.; Fardoux, J.; Giraud, E.; Dreyfus, B.; Bailly, X. Nickel resistance determinants in *Bradyrhizobium* strains from nodules of the endemic New Caledonia legume *Serianthes calycina*. *Appl. Environ. Microbiol.* **2007**, *73*, 8018–8022. [CrossRef] [PubMed]
43. Rajkumar, M.; Narasimha, M.; Prasad, V.; Freitas, H.; Ae, N. Biotechnological applications of serpentine soil bacteria for phytoremediation of trace metals. *Crit. Rev. Biotechnol.* **2009**, *29*, 120–130. [CrossRef] [PubMed]
44. Friesen, M.L. Widespread fitness alignment in the legume—Rhizobium symbiosis. *New Phytol.* **2012**, *194*, 1096–1111. [CrossRef] [PubMed]
45. Grison, C.M.; Jackson, S.; Merlot, S.; Dobson, A.; Grison, C. *Rhizobium metallidurans* sp. nov., a symbiotic heavy metal resistant bacterium isolated from the anthyllis vulneraria Zn-hyperaccumulator. *Int. J. Syst. Evol. Microbiol.* **2015**, *65*, 1525–1530. [CrossRef] [PubMed]
46. Ye, M.; Liao, B.; Li, J.T.; Mengoni, A.; Hu, M.; Luo, W.C.; Shu, W.S. Contrasting patterns of genetic divergence in two sympatric pseudo-metallophytes: *Rumex acetosa* L. and *Commelina communis* L. *BMC Evol. Biol.* **2012**, *12*, 84. [CrossRef] [PubMed]
47. Mohamad, R.; Maynaud, G.; Le Quéré, A.; Vidal, C.; Klonowska, A.; Yashiro, E.; Cleyet-Marel, J.-C.; Brunel, B. Ancient heavy metal contamination in soils as a driver of tolerant *Anthyllis vulneraria* rhizobial communities. *Appl. Environ. Microbiol.* **2016**, *83*, e01735-16. [CrossRef] [PubMed]
48. Vidal, C.; Chantreuil, C.; Berge, O.; Mauré, L.; Escarré, J.; Béna, G.; Brunel, B.; Cleyet-Marel, J.C. *Mesorhizobium metallidurans* sp. nov., a metal-resistant symbiont of *Anthyllis vulneraria* growing on metallicolous soil in Languedoc, France. *Int. J. Syst. Evol. Microbiol.* **2009**, *59*, 850–855. [CrossRef] [PubMed]
49. Sujkowska-Rybkowska, M.; Ważny, R. Metal resistant rhizobia and ultrastructure of *Anthyllis vulneraria* nodules from zinc and lead contaminated tailing in Poland. *Int. J. Phytoremediat.* **2018**, *20*, 709–720. [CrossRef] [PubMed]
50. El Aafi, N.; Saidi, N.; Maltouf, A.F.; Perez-Palacios, P.; Dary, M.; Brhada, F.; Pajuelo, E. Prospecting metal-tolerant rhizobia for phytoremediation of mining soils from Morocco using *Anthyllis vulneraria* L. *Environ. Sci. Pollut. Res.* **2015**, *22*, 4500–4512. [CrossRef] [PubMed]
51. Maynaud, G.; Willems, A.; Soussou, S.; Vidal, C.; Mauré, L.; Moulin, L.; Cleyet-Marel, J.-C.; Brunel, B. Molecular and phenotypic characterization of strains nodulating *Anthyllis vulneraria* in mine tailings, and proposal of *Aminobacter anthyllidis* sp. nov., the first definition of Aminobacter as legume-nodulating bacteria. *Syst. Appl. Microbiol.* **2012**, *35*, 65–72. [CrossRef] [PubMed]
52. Gilbert, L.B.; Heeb, P.; Gris, C.; Timmers, T.; Batut, J.; Masson-boivin, C. Experimental evolution of a plant pathogen into a legume symbiont. *PLoS Biol.* **2010**, *8*. [CrossRef]
53. Vamerali, T.; Bandiera, M.; Mosca, G. Field crops for phytoremediation of metal-contaminated land. A review. *Environ. Chem. Lett.* **2010**, *8*, 1–17. [CrossRef]

54. Delgadillo, J.; Lafuente, A.; Doukkali, B.; Redondo-Gómez, S.; Mateos-Naranjo, E.; Caviedes, M.A.; Pajuelo, E.; Rodríguez-Llorente, I.D. Improving legume nodulation and Cu rhizostabilization using a genetically modified rhizobia. *Environ. Technol.* **2015**, *36*, 1237–1245. [CrossRef] [PubMed]
55. Kong, Z.; Glick, B.R.; Duan, J.; Ding, S.; Tian, J.; McConkey, B.J.; Wei, G. Effects of 1-aminocyclopropane-1-carboxylate (ACC) deaminase-overproducing *Sinorhizobium meliloti* on plant growth and copper tolerance of *Medicago lupulina*. *Plant Soil* **2015**, *70*, 5891–5897. [CrossRef]
56. Ghnaya, T.; Mnassri, M.; Ghabriche, R.; Wali, M.; Poschenrieder, C.; Lutts, S.; Abdelly, C. Nodulation by *Sinorhizobium meliloti* originated from a mining soil alleviates Cd toxicity and increases Cd-phytoextraction in *Medicago sativa* L. *Front. Plant Sci.* **2015**, *6*, 1–10. [CrossRef] [PubMed]
57. Zribi, K.; Nouairi, I.; Slama, I.; Talbi-Zribi, O.; Mhadhbi, H. *Medicago sativa—Sinorhizobium meliloti* Symbiosis Promotes the Bioaccumulation of Zinc in Nodulated Roots. *Int. J. Phytoremediat.* **2015**, *17*, 49–55. [CrossRef] [PubMed]
58. Saadani, O.; Fatnassi, I.C.; Chiboub, M.; Abdelkrim, S.; Barhoumi, F.; Jebara, M.; Jebara, S.H. In situ phytostabilisation capacity of three legumes and their associated Plant Growth Promoting Bacteria (PGPBs) in mine tailings of northern Tunisia. *Ecotoxicol. Environ. Saf.* **2016**, *130*, 263–269. [CrossRef] [PubMed]
59. Bianucci, E.; Godoy, A.; Furlan, A.; Peralta, J.M.; Hernández, L.E.; Carpena-Ruiz, R.O.; Castro, S. Arsenic toxicity in soybean alleviated by a symbiotic species of *Bradyrhizobium*. *Symbiosis* **2018**, *74*, 167–176. [CrossRef]
60. Chen, J.; Liu, Y.Q.; Yan, X.W.; Wei, G.H.; Zhang, J.H.; Fang, L.C. *Rhizobium* inoculation enhances copper tolerance by affecting copper uptake and regulating the ascorbate-glutathione cycle and phytochelatin biosynthesis-related gene expression in *Medicago sativa* seedlings. *Ecotoxicol. Environ. Saf.* **2018**, *162*, 312–323. [CrossRef] [PubMed]
61. Zribi, K.; Djébali, N.; Mrabet, M.; Khayat, N.; Smaoui, A.; Mlayah, A.; Aouani, M.E. Physiological responses to cadmium, copper, lead, and zinc of *Sinorhizobium* sp. strains nodulating *Medicago sativa* grown in Tunisian mining soils. *Ann. Microbiol.* **2012**, *62*, 1181–1188. [CrossRef]
62. Fan, M.; Xiao, X.; Guo, Y.; Zhang, J.; Wang, E.; Chen, W.; Lin, Y.; Wei, G. Enhanced phytoremdiation of *Robinia pseudoacacia* in heavy metal-contaminated soils with rhizobia and the associated bacterial community structure and function. *Chemosphere* **2018**, *197*, 729–740. [CrossRef] [PubMed]
63. Van Houdt, R.; Mergeay, M. Genomic context of metal response genes in *Cupriavidus metallidurans* with a focus on strain CH34. In *Metal Response in Cupriavidus Metallidurans*; Springer: Heidelberg, Germany, 2015; pp. 21–44. ISBN 978-3-319-20594-6.
64. Rozycki, T. Von; Nies, Æ.D.H.; Alcaligenes, W.Á.; Ch, Á.Á.H. *Cupriavidus metallidurans*: Evolution of a metal-resistant bacterium. *Anton. Leeuwenhoek* **2008**, *96*, 115–139. [CrossRef] [PubMed]
65. Rubio-Sanz, L.; Prieto, R.I.; Imperial, J.; Palacios, J.M.; Brito, B. Functional and expression analysis of the metal-inducible *dmeRF* system from *Rhizobium leguminosarum* bv. *viciae*. *Appl. Environ. Microbiol.* **2013**, *79*, 6414–6422. [CrossRef] [PubMed]
66. Haney, C.J.; Grass, G.; Franke, S.; Rensing, C. New developments in the understanding of the cation diffusion facilitator family. *J. Ind. Microbiol. Biotechnol.* **2005**, *32*, 215–226. [CrossRef] [PubMed]
67. Zielazinski, E.L.; González-Guerrero, M.; Subramanian, P.; Stemmler, T.L.; Argüello, J.M.; Rosenzweig, A.C. *Sinorhizobium meliloti* Nia is a P1B-5-ATPase expressed in the nodule during plant symbiosis and is involved in Ni and Fe transport. *Metallomics* **2013**, *5*, 1614–1623. [CrossRef] [PubMed]
68. Li, Z.; Lu, M.; Wei, G. An omp gene enhances cell tolerance of Cu(II) in Sinorhizobium meliloti CCNWSX0020. *World J. Microbiol. Biotechnol.* **2013**, *29*, 1655–1660. [CrossRef] [PubMed]
69. Lu, M.; Li, Z.; Liang, J.; Wei, Y.; Rensing, C.; Wei, G. Zinc resistance mechanisms of P 1B-type ATPases in *Sinorhizobium meliloti* CCNWSX0020. *Sci. Rep.* **2016**, *6*, 1–12. [CrossRef]
70. Lu, M.; Jiao, S.; Gao, E.; Song, X.; Li, Z.; Hao, X.; Rensing, C.; Wei, G. Transcriptome response to heavy metals in *Sinorhizobium meliloti* CCNWSX0020 reveals new metal resistance determinants that also promote bioremediation by *Medicago lupulina* in metal contaminated soil. *Appl. Environ. Microbiol.* **2017**, *83*. [CrossRef] [PubMed]
71. Grass, G.; Rensing, C. CueO is a multi-copper oxidase that confers copper tolerance in *Escherichia coli*. *Biochem. Biophys. Res. Commun.* **2001**, *286*, 902–908. [CrossRef] [PubMed]
72. Franke, S.; Grass, G.; Rensing, C.; Nies, D.H. Molecular analysis of the copper-transporting efflux system CusCFBA of *Escherichia coli*. *J. Bacteriol.* **2003**, *185*, 3804–3812. [CrossRef] [PubMed]

73. Long, F.; Su, C.-C.; Lei, H.-T.; Bolla, J.R.; Do, S.V.; Yu, E.W. Structure and mechanism of the tripartite CusCBA heavy-metal efflux complex. *Philos. Trans. R. Soc. B Biol. Sci.* **2012**, *367*, 1047–1058. [CrossRef] [PubMed]
74. Kahn, D.; David, M.; Domergue, O.; Daveran, M.L.; Ghai, J.; Hirsch, P.R.; Batut, J. Rhizobium meliloti fixGHI sequence predicts involvement of a specific cation pump in symbiotic nitrogen fixation. *J. Bacteriol.* **1989**, *171*, 929–939. [CrossRef] [PubMed]
75. Batut, J.; Terzaghi, B.; Gherardi, M.; Huguet, M.; Terzaghi, E.; Garnerone, A.M.; Boistard, P.; Huguet, T. Localization of a symbiotic *fix* region on *Rhizobium meliloti* pSym megaplasmid more than 200 kilobases from the *nod-nif* region. *Mol. Gen. Genet. MGG* **1985**, *199*, 232–239. [CrossRef]
76. Romaniuk, K.; Dziewit, L.; Decewicz, P.; Mielnicki, S.; Radlinska, M.; Drewniak, L. Molecular characterization of the pSinB plasmid of the arsenite oxidizing, metallotolerant *Sinorhizobium* sp. M14—Insight into the heavy metal resistome of sinorhizobial extrachromosomal replicons. *FEMS Microbiol. Ecol.* **2017**, *93*. [CrossRef] [PubMed]
77. Brokx, S.J.; Rothery, R.A.; Zhang, G.; Ng, D.P.; Weiner, J.H. Characterization of an *Escherichia coli* sulfite oxidase homologue reveals the role of a conserved active site cysteine in assembly and function. *Biochemistry* **2005**, *44*, 10339–10348. [CrossRef] [PubMed]
78. Wilson, J.J.; Kappler, U. Sulfite oxidation in *Sinorhizobium meliloti*. *Biochim. Biophys. Acta (BBA) Bioenerg.* **2009**, *1787*, 1516–1525. [CrossRef] [PubMed]
79. Porter, S.S.; Chang, P.L.; Conow, C.A.; Dunham, J.P.; Friesen, M.L. Association mapping reveals novel serpentine adaptation gene clusters in a population of symbiotic *Mesorhizobium*. *ISME J.* **2017**, *11*, 248–262. [CrossRef] [PubMed]
80. Li, Z.; Ma, Z.; Hao, X.; Rensing, C.; Wei, G. Genes conferring copper resistance in *Sinorhizobium meliloti* CCNWSX0020 also promote the growth of *Medicago lupulina* in copper-contaminated soil. *Appl. Environ. Microbiol.* **2014**, *80*, 1961–1971. [CrossRef] [PubMed]
81. Hao, X.; Xie, P.; Zhu, Y.-G.; Taghavi, S.; Wei, G.; Rensing, C. Copper tolerance mechanisms of *Mesorhizobium amorphae* and its role in aiding phytostabilization by *Robinia pseudoacacia* in copper contaminated soil. *Environ. Sci. Technol.* **2015**, *49*, 2328–2340. [CrossRef] [PubMed]
82. Maynaud, G.; Brunel, B.; Yashiro, E.; Mergeay, M.; Cleyet-Marel, J.C.; Le Quéré, A. CadA of *Mesorhizobium metallidurans* isolated from a zinc-rich mining soil is a PIB-2-type ATPase involved in cadmium and zinc resistance. *Res. Microbiol.* **2014**, *165*, 175–189. [CrossRef] [PubMed]
83. Rubio-Sanz, L.; Brito, B.; Palacios, J. Analysis of metal tolerance in *Rhizobium leguminosarum* strains isolated from an ultramafic soil. *FEMS Microbiol. Lett.* **2018**, *365*, fny010. [CrossRef] [PubMed]
84. Pérez-Palacios, P.; Romero-Aguilar, A.; Delgadillo, J.; Doukkali, B.; Caviedes, M.A.; Rodríguez-Llorente, I.D.; Pajuelo, E. Double genetically modified symbiotic system for improved Cu phytostabilization in legume roots. *Environ. Sci. Pollut. Res.* **2017**, *24*, 14910–14923. [CrossRef] [PubMed]
85. Zhang, J.; Xu, Y.; Cao, T.; Chen, J.; Rosen, B.P.; Zhao, F.-J. Arsenic methylation by a genetically engineered Rhizobium-legume symbiont. *Plant Soil* **2017**, *416*, 259–269. [CrossRef] [PubMed]
86. Kong, Z.; Mohamad, O.A.; Deng, Z.; Liu, X.; Glick, B.R.; Wei, G. Rhizobial symbiosis effect on the growth, metal uptake, and antioxidant responses of *Medicago lupulina* under copper stress. *Environ. Sci. Pollut. Res.* **2015**, *22*, 12479–12489. [CrossRef] [PubMed]
87. Rajkumar, M.; Ae, N.; Prasad, M.N.V.; Freitas, H. Potential of siderophore-producing bacteria for improving heavy metal phytoextraction. *Trends Biotechnol.* **2010**, *28*, 142–149. [CrossRef] [PubMed]
88. Valls, M.; De Lorenzo, V. Exploiting the genetic and biochemical capacities of bacteria for the remediation of heavy metal pollution. *FEMS Microbiol. Rev.* **2002**, *26*, 327–338. [CrossRef] [PubMed]
89. Nies, D.H. Efflux-mediated heavy metal resistance in prokaryotes. *FEMS Microbiol. Rev.* **2003**, *27*, 313–339. [CrossRef]
90. Drewniak, L.; Dziewit, L.; Ciezkowska, M.; Gawor, J.; Gromadka, R.; Sklodowska, A. Structural and functional genomics of plasmid pSinA of *Sinorhizobium* sp. M14 encoding genes for the arsenite oxidation and arsenic resistance. *J. Biotechnol.* **2013**, *164*, 479–488. [CrossRef] [PubMed]
91. Drewniak, L.; Matlakowska, R.; Sklodowska, A. Arsenite and arsenate metabolism of *Sinorhizobium* sp. M14 living in the extreme environment of the Zloty Stok gold mine. *Geomicrobiol. J.* **2008**, *25*, 363–370. [CrossRef]
92. DiCenzo, G.C.; Finan, T.M. The divided bacterial genome: Structure, function, and evolution. *Microbiol. Mol. Biol. Rev.* **2017**, *81*, e00019-17. [CrossRef] [PubMed]

93. DiCenzo, G.C.; Benedict, A.B.; Fondi, M.; Walker, G.C.; Finan, T.M.; Mengoni, A.; Griffitts, J.S. Robustness encoded across essential and accessory replicons of the ecologically versatile bacterium *Sinorhizobium meliloti*. *PLoS Genet.* **2018**, *14*, e1007357. [CrossRef] [PubMed]
94. DiCenzo, G.C.; Wellappili, D.; Golding, G.B.; Finan, T.M. Inter-replicon gene flow contributes to transcriptional integration in the *Sinorhizobium meliloti* multipartite genome. *G3 Genes Genomes Genet.* **2018**, *8*, 1711–1720. [CrossRef] [PubMed]
95. Landeta, C.; Dávalos, A.; Cevallos, M.Á.; Geiger, O.; Brom, S.; Romero, D. Plasmids with a chromosome-like role in Rhizobia. *J. Bacteriol.* **2011**, *193*, 1317–1326. [CrossRef] [PubMed]
96. DiCenzo, G.C.; Checcucci, A.; Bazzicalupo, M.; Mengoni, A.; Viti, C.; Dziewit, L.; Finan, T.M.; Galardini, M.; Fondi, M. Metabolic modelling reveals the specialization of secondary replicons for niche adaptation in *Sinorhizobium meliloti*. *Nat. Commun.* **2016**, *7*, 12219. [CrossRef] [PubMed]
97. Galardini, M.; Brilli, M.; Spini, G.; Rossi, M.; Roncaglia, B.; Bani, A.; Chiancianesi, M.; Moretto, M.; Engelen, K.; Bacci, G.; et al. Evolution of intra-specific regulatory networks in a multipartite bacterial genome. *PLoS Comput. Biol.* **2015**, *11*, e1004478. [CrossRef] [PubMed]
98. Galardini, M.; Pini, F.; Bazzicalupo, M.; Biondi, E.G.; Mengoni, A. Replicon-dependent bacterial genome evolution: The case of *Sinorhizobium meliloti*. *Mol. Biol.* **2013**, *5*, 542–558. [CrossRef] [PubMed]
99. Checcucci, A.; diCenzo, G.C.; Ghini, V.; Bazzicalupo, M.; Beker, A.; Decorosi, F.; Dohlemann, J.; Fagorzi, C.; Finan, T.M.; Fondi, M.; et al. Creation and multi-omics characterization of a genomically hybrid strain in the nitrogen-fixing symbiotic bacterium *Sinorhizobium meliloti*. *bioRxiv* **2018**. [CrossRef]
100. Checcucci, A.; DiCenzo, G.C.; Bazzicalupo, M.; Mengoni, A. Trade, diplomacy, and warfare: The quest for elite rhizobia inoculant strains. *Front. Microbiol.* **2017**, *8*, 2207. [CrossRef] [PubMed]
101. Van Opijnen, T.; Camilli, A. Transposon insertion sequencing: A new tool for systems-level analysis of microorganisms. *Nat. Rev. Microbiol.* **2013**, *11*, 435–442. [CrossRef] [PubMed]

© 2018 by the authors. Licensee MDPI, Basel, Switzerland. This article is an open access article distributed under the terms and conditions of the Creative Commons Attribution (CC BY) license (http://creativecommons.org/licenses/by/4.0/).

Article

Genomic and Biotechnological Characterization of the Heavy-Metal Resistant, Arsenic-Oxidizing Bacterium *Ensifer* sp. M14

George C diCenzo [1,*,†], Klaudia Debiec [2,*,†], Jan Krzysztoforski [3], Witold Uhrynowski [2], Alessio Mengoni [1], Camilla Fagorzi [1], Adrian Gorecki [4], Lukasz Dziewit [4], Tomasz Bajda [5], Grzegorz Rzepa [5] and Lukasz Drewniak [2]

1. Laboratory of Microbial Genetics, Department of Biology, University of Florence, via Madonna del Piano 6, 50019 Sesto Fiorentino, Italy; alessio.mengoni@unifi.it (A.M.); camilla.fagorzi@unifi.it (C.F.)
2. Laboratory of Environmental Pollution Analysis, Faculty of Biology, University of Warsaw, Miecznikowa 1, 02-096 Warsaw, Poland; w.uhrynowski@biol.uw.edu.pl (W.U.); ldrewniak@biol.uw.edu.pl (L.Dr.)
3. Faculty of Chemical and Process Engineering, Warsaw University of Technology, Warynskiego 1, 00-645 Warsaw, Poland; jan.krzysztoforski@pw.edu.pl
4. Department of Bacterial Genetics, Institute of Microbiology, Faculty of Biology, University of Warsaw, Miecznikowa 1, 02-096 Warsaw, Poland; agorecki@biol.uw.edu.pl (A.G.); ldziewit@biol.uw.edu.pl (L.Dz.)
5. Department of Mineralogy, Petrography and Geochemistry, Faculty of Geology, Geophysics and Environmental Protection, AGH University of Science and Technology, Mickiewicza 30, 30-059 Krakow, Poland; bajda@agh.edu.pl (T.B.); gprzepa@cyf-kr.edu.pl (G.R.)
* Correspondence: georgecolin.dicenzo@unifi.it (G.C.d.); k.debiec@biol.uw.edu.pl (K.D.)
† These authors contributed equally to this paper.

Received: 14 June 2018; Accepted: 25 July 2018; Published: 27 July 2018

Abstract: *Ensifer* (*Sinorhizobium*) sp. M14 is an efficient arsenic-oxidizing bacterium (AOB) that displays high resistance to numerous metals and various stressors. Here, we report the draft genome sequence and genome-guided characterization of *Ensifer* sp. M14, and we describe a pilot-scale installation applying the M14 strain for remediation of arsenic-contaminated waters. The M14 genome contains 6874 protein coding sequences, including hundreds not found in related strains. Nearly all unique genes that are associated with metal resistance and arsenic oxidation are localized within the pSinA and pSinB megaplasmids. Comparative genomics revealed that multiple copies of high-affinity phosphate transport systems are common in AOBs, possibly as an As-resistance mechanism. Genome and antibiotic sensitivity analyses further suggested that the use of *Ensifer* sp. M14 in biotechnology does not pose serious biosafety risks. Therefore, a novel two-stage installation for remediation of arsenic-contaminated waters was developed. It consists of a microbiological module, where M14 oxidizes As(III) to As(V) ion, followed by an adsorption module for As(V) removal using granulated bog iron ores. During a 40-day pilot-scale test in an abandoned gold mine in Zloty Stok (Poland), water leaving the microbiological module generally contained trace amounts of As(III), and dramatic decreases in total arsenic concentrations were observed after passage through the adsorption module. These results demonstrate the usefulness of *Ensifer* sp. M14 in arsenic removal performed in environmental settings.

Keywords: *Ensifer* (*Sinorhizobium*) sp. M14; arsenic-oxidizing bacteria; heavy metal resistance; draft genome sequence; comparative genomic analysis; biosafety; biotechnology for arsenic removal; adsorption; water treatment; in situ (bio)remediation

1. Introduction

The development and implementation of bioremediation technologies based on bioaugmentation requires the selection of appropriate microbial strains. A basic requirement of strains used as bioaugmentation agents is their ability to survive in the environment into which they are introduced. Thus, such strains are usually characterized by high tolerance to heavy metals [1,2], resistance and ability to use organic (sometimes toxic) compounds [3,4], resistance to antibiotics [5], and an ability to thrive in the presence of local bacteriophages and microorganisms. Another important feature of strains used in bioaugmentation is their ability to perform effective transformation of the particular compound under changing environmental conditions (e.g., temperature, humidity, and pH). This is always the critical limitation, as many strains effective under laboratory conditions are, in fact, ineffective in field applications. Microorganisms suitable in bioremediation should maintain their activity in various seasons and under variable substrate inflow. A very important factor influencing the decision to apply a given microorganism in practice is also its interaction with the environment [6]. Strains that contribute to the uncontrolled release of contaminants, dissemination of antibiotic resistance genes, or disrupt the functioning of the ecosystem (e.g., by eliminating key microorganisms) should not be applied in open (uncontrolled) usage.

In this study, we provide a detailed characterization of *Sinorhizobium* sp. M14 (renamed here to *Ensifer* sp. M14 due to its phylogenetic positioning within the *Ensifer* clade), which is a strain with high potential to be used in bioremediation technologies for the removal of arsenic from contaminated waters and wastewaters. *Ensifer* sp. M14 is a psychrotolerant strain that was isolated from the microbial mats present in the arsenic-rich bottom sediments of an abandoned gold mine in Zloty Stok (Poland) [7]. The arsenic concentration in the mine waters reaches ~6 mg L^{-1}, while in the microbial mats the level of accumulated arsenic is close to 20 g L^{-1} [8]. Previous physiological studies showed that *Ensifer* sp. M14 tolerates extremely high concentrations of arsenate [As(V)—up to 250 mM] and arsenite [As(III)—up to 20 mM], and is able to oxidize As(III) both chemolithoautotrophically [using arsenite or arsenopyrite (FeAsS) as a source of energy] and heterotrophically [7]. Batch experiments performed under various conditions of pH, temperature, and arsenic concentration confirmed the high adaptive potential of *Ensifer* sp. M14 [9]. The strain was capable of intensive growth and efficient biooxidation in a wide range of conditions, including low temperature [As(III) oxidation rate = 0.533 mg L^{-1} h^{-1} at 10 °C]. Continuous flow experiments under environment-like conditions (2 L flow bioreactor) showed that *Ensifer* sp. M14 efficiently transforms As(III) into As(V) [24 h of residence time was sufficient to oxidize 5 mg L^{-1} of As(III)], but its activity depended mainly on the retention time in the bioreactor, which may be accelerated by stimulation with yeast extract as a source of nutrients [9].

Analysis of the extrachromosomal replicons of *Ensifer* sp. M14 revealed that its arsenic metabolism properties are linked with the presence of the mega-sized plasmid pSinA (109 kbp) [10]. The loss of the pSinA plasmid from *Ensifer* sp. M14 cells (using a target-oriented replicon curing technique [11]) eliminated the ability to oxidize As(III), and caused deficiencies in resistance to arsenic and heavy metals (Cd, Co, Zn, and Hg). In turn, the introduction of this plasmid into other representatives of the *Alphaproteobacteria* showed that cells with pSinA acquired the ability to oxidize arsenite and exhibited higher tolerance to arsenite than their parental, pSinA-less, wild-type strains. Horizontal transfer of arsenic metabolism genes by *Ensifer* sp. M14 was also confirmed in microcosm experiments [10]. The plasmid pSinA was successfully transferred via conjugation into indigenous bacteria of *Alpha*- and *Gammaproteobacteria* classes from the microbial community of As-contaminated soils. Transconjugants carrying plasmid pSinA expressed arsenite oxidase and stably maintained pSinA in their cells after approximately 60 generations of growth under nonselective conditions [10].

The second mega-sized replicon of *Ensifer* sp. M14—plasmid pSinB (300 kbp)—also plays an important role in the adaptation of the host to the mine environment. Structural and functional analysis of this plasmid showed that it carries gene clusters involved in heavy metals resistance. Among these are genes encoding efflux pumps, permeases, transporters, and copper oxidases, which are responsible for resistance to arsenic, cobalt, zinc, cadmium, iron, mercury, nickel, copper, and silver [12].

In this paper, we obtained a draft genomic sequence of *Ensifer* sp. M14 and performed complex genome-guided characterization of this bacterium. Special considerations were given to (i) determination of the metabolism of phosphate, sulfur, iron, and one-carbon substrates, and (ii) investigation of the biosafety of *Ensifer* sp. M14 in the context of its release to the environment (e.g., determination of the presence of virulence and antibiotic resistance genes). These analyses revealed hints about the potential application of this strain in biotechnological applications; for example, the ability of it to survive environmental stresses, and whether it is likely to pose a safety risk. As the genomic analyses were consistent with *Ensifer* sp. M14 having potential application in biotechnology, we performed a large-scale simulation of the usage of M14 in the biological and chemical removal of arsenic from contaminated waters. The results support that the developed low-cost approach is an efficient method for the removal of arsenic from contaminated water.

2. Materials and Methods

2.1. Genome Sequencing, Assembly, and Annotation

Ensifer sp. M14 (available on request from the authors) was grown at 30 °C to stationary phase in TY medium (5 g L^{-1} tryptone, 3 g L^{-1} yeast extract, and 0.4 g L^{-1} calcium chloride). Genomic DNA was isolated from the culture using a cetyltrimethylammonium bromide (CTAB) method [13] modified for bacterial DNA isolation as described by the Joint Genome Institute [14]. Sequencing was performed at IGATech (Udine, Italy) using an Illumina HiSeq2500 instrument with 125-bp paired-end reads. Two independent sequencing runs were performed. Reads were assembled into scaffolds using SPAdes v3.9.0 [15,16]. The scaffolds returned by SPAdes were parsed to remove those with less than 10× coverage or with a length below 200 nucleotides. Using FastANI [17], one-way average nucleotide identity (ANI) of the *Ensifer* sp. M14 assembly was calculated against the 887 alpha-proteobacterial genomes available through the National Center for Biotechnological Information (NCBI) with an assembly level of 'complete' or 'chromosome'. The 10 genomes most closely related to *Ensifer* sp. M14 were identified on the basis of the ANI results. These 10 genomes, together with the complete pSinA and pSinB plasmid sequences [10,12], were used as reference genomes for further scaffolding of the assembly using MeDuSa [18]. The *Ensifer* sp. M14 assembly was then annotated using prokka version 1.12-beta [19], annotating coding regions with Prodigal [20], tRNA with Aragon [21], rRNA with Barrnap (github.com/tseemann/barrnap), and ncRNA with Infernal [22] and Rfam [23]. The predicted coding sequences were associated with Cluster of Orthologous Genes (COG) categories, Gene Ontology (GO) terms, Kyoto Encyclopedia of Genes and Genomes (KEGG) pathway terms, and eggNOG annotations using eggNOG-mapper version 0.99.2-3-g41823b2 [24]. The assembly was deposited to NCBI with the GenBank accession QJNR00000000 (the version described in this paper is version QJNR01000000) and the BioSample accession SAMN09254189.

2.2. Phylogenetic Analysis

Initially, all 133 *Sinorhizobium*/*Ensifer* genomes available through NCBI, regardless of assembly level, were downloaded. FastANI [17] was used to calculate one-way ANI values between *Ensifer* sp. M14 and each of the 133 downloaded genomes. Only the strains meeting at least one of the following two requirements were kept for further analyses: (i) had a genome assembly level of 'complete' or 'chromosome', or (ii) had an ANI value of at least 85% compared to *Ensifer* sp. M14. This resulted in a final set of 46 strains, when including *Ensifer* sp. M14.

The pangenome of the 46 strains was calculated using Roary version 3.11.3 [25], as described below, following re-annotation with prokka version 1.12-beta [19]. Included in the Roary output was a concatenated nucleotide alignment of the 1652 core genes, each individually aligned with PRANK [26]. The core gene alignment was used to build a maximum likelihood phylogeny with RAxML version 8.2.9 [27] using the following command:

raxmlHPC-HYBRID-SSE3-T 5-s input.fasta-N autoMRE-n output-f a-p 12345-x 12345-m GTRCAT.

The final tree is the bootstrap best tree following 50 bootstrap replicates, and was visualized using the online iTOL (Interactive Tree of Life) webserver [28].

Strains were grouped into putative species on the basis of ANI and average amino acid identity (AAI) values, using thresholds of 96% for both measures. Groupings for ANI were the same at thresholds of 96% and 94%. Pairwise ANI values were calculated between each strain using FastANI [17], and the values in both directions were averaged. The CompareM workflow (github.com/dparks1134/CompareM) was used for calculating the AAI values. In the CompareM workflow, orthologous proteins were first identified using DIAMOND with the sensitive setting [29], and thresholds of 40% identity over 70% the length of the protein and a maximum e-value of $1e^{-12}$ were applied, as these are the thresholds used in the myTaxa program [30].

2.3. Sinorhizobium/Ensifer Pangenome Calculation

All 46 strains included in the phylogenetic analyses were reannotated using prokka version 1.12-beta [19], to ensure consistent annotation. The pangenome of the 46 reannotated strains was then determined with Roary version 3.11.3 [25], using an amino acid identity threshold of 80% and the following command:

roary-p 20-f Output-e-I 80-g 100,000 Input/*.gff.

For comparison of the gene content of *Ensifer* sp. M14, *Ensifer* sp. A49, *Ensifer adhaerens* OV14, and *Ensifer adhaerens* Casida A, the data was extracted from the full 46-strain pangenome. The complete gene presence/absence output from Roary is provided as Data Set S1. Several short proteins of *Ensifer* sp. M14 were not present in the output of the Roary analysis; these proteins were not considered when identifying unique genes.

2.4. Comparative Genomics of Arsenic Oxidizing Bacteria

The genomes of *Agrobacterium tumefaciens* 5A [31], *Agrobacterium tumefaciens* Ach5 [32], *Ensifer adhaerens* OV14 [33], *Neorhizobium galegae* HAMBI 540 [34], and *Rhizobium* sp. NT-26 [35] were downloaded from NCBI GenBank and reannotated using prokka, as described above for *Ensifer* sp. M14. The GenBank files of the re-annotated genomes, and the *Ensifer* sp. M14 genome, were uploaded to the KBase webserver [36], and OrthoMCL [37] was run on the KBase server using an e-value threshold of $1e^{-12}$. Identification of phosphate transport and arsenic resistance genes in other bacterial genomes (*Achromobacter arsenitoxydans* SY8 [38], *Herminiimonas arsenicoxydans* ULPAs1 [39], and *Pseudomonas stutzeri* TS44 [40]) was accomplished by manually searching the GenBank file of the RefSeq annotated genomes [41].

2.5. Identification of Prophage Loci

PhiSpy version 3.2 [42], implemented in Python, was used to predict phage genes. The *Ensifer* sp. M14 GenBank file produced with prokka was converted to SEED format using the genbank_to_seed.py script. The converted file was then used as input for the PhiSpy.py script, using the generic test set for training.

2.6. Identification of Putative Antibiotic Resistance Genes

To identify putative antibiotic resistance genes, the Resistance Gene Identifier (RGI) in the Comprehensive Antibiotic Resistance Database (CARD) software was used [43]. Hits showing at least 50% identity with the reference protein were considered significant. Each hit was verified manually using BLASTp analysis.

2.7. Analysis of the Antimicrobial Susceptibility Patterns

To determine the antimicrobial susceptibility patterns of *Ensifer* sp. M14, minimum inhibitory concentrations (MICs) of 11 antimicrobial agents were assessed using Etest™ (Liofilchem, Roseto degli Abruzzi, Italy). The analysis was conducted according to the European Committee on Antimicrobial Susceptibility Testing (EUCAST) recommendations [44]. The following antibiotics (selected based on the bioinformatic analyses that identified putative antibiotic resistance genes) were used:

(i) aminoglycosides–gentamicin (GN; concentration of antibiotic: 0.064–1024 µg mL^{-1} Roseto degli Abruzzi[1]); (ii) β-lactams (penicillin derivatives)–ampicillin (AMP; 0.016–256 µg mL^{-1}); (iii) β-lactams (cephalosporins)–cefixime (CFM; 0.016–256 µg mL^{-1}); (iv) β-lactams (cephalosporins)–cefotaxime (CTX; 0.016–256 µg mL^{-1}); (v) β-lactams (cephalosporins)–ceftriaxone (CRO; 0.016–256 µg mL^{-1}); (vi) fluroquinolones–ciprofloxacin (CIP; 0.002–32 µg mL^{-1}); (vii) fluroquinolones–moxifloxacin (MXF; 0.002–32 µg mL^{-1}); (viii) phenicols–chloramphenicol (C; 0.016–256 µg mL^{-1}); (ix) rifamicyns–rifampicin (RD; 0.016–256 µg mL^{-1}); (x) sulfonamides–trimethoprim (TM; 0.002–32 µg mL^{-1}); and (xi) tetracyclines–tetracycline (TE; 0.016–256 µg mL^{-1}). The susceptibility testing was performed at 30 °C for 20 h. After incubation, plates were photographed and MICs were defined. Antimicrobial susceptibility data were interpreted according to the EUCAST breakpoint table version 8.0 [45].

2.8. Search for Symbiotic Proteins

A custom pipeline based on the use of hidden Markov models (HMM) was used to search the proteomes of all 46 *Sinorhizobium/Ensifer* strains for the presence of the nodulation proteins NodA, NodB, and NodC, as well as for the nitrogenase proteins NifH, NifD, and NifK. This pipeline is dependent on HMMER version 3.1b2 [46], and the complete Pfam-A version 31.0 (16,712 HMMs) and TIGERFAM version 15.0 (4488 HMMs) databases [47,48]. After downloading the HMM databases, hmmconvert was used to ensure consistent formatting. The two databases were combined into a single HMM database, and then converted into a searchable database with hmmpress. Additionally, the HMM seed alignments for NodA (TIGR04245), NodB (TIGR04243), NodC (TIGR04242), NifH (TIGR01287), NifD (TIGR01282), and NifK (TIGR01286) were downloaded from the TIGRFAM database [47].

For each HMM seed alignment, a HMM was built using hmmbuild, and the output was then searched against the complete set of *Sinorhizobium/Ensifer* proteins using hmmsearch. The output was parsed, and the amino acid sequences for each of the hits (regardless of e-value) were collected. Each set of sequences were then searched against the combined HMM database using hmmscan, and the output parsed to identify the top scoring HMM hit for each query protein. Proteins were annotated as follows: NodA if the top hit was TIGR04245 (TIGRFAM) or NodA (Pfam); NodB if the top hit was TIGR04243 (TIGRFAM); NodC if the top hit was TIGR04242 (TIGRFAM); NifH if the top hit was TIGR01287 (TIGRFAM) or Fer4_NifH (Pfam); NifD if the top hit was TIGR01282 (TIGRFAM), TIGR01860 (TIGRFAM), or TIGR01861 (TIGRFAM); NifK if the top hit was TIGR02932 (TIGRFAM), TIGR02931 (TIGRFAM), or TIGR01286 (TIGRFAM).

2.9. Cluster of Orthologous Genes Functional Annotation

Proteomes were annotated with COG functional categories using eggNOG-mapper version 0.99.2-3-g41823b2 [24]. The output of eggNOG-mapper was parsed with a custom Perl script to count the percentage of proteins annotated with each functional category. Fisher exact tests, performed using MATLAB R2016b (www.mathworks.com), were performed to identify statistically significant differences ($p < 0.05$) between *Ensifer* sp. M14 and the other strains.

2.10. In Silico Metabolic Reconstruction and Constraint-Based Modelling

Metabolic reconstruction steps and constraint-based metabolic modeling were performed in MATLAB 2017a (Mathworks, Natick, MA, USA), using the Gurobi 7.0.2 solver (gurobi.com), SBMLToolbox 4.1.0 [49], libSBML 5.13.0 [50], and scripts from the COBRA Toolbox [51] and the Tn-Core Toolbox [52]. The ability of the *Ensifer* sp. M14 model to grow when individually provided with 163 carbon sources was tested using flux balance analysis (FBA) as implemented in the 'optimizeCbModel' function of the COBRA Toolbox.

An initial draft metabolic reconstruction was prepared using the online KBase webserver [36]. The *Ensifer* sp. M14 genome was uploaded and re-annotated with RAST functions using the 'annotate microbial genome' function. The re-annotated genome was used to build a draft model with the 'build metabolic model' function, performing gap-filling on a glucose minimal medium, and with automatic biomass template selection. This reconstruction was downloaded in SBML format, and then imported into MATLAB as a COBRA formatted metabolic model for further manipulation. After removing duplicate genes from the gene list and updating the gene-reaction rules appropriately, the model was expanded based on the reaction content of the curated iGD1575 and iGD726 metabolic reconstructions of the closely related species *Sinorhizobium meliloti* [53,54]. First, a BLAST bidirectional best hit approach was used to identify putative orthologs (at least 70% identity over at least 70% the protein length) between *S. meliloti* Rm1021 and *Ensifer* sp. M14. All *S. meliloti* genes without a putative ortholog in *Ensifer* sp. M14 were deleted from the iGD1575 and iGD726 models, and the constrained reactions removed. Next, the reactions of iGD726 and the draft *Ensifer* sp. M14 model were compared based on their equations, and all reactions unique to iGD726 were identified and transferred to the *Ensifer* sp. M14 model. Exceptions were iGD726 reactions that differed from a reaction in the *Ensifer* sp. M14 model only in the presence/absence of a proton or in metabolite stoichiometry. This process was then repeated, transferring the unique reactions of iGD1575 to the partially expanded model. When transferring reactions, associated genes were also transferred and changed to the name of the *Ensifer* sp. M14 orthologs. Following the expansions, all reactions producing dead-end metabolites were iteratively removed from the model. The final model contained 1491 genes, 1561 reactions, and 1105 metabolites, and is available in Data Set S2.

2.11. Prediction of Secondary Metabolism

Loci encoding secondary metabolic pathways were predicted in the *Ensifer* sp. M14 genome using the antiSMASH webserver [55]. The *Ensifer* sp. M14 GenBank file was uploaded to the bacterial version of antiSMASH, and the analysis was run with all options selected with default parameters.

2.12. Construction of a Pilot-Scale Installation for Arsenic Bioremediation

A pilot-scale installation for the removal of arsenic from contaminated waters was developed. The installation was operated using water from a dewatering system of a former gold mine located in the Zloty Stok area (SW Poland), which is highly polluted with arsenic. The total arsenic concentration, arsenic speciation, as well as detailed chemical and physical characteristics of the water are presented elsewhere [56]. The installation consisted of two modules: the microbiological module and the adsorption module (Figure 1).

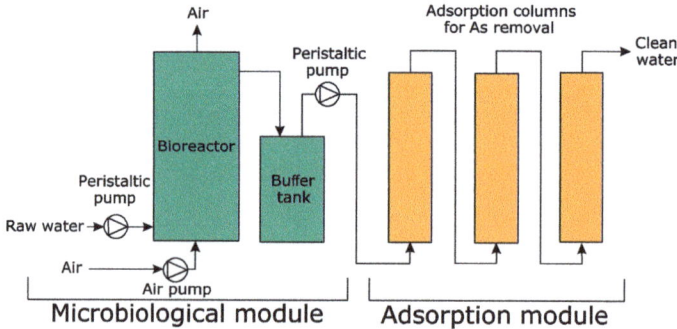

Figure 1. The pilot-scale installation used for remediation of arsenic contaminated water. The image is a schematic representation of the pilot-scale installation developed as part of this work. Both the microbiological and adsorption modules are shown.

The microbiological module was based on the activity of *Ensifer* sp. M14, which was used as an arsenite biooxidizer. This module included a 200 L bioreactor with an electric heater. The contaminated water flowing out from the gold mine was fed into the bioreactor through a pressure reducer and a peristaltic pump at a volume flow rate of 8.33 L h^{-1}, corresponding to a residence time of 24 h in the bioreactor. Outflow of the water occurred as overflow in the upper part of the bioreactor. To increase the effectiveness of the arsenite biooxidition, the bioreactor was equipped with an additional aeration system that consisted of an air pump producing compressed air. The additional aeration system was included in our previous study and showed that the arsenite oxidation efficiency of *Ensifer* sp. M14 is higher in the presence of additional aeration during continuous culturing [9]. Moreover, yeast extract was added to the bioreactor as a source of vitamins (growth supplements). Fifty grams of powdered yeast extract (Sigma-Aldrich, St. Louis, MO, USA) was added to the bioreactor twice a week. This was done as we previously observed that the presence of yeast extract led to an increase in the growth and efficiency of arsenite biooxidation of *Ensifer* sp. M14 during continuous culturing [9]. This relationship was also confirmed in other papers [10,57]. The supply of air also contributed to the mixing of the bioreactor content. The bioreactor was equipped with a multifunctional electrode dedicated to controlling the chemical and physical parameters of water, specifically, to monitor pH, redox potential, and temperature (Hydrolab HL4, OTT Hydromet, Kempten, Germany). The water leaving the bioreactor was fed into a 60 L buffer tank, which functioned as the connecting element between the bioreactor and the adsorption module. The inclusion of the buffer tank helped maintain a constant water level in the adsorption columns and ensured a constant flow of water from the bioreactor to the adsorption columns.

The adsorption module consisted of three columns (17 L volume each) filled with granulated bog iron ores (about 15 kg per column) and connected in series (Figure 1). The detailed chemical and physical parameters, chemical composition, and stability of the adsorbent were presented previously [56]. Contaminated water from the buffer tank (after passing through the microbiological module) was fed into the first column using a second peristaltic pump at a volume flow rate of 8.33 L h^{-1}, which corresponded to approximately one hour of residence time per column.

The installation was also equipped with a process control system (operated at the location of the pilot plant or remotely via a Global System for Mobile Communications (GSM)) that monitored and controlled key process parameters including the volume flow rate of the water, the water temperature at the inlet, in the bioreactor, and at the outlet of the pilot plant, as well as the ambient temperature.

2.13. Installation Start-Up

Scale-up of the installation (from laboratory scale to pilot scale) required the development of procedures for successful start-up based on the results of our previous study [9]. The first step of the start-up of the microbiological module was inoculation of the bioreactor with an appropriate amount of *Ensifer* sp. M14. The bioreactor filled with arsenic contaminated water was inoculated with 200 mL of a highly concentrated overnight culture of *Ensifer* sp. M14 suspended in 0.85% NaCl solution. The initial OD$_{600}$ in the bioreactor was 0.01. In earlier experiments, it was determined that a starting cell density of 10^8 CFU mL^{-1} (which corresponds to an OD$_{600}$ of 0.1) is required for the installation to work properly [9]. To increase the density of the *Ensifer* sp. M14, the water in the bioreactor was supplemented with powdered yeast extract to a final concentration of 0.04%. Additionally, aeration was applied. Finally, the temperature of the water was increased (from 10 to 22 °C) with the use of an electric heater placed in the bioreactor. Application of all these treatments led to an OD$_{600}$ value of 0.1 within 24 h.

Start-up procedures related to the adsorption module mainly concerned the preparation of the adsorbent for its usage. After filling the columns with granulated bog iron ores, it was necessary to condition the adsorbent (rinsing the adsorbent with the tap water without arsenic) to remove all the loosely bound fractions.

2.14. Biological and Chemical Analyses

Arsenic speciation was investigated with the use of ion chromatography on an IonPac AS18 (2 mm, Dionex, Lübeck, Germany) column on an ICS Dionex 3000 (Lübeck, Germany) instrument equipped with an ASRS® 2 mm suppressor, which was coupled to a ZQ 2000 mass spectrometer via an electrospray source (Waters, Milford, MA, USA) according to the method described by Debiec et al. [9]. In the adsorption module, the total arsenic concentration was investigated. Total arsenic concentration was measured using a Graphite Furnace Atomic Absorption Spectrometry (GFAAS; AA Solaar M6 Spectrometer, TJA Solutions, Waltham, MA, USA). Arsenic standard solutions (Merck, Darmstadt, Germany) were prepared in 3% HNO_3. The pH and redox potential were measured only in the microbiological module. Samples of raw water, water from the bioreactor, as well as water at the inflow and outflow of each adsorption column were collected once a day during the first 8 days, and then three times a week up to day 40. Samples taken from the bioreactor were stored at −20 °C, while samples collected from the adsorption module were stored at 4 °C. This experiment was repeated twice.

3. Results and Discussion

3.1. Sequencing of the Ensifer sp. M14 Genome

The draft genome sequence of *Ensifer* sp. M14 was obtained as described in the Materials and Methods, and the general genomic features are described in Table 1.

Table 1. Features of the *Ensifer* sp. M14 genome assembly.

Length	7,345,249 bp
G + C content	61.47%
CDS	6874
rRNA	3
tRNA	53
Miscellaneous RNA	33
Scaffolds	45
Scaffold N50 (L50)	4400,487 (1)
CDS with COG terms *,†	64.00%
CDS with GO terms *	28.70%
CDS with KEGG pathway terms *	35.50%
CDS with eggNOG annotations *,¥	80.50%
CDS with no similarity *	9.40%

* As determined using eggnog-mapper [24]. Those genes not returned in the eggNOG-mapper output were said to have no similarity; † Excluding those annotated with COG category S (unknown function); ¥ Excluding those annotated as protein/domain of unknown/uncharacterized function. CDS (Coding Sequences); COG (Cluster of Orthologous Genes); KEGG (Kyoto Encyclopedia of Genes and Genomes); GO (Gene Ontology).

The assembly consists of 7,345,249 bp spread over 45 scaffolds at an average coverage of 118×. Of the 45 scaffolds, 12 are over 40 kbp in size and account for 98.7% of the assembly. Based on similarity searches of the scaffolds, previous plasmid profiling of *Ensifer* sp. M14 [10,12], and the finished genomes of related strains [33,58], we predict that the *Ensifer* sp. M14 genome consists of one chromosome (at least 4.4 Mbp in size), two additional large replicons (chromids and/or large megaplasmids, at least 1.6 Mbp and 0.6 Mbp in size), and the two previously reported smaller megaplasmids (pSinA and pSinB, 109 kbp and 300 kbp, respectively, based on previous papers [10,12]). A total of 6874 coding sequences were predicted, which is more than the 6218 predicted in *S. meliloti* Rm1021 and the 6641 of *E. adhaerens* Casida A, but less than the 7033 predicted in *E. adhaerens* OV14 [33,58,59]. Six putative prophages were identified on Scaffold 4 (the chromosome) using PhiSpy [42]; these ranged in size from 21 to 65 genes, and accounted for a total of 292 genes (Data Set S3). However, no CRISPR loci were detected during annotation with prokka [19]; a questionable, short CRIPSR with one spacer was detected with CRISPRfinder [59], but its location within a predicted coding region suggests it is unlikely to be a true CRISPR locus. No evidence for the presence of the common nodulation genes *nodABC* or the nitrogenase genes *nifHDK* was found using a hidden Markov model based approach. The *Ensifer* sp. M14 assembly has been deposited in GenBank under the accession QJNR00000000, as part of the BioSample SAMN09254189.

3.2. Taxonomic Analysis of Ensifer sp. M14

Phylogenetic analyses were performed to identify the relationships between *Ensifer* sp. M14 and previously sequenced *Sinorhizobium/Ensifer* strains. Forty-five *Sinorhizobium/Ensifer* genomes were downloaded from the NCBI database (see Materials and Methods for criteria for strain inclusion), and a maximum likelihood phylogeny of these strains plus *Ensifer* sp. M14 was built based on 1652 core genes (Figure 2).

The 46 strains were grouped into putative species on the basis of whole genome ANI and AAI values (Figures S1 and S2). The results revealed that *Ensifer* sp. M14 is closely related to *Ensifer* (*Sinorhizobium*) sp. A49 (98.5% ANI and 98.9% AAI), and that these strains likely belong to a new species. *Ensifer* sp. A49 was previously isolated from soil of the Fureneset Rural Development Centre of Fjaler, Norway [60]. However, the pSinA and pSinB plasmids, carrying genes involved in arsenic oxidation and heavy metal resistance [10,12], appear to be specific to *Ensifer* sp. M14 and may therefore have been gained during growth in the Zloty Stok gold mine [7]. The most closely related named species is *Ensifer adhaerens*, which includes bacterial predators capable of feeding on organisms such as *Micrococcus luteus* [58,61].

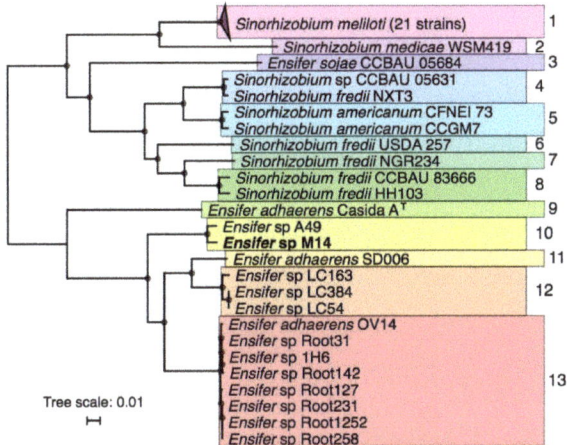

Figure 2. Phylogeny of a selected 46 *Sinorhizobium/Ensifer* strains with a publicly available whole genome sequence. An unrooted RAxML maximum likelihood phylogeny of 46 *Sinorhizobium/Ensifer* strains was prepared on the basis of the concatenated nucleotide alignments of 1652 core genes. The presented tree is the bootstrap best tree following 50 bootstrap replicates, and the scale represents the mean number of nucleotide substitutions per site. Nodes with 100% bootstrap support are indicated by the black circles. The colors and numbers to the right of the tree are used to indicate strains that group into putative species on the basis of average nucleotide identity (>96% ANI; same results were obtained with >94% ANI) and average amino acid identity (>96% AAI), as described in the Materials and Methods. Type strains are indicated by the 'T'. The accessions for all strains included in this figure are provided in Table S1.

3.3. Identification of Unique Features of the Ensifer sp. M14 Genome

A global, functional analysis of the *Ensifer* sp. M14 proteome was performed using COG categories, and the proteome was compared with closely related species to identify general functional biases. This analysis was performed with the goal of identifying recently acquired genomic islands that may contribute to the adaptation of *Ensifer* sp. M14 to the gold mine environment. When compared with *Ensifer* sp. A49, *E. adhaerens* OV14, and *E. adhaerens* Casida A, no statistically significant biases (pairwise Fisher's exact tests, $p > 0.05$ in all cases) in COG category abundances were detected in the *Ensifer* sp. M14 proteome (Figure 3A). However, there was a slight, but statistically insignificant (pairwise Fisher's exact tests, $p > 0.05$), enrichment in inorganic ion transport and metabolism (COG P) in the proteomes of *Ensifer* sp. M14 and *Ensifer* sp. A49 compared to the other two strains (Figure 3A). These results suggest no gross functional changes in the *Ensifer* sp. M14 genome occurred during adaptation to growth in the Zloty Stok gold mine, at least at the general level of COG categories.

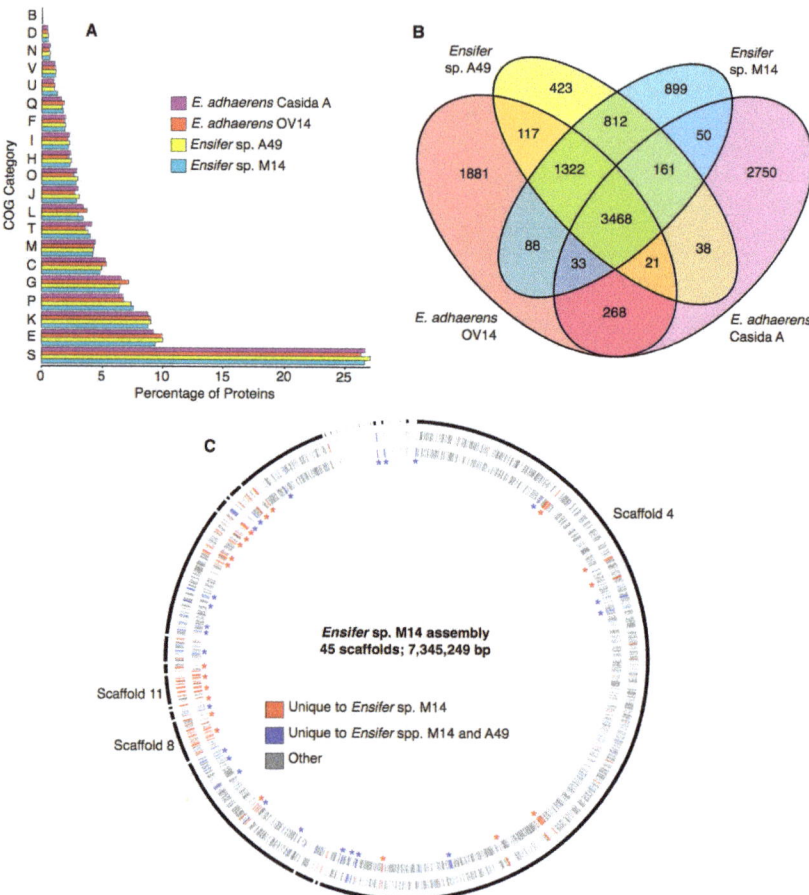

Figure 3. General features of the genome of *Ensifer* sp. M14 and related strains. (**A**) The percentage of proteins encoded by each strain annotated with each COG (Cluster of Orthologous Genes) functional category. COG categories not represented in the proteome are excluded from the graph. COG category definitions are provided in Table S4. (**B**) A Venn diagram indicating the number of genes shared among these four strains, as extracted from the pangenome of the 46 strains shown in Figure 2. (**C**) A circular plot, prepared with Circos version 0.67-7 [62], showing the scaffolds of the *Ensifer* sp. M14 assembly (outer black curved lines) including the plasmids, and the predicted coding sequences on the positive strand (outer ring) and negative strand (inner ring). Scaffolds are drawn proportional to their size, and they are presented in the order they are numbered. Scaffold 4 (chromosome), 8 (pSinB), and 11 (pSinA) are labelled. The coding regions are colored according to their conservation level, with red indicating genes unique to *Ensifer* sp. M14, and yellow indicating species common and unique to *Ensifer* spp. M14 and A49. Some multi-gene loci unique to M14 (red asterisks) or unique to M14 and A49 (blue asterisks) are indicated.

Despite the similarity in COG abundances, the *Ensifer* sp. M14 genome contains a large number of unique genes. There are 899 genes found in *Ensifer* sp. M14 but not in *Ensifer* sp. A49, *E. adhaerens* OV14, or *E. adhaerens* Casida A, while an additional 812 are found in M14 and A49, but not OV14 or Casida A (Figure 3B). Of the 899 genes specific to *Ensifer* sp. M14, 656 (9.4% of the genome) were not detected in any of the other 45 *Sinorhizobium / Ensifer* strains included in the phylogenetic analysis (Data Sets S1 and S4). Five hundred and ninety of the 656 unique proteins had a blast hit (e-value $\leq 1e^{-10}$) when queried against the NCBI non-redundant protein database, consistent with the corresponding genes being real genes that were likely acquired from other organisms through horizontal gene transfer (HGT). Mapping the location of the 656 unique genes across the assembly revealed the presence of several putative genomic islands (GIs) likely acquired through recent HGT since the divergence of *Ensifer* sp. M14 from *Ensifer* sp. A49 (Figure 3C, Data Set S4). Scaffolds 11 and 8, which correspond to the pSinA and pSinB plasmids, respectively, were not surprisingly enriched in unique genes, and together account for 217 (33%) of the unique genes. As described in detail elsewhere, these plasmids carry numerous functions associated with arsenic oxidation [10] and heavy metal resistance [10,12]. Of the 439 unique genes spread among the other scaffolds, 309 (70.4%) were annotated as hypothetical genes. Little else of interest was detected among the unique genes (Data Set S4); however, scaffold 36 was predicted to encode a zinc transporting ATPase, and a few genes related to stress resistance or drug resistance were found (discussed later). Overall, these results suggest that essentially all of the recently acquired traits associated with heavy metal resistance, arsenic oxidation, and adaptation to the stressful conditions of the Zloty Stock gold mine are associated with the pSinA and pSinB plasmids.

3.4. Metabolism of Ensifer sp. M14

Detailed phenotypic characterization of *Ensifer* sp. M14 was previously reported [7]. To further evaluate (in silico) the metabolic and transport potential of *Ensifer* sp. M14, a draft metabolic reconstruction was prepared encompassing 1491 genes and 1289 gene-associated reactions (Data Set S2). As expected based on the metabolism of related organisms [63], glycolysis in *Ensifer* sp. M14 is predicted to proceed through the Entner–Duodoroff pathway (Figures S3–S5). Growth simulations using Flux Balance Analysis suggested that *Ensifer* sp. M14 has a broad metabolic capacity, with a predicted ability to catabolize 72 carbon sources, including a variety of sugars, sugar alcohols, and organic acids (Table S2). This is consistent with previous work, which found that *Ensifer* sp. M14 could grow on 12 of 16 tested carbon substrates, including glucose, xylose, and lactate [7]. The following paragraphs provide a description of several metabolic capabilities that may be relevant to survival in the stressful environment of the Zloty Stok gold mine, and/or to resistance to elevated arsenic concentrations.

3.4.1. Phosphate Transport

The metabolic reconstruction indicated that *Ensifer* sp. M14 encodes two copies of the PstSCAB-PhoU high-affinity phosphate transporter (*BLJAPNOD_00112* through *BLJAPNOD_00116*; and *BLJAPNOD_05453* through *BLJAPNOD_05457*). Further examination of the *Ensifer* sp. M14 genome additionally revealed two copies of the PhnCDE(T) high-affinity phosphate and phosphonate transport system (*BLJAPNOD_04783* through *BLJAPNOD_04786*; and *BLJAPNOD_05447* through *BLJAPNOD_05450*). Notably, one copy of PstSCAB-PhoU and one copy of PhnCDE(T) were adjacent to the arsenic oxidation gene cluster within pSinA. This led us to explore the presence of phosphate transport systems in other arsenic-oxidizing bacteria (AOB). Using OrthoMCL [37], orthologous proteins were identified among six strains from the family *Rhizobiaceae* (Table S3): these included three AOB (*Ensifer* sp. M14, *A. tumefaciens* 5A, and *Rhizobium* sp. NT-26), as well as three related strains that are not AOB (*E. adhaerens* OV14, *N. galegae* HAMBI 540, and *A. tumefaciens* Ach5). Thirteen proteins were found to be common and specific to the three AOB, which not surprisingly included the arsenic oxidation gene cluster [10]. Notably, included within these 13 proteins were subunits of the PstSCAB-PhoU and PhnCDE(T) transporters. While all six strains encoded orthologous versions of PstSCAB-PhoU and PhnCDE(T), all three AOB encoded additional copies adjacent to their arsenic oxidation loci. Examining the genomes of three additional diverse AOB (*H. arsenicoxydans* ULPAs1, *A. arsenitoxydans* SY8, and *P. stutzeri* TS44) revealed that the first two also contained a second copy of the PstSCAB transporter in close proximity to arsenite related genes.

Based on the above results, we predict that phosphate transport genes are commonly associated with arsenite resistance loci [64]. Arsenates and phosphate are chemical analogs, with the toxicity of arsenic being a result of arsenic replacing phosphate in key biological molecules [65]. Similarly, arsenic competes with phosphate for transport through phosphate transport systems, including the PstSCAB and PhnCDE(T) systems [66–68], potentially resulting in phosphate starvation. However, the phosphate periplasmic binding proteins of at least some PstSCAB-PhoU systems, such as from the arsenic-resistant strain *Halomonas* strain GFAJ-1, display a strong preference for binding phosphate over arsenic [68]. Thus, the presence of additional high-affinity phosphate systems in AOB may be a mechanism to increase the rate (and selectivity) of phosphate import, thereby reducing the toxic effects of elevated environmental arsenic concentrations.

3.4.2. Sulfur Metabolism

We evaluated sulfur metabolism by *Ensifer* sp. M14, as sulfur compounds, such as sulfide, can be abundant in gold mines, and the arsenic oxidase enzyme contains an iron-sulfur subunit [64]. *Ensifer* sp. M14 appears to have a variety of mechanisms for sulfate assimilation. Based on the metabolic reconstruction, the genome is predicted to encode multiple sulfate and thiosulfate transporters. It is further predicted to encode several putative thiosulfate sulfurtransferases and a hydrogen sulfide oxidoreductase (*BLJAPNOD_03089*); in contrast, a sulfite oxidoreductase was not identified. Genes *BLJAPNOD_05764* through *BLJAPNOD_05768* may encode for the transport and metabolism of taurine, while *BLJAPNOD_05769* may encode the TauR taurine transcriptional regulator. *Ensifer* sp. M14 is also predicted to encode an alkanesulfonate monooxygenase (*BLJAPNOD_06609*). At least one copy of each of the subunits of the SsuABC alkanesufonate ABC-type transporter are also predicted to be encoded in the genome; however, no locus appeared to contain all three.

3.4.3. One-Carbon Metabolism

Ensifer sp. M14 is capable of growing with carbon dioxide or bicarbonate as the sole source of carbon [7], although the underlying metabolic pathway for this capability has not been examined. The metabolic reconstruction identified a putative formamide amidohydrolase (*BLJAPNOD_04973*) and putative formate dehydrogenases (*BLJAPNOD_00952 and BLJAPNOD_03433*), suggestive of the utilization of these one-carbon compounds. No clear evidence for genes associated with methanol or methylamine metabolism were found. However, the mechanism underlying one-carbon metabolism remains unclear. Unlike *S. meliloti* [69], *Ensifer* sp. M14 does not appear to encode the Calvin–Benson–Bassham cycle, nor were we able to identify any of the complete carbon-fixation pathways [70]. However, multiple enzymes potentially involved in the incorporation of bicarbonate were identified. These include putative acetyl-CoA carboxylases (*BLJAPNOD_03269, BLJAPNOD_04937, BLJAPNOD_04938*), a putative 3-oxopropanoate oxidoreductase (*BLJAPNOD_03990*), putative propanoyl-CoA carboxylases (*BLJAPNOD_06206, BLJAPNOD_06208*), a putative pyruvate carboxylase (*BLJAPNOD_00700*), and a phosphoenolpyruvate carboxylase (*BLJAPNOD_01050*).

3.4.4. Iron Transport and Metabolism

Due to the involvement of iron in arsenic oxidation, the transport and metabolism of this metal was examined. *Ensifer* sp. M14 is predicted to encode several transporters of iron or iron containing compounds. The genes *BLJAPNOD_01755* and *BLJAPNOD_01831* are predicted to encode a ferrous iron (Fe^{2+}) permease (EfeU) and a ferrous iron efflux pump (FieF), respectively. Genes *BLJAPNOD_05889* through *BLJAPNOD_05891* may encode a FecBDE ferric dicitrate transporter, while BLJAPNOD_05888 may encode the FecA ferric dicitrate outer membrane receptor protein. The genes *BLJAPNOD_00861* through *BLJAPNOD_00863* may encode a second ferric dicitrate transporter. Additionally, the genes *BLJAPNOD_05777*, *BLJAPNOD_05780*, and *BLJAPNOD_05781* may form an ABC-type transport system for iron or an iron complexes. Moreover, three putative FhuA ferrichrome (iron containing siderophore) transporting outer membrane proteins (*BLJAPNOD_04144*, *BLJAPNOD_04445*, *BLJAPNOD_05778*), and a FcuA ferrichrome receptor (*BLJAPNOD_05962*) are predicted to be encoded in the genome. A putative FepCDG ferric enterobactin transporter (*BLJAPNOD_04147*, *BLJAPNOD_04148*, *BLJAPNOD_04149*) and a PfeA enterobactin receptor (*BLJAPNOD_05560*) are also annotated. Aside from transport, *Ensifer* sp. M14 is predicted to encode a ferric reductase (*BLJAPNOD_01976–fhuF*), a ferrous oxidoreductase (*BLJAPNOD_01631*), and a ferric-chelate reductase (*BLJAPNOD_02273*). Additionally, the five gene operon (*BLJAPNOD_05798-BLJAPNOD_05802*) was predicted (using antiSMASH [55]) to encode a siderophore (aerobacin-like) biosynthetic pathway. Finally, the ferric uptake regulator (Fur) is predicted to be encoded by *BLJAPNOD_00930*.

3.4.5. Halotolerance

The *Ensifer* sp. M14 genome was searched for genes relevant to halotolerance as *Ensifer* sp. M14 has been shown to grow in highly saline environments with up to 20 mg L^{-1} NaCl [10]. Examination of the *Ensifer* sp. M14 genome with antiSMASH [55] identified a 13 gene locus (*BLJAPNOD_06859* to *BLJAPNOD_06872*) in which 12 of the genes showed similarity to 12 of the 15 genes of a known salecan biosynthetic cluster. Salecan is a water-soluble β-glucan also produced by the salt tolerant strain *Agrobacterium* sp. ZX09 [71]. Thus, this locus in *Ensifer* sp. M14 may encode for the biosynthesis of salecan, or another carbohydrate, that contributes to halotolerance. Additionally, *Ensifer* sp. M14 is predicted to be capable of synthesizing the compatible solute betaine from choline using the BetA (*BLJAPNOD_01468*, *BLJAPNOD_03726*, *BLJAPNOD_06536*) and BetB (*BLJAPNOD_00678*, *BLJAPNOD_03725*, *BLJAPNOD_05671*) pathway, as well as from choline-O-sulfate with BetC (*BLJAPNOD_02271*, *BLJAPNOD_03724*). The genome is further predicted to encode numerous proteins related to glycine betaine and proline betaine transport. Finally, as previously reported [10], pSinA encodes a putative NhaA pH-dependent sodium/proton antiporter (*BLJAPNOD_05431*), which may contribute to adaptation to high salinity [72].

3.4.6. Heavy Metal Resistance

Ensifer sp. M14 displays high resistance to numerous heavy metals [7]. Previous work identified eight modules related to heavy metal resistance on the pSinB replicon of *Ensifer* sp. M14 [12]. These modules were involved in resistance to arsenic, cadmium, cobalt, copper, iron, mercury, nickel, silver, and zinc [12]. Additionally, pSinA contains a locus involved in resistance to cadmium, zinc, cobalt, and mercury [10]. Our analyses reported above suggested that the majority, if not all, genes relevant to adaptation to the heavy metal-rich environment in the Zloty Stok gold mine are located on the pSinA and pSinB plasmids [10,12].

3.5. Biosafety Considerations of Ensifer sp. M14

The *Sinorhizobium/Ensifer* group of bacteria contain numerous plant symbionts and other biotechnologically relevant strains, but it lacks known pathogens. Considering this, and the observation that none of the genomic islands detected in *Ensifer* sp. M14 appear to be pathogenicity islands, it is unlikely that *Ensifer* sp. M14 is pathogenic. Therefore, the environmental release of *Ensifer* sp. M14 is not expected to pose a biosafety risk from that perspective. Additionally, analysis of the secondary metabolism of *Ensifer* sp. M14 with antiSMASH [55] did not identify antibiotic synthesis loci. However, *Ensifer* sp. M14 may carry several antimicrobial resistance (AMR) genes. The analysis applying the RGI analyzer revealed the presence of 12 putative antibiotic resistance genes/gene clusters (Table 2). It is worth mentioning that the best hits were found for four *acrAB(-TolC)* modules encoding resistance-nodulation-cell division (RND) type multidrug efflux systems, while the remaining eight genes were much more divergent compared with the reference proteins (they were detected only when applying the LOOSE algorithm of the RGI analyzer). This may suggest that these hits are accidental, and that the identified genes are not truly AMR genes, or that these are novel, emergent threats and more distant homologs of known reference genes.

Table 2. Putative antimicrobial resistance genes found in the *Ensifer* sp. M14 genome.

Scaffold	Gene ID	CARD Database Hit	Predicted Resistance to	Tested Antibiotics
Scaffold_4	BLJAPNOD_00187-BLJAPNOD_00188	acrAB	Fluoroquinolone Tetracyclines	CIP (S); MXF (S) TE (S/R)
Scaffold_4	BLJAPNOD_00458	cmlA/floR	Chloramphenicol	C (R)
Scaffold_4	BLJAPNOD_00485-BLJAPNOD_00487	acrAB-TolC	Tetracyclines Cephalosporins Penams Phenicols Rifamycins Fluoroquinolones	TE (S/R) CFM (S); CRO (S); CTX (S) AMP (R) C (R) RD (R) CIP (S); MXF (S)
Scaffold_4	BLJAPNOD_00960	aph(3′)-IIa	Aminoglycosides	CN (S)
Scaffold_4	BLJAPNOD_01284	adeF	Fluoroquinolones Tetracyclines	CIP (S); MXF (S) TE (S/R)
Scaffold_4	BLJAPNOD_02256	bla$_{OXA}$	Cephalosporins Penams	CFM (S); CRO (S); CTX (S) AMP (R)
Scaffold_4	BLJAPNOD_02798	aph(6)-Ic	Aminoglycosides	CN (S)
Scaffold_7	BLJAPNOD_04982	aph(3″)-Ib	Aminoglycosides	CN (S)
Scaffold_8	BLJAPNOD_05149-BLJAPNOD_05151	acrAB-TolC	Tetracyclines Cephalosporins Penams Phenicols Rifamycins Fluoroquinolones	TE (S/R) CFM (S); CRO (S); CTX (S) AMP (R) C (R) RD (R) CIP (S); MXF (S)
Scaffold_14	BLJAPNOD_05841-BLJAPNOD_05842	acrAB	Fluoroquinolone Tetracyclines	CIP (S); MXF (S) TE (S/R)
Scaffold_17	BLJAPNOD_06442	dfrA12	Trimethoprim	TM (S)
Scaffold_18	BLJAPNOD_06615	aph(6)-Ic	Aminoglycosides	CN (S)

The most significant hits, defined with the usage of the STRICT algorithm of the RGI analyzer, are bolded. Abbreviations: AMP—ampicilin; C—chloramphenicol; CN—gentamicin; CFM—cefixime; CTX—cefotaxime; CRO—ceftriaxone; CIP—ciprofloxacin; TE—tetracycline; TM—trimethoprim; MXF—moxifloxacin; RIF—rifampicin; R—resistant; S—susceptibility; S/R—inability of interpretation of the result (threshold value).

Previous analyses revealed that the closely related organism *E. adhaerens* OV14 displays resistance to numerous antibiotics, including, among others, ampicillin, spectinomycin, kanamycin, and carbenicillin [73]. Therefore, to check whether the predicted antibiotic resistance genes truly associated with antibiotic resistance in *Ensifer* sp. M14, the MICs of 11 antibiotics were determined using Etests. Results from the Etests showed that *Ensifer* sp. M14 is resistant to ampicillin (MIC: 12.0 mg L^{-1}), chloramphenicol (MIC: 8.0 mg L^{-1}), and rifampicin (MIC 4.0 mg L^{-1}), while it is susceptible to cefixime, cefotaxime, ceftriaxone, ciprofloxacin, gentamicin, moxifloxacin, and trimethoprim. In the case of tetracycline, the MIC values fluctuated around the threshold for classification as resistant (1–4 mg L^{-1}); hence, precise interpretation of this result is not possible. Resistance to antibiotics belonging to the penams, phenicols, and rifamicyns families may be explained by the presence of efflux pumps belonging to the RND family. These multidrug resistance systems are highly prevalent in Gram-negative bacteria, and play an important role in resistance to various types of stress factors, including antibiotics [74]. It is also worth mentioning, that environmental isolates of *Alphaproteobacteria* usually possess several copies of genetic modules encoding RND type multidrug efflux systems, which may be linked with their adaptation to the heterogeneity of the soil habitat [75,76]. Therefore, we think that the environmental release of *Ensfier* sp. M14 is unlikely to pose a biosafety risk.

3.6. Development of a Pilot-Scale Installation for Arsenic Bioremediation

The genomic analyses suggested that *Ensifer* sp. M14 contains several genetic features that may allow it to be successfully used in environmental bioremediation applications. In addition, previous experimental studies demonstrated that this strain can efficiently transform As(III) into As(V) (24 h of residence time was sufficient to oxidize 5 mg L^{-1} of As(III) in the laboratory) [9]. We therefore attempted to prepare an installation for environmental bioremediation of arsenic contaminated water involving *Ensifer* sp. M14. The purification of arsenic contaminated waters constitutes a serious environmental challenge, as most of the available chemical and physical methods are dedicated to the selective removal of As(V), and are inefficient with regard to As(III). Thus, the aim of the microbiological module of the installation was to harness the arsenite oxidation capabilities of *Ensifer* sp. M14 to ensure efficient oxidation of As(III) to facilitate its subsequent removal. We reasoned that combining a biological approach with an appropriate physicochemical process (i.e., adsorption) could overcome the constraints and reservations of the conventional methods dedicated to the removal of arsenic from contaminated waters [77,78].

3.7. The Activity and Characterization of the Microbiological Module of the Pilot-Scale Installation

In our preliminary study [9], we observed that efficient functioning of the laboratory-scale installation required a high density of *Ensifer* sp. M14 (OD$_{600}$ between 0.1 and 0.2). This is in part because the quantity of *Ensifer* sp. M14 usually decreases quite intensively during the first hours/days of continuous culturing in the bioreactor [9]. Although appropriate growth conditions and length of residence time during continuous cultures were previously determined [9], the move from the laboratory-scale to pilot-scale installation meant it was necessary to re-evaluate them. In particular, replacement of the synthetic medium by natural arsenic contaminated water, as well as increasing the scale of application, may result in a deceleration of bacterial growth and a decrease in the efficiency of the biooxidation processes [79].

3.7.1. Microbial Growth and Efficiency of Arsenic Biooxidation in the Bioreactor

Using the start-up procedures described in the Materials and Methods, the initial quantity of bacteria in the bioreactor after yeast extract augmentation was about 10^8 CFU mL^{-1} (Figure 4). The value was almost nine orders of magnitude higher compared to raw arsenic-contaminated water, where the CFU mL^{-1} (when plated on Luria-Bertani agar medium) was about 10^0.

As expected based on our preliminary study [9], the density of bacteria decreased systematically during the first few days of operation, reaching a density on the magnitude of 10^3 CFU mL^{-1} on day seven (Figure 4A). After this point, the density of bacteria largely stabilized, with the exception of a few days (days 17–20), when an ~100-fold drop in bacterial density was observed (Figure 4A). A bacterial concentration of 10^3 CFU mL^{-1} in the bioreactor generally appeared sufficient for efficient biooxidation of the arsenite in the contaminated water, as there was generally little to no arsenite detected in the water following passage through the bioreactor (Figure 4D). The exceptions were five of the nine measurements taken between days 15 and 31, inclusive, when arsenite accounted for up to 62.86% of the total arsenic concentration; this corresponded with the drop in the density of bacteria within the bioreactor (Figure 4A). Thus, the low arsenite concentration throughout the majority of the experiment suggests that the microbiological module efficiently converted the As(III) to As(V).

Recently, Tardy et al. [52] showed that efficient arsenite biooxidation in environmental samples of water at 20 °C occurred after eight days of culture (batch experiment), and the quantity of bacteria at the end of their experiment was about 10^5 CFU mL^{-1}. On the other hand, Kamde et al. [80] reported that arsenic removal was most intensive when the quantity of bacteria was about 28 CFU mL^{-1} (batch cultures with the use on synthetic medium). The higher quantity of bacteria in the abovementioned papers in comparison with our study is presumably related to differences in culture conditions (various media and/or culturing methods).

Our data (Figure 4A,D) is also consistent with a relationship between the quantity of *Ensifer* sp. M14 and the efficiency of arsenic biooxidation, as were our preliminary experiments in batch cultures (data not shown). Indeed, many studies have observed a positive correlation between the density of bacteria and the rate of metal metabolism or biotransformation for arsenic compounds and other elements [80–82].

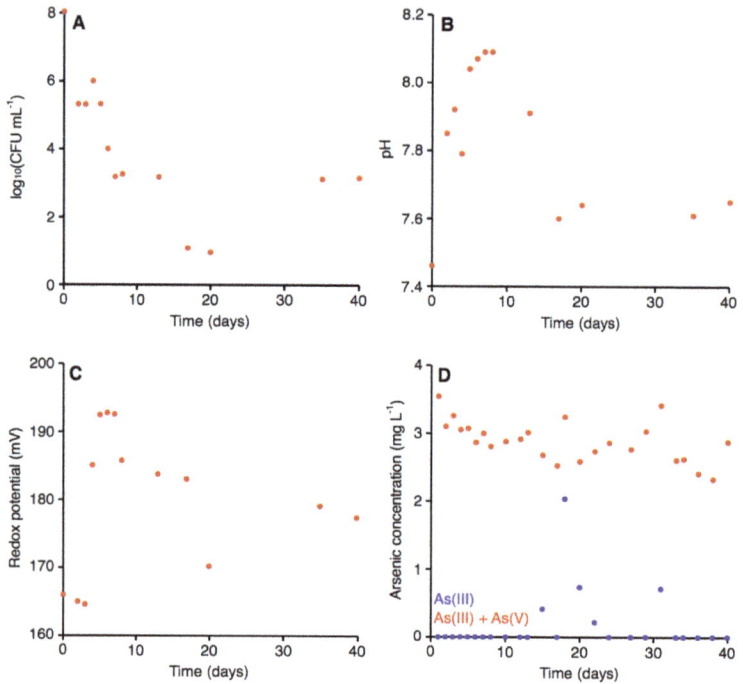

Figure 4. Parameters of the water in the bioreactor of the microbiological module. The graphs display (**A**) the quantity of bacteria, (**B**) the pH of the water, (**C**) the redox potential of the water, and (**D**) the concentration of As(III) (blue) and total arsenic (red) in the water.

3.7.2. Physical and Chemical Characterization of the Bioreactor

Previous studies have observed that there is a relationship between pH and redox potential with the arsenite/arsenate ratio; arsenites are the predominant form in reducing conditions and lower pH values, as the concentration of the arsenate form increases, both pH and redox potential also increase [83,84]. We therefore evaluated the pH and the redox potential in the treated water. For both parameters, the biological treatment had a small but noticeable effect. In the raw water, the pH and the redox potential were 7.48 and 170.90 mV, respectively [56]. In the case of the pH, the value in the bioreactor systematically increased up to the eighth day, with the treated water reaching a pH of 8.09 (Figure 4B). The pH returned to 7.60 by day 17, following which the pH stabilized in the range of 7.60 to 7.65 until the end of the experiment (day 40). In general, the redox potential remained relatively stable (Figure 4C). For the first three days, a value around 155.00 mV was observed, following which the redox potential increased and stabilized (with a slight, gradual decrease) within a range from 177.00 mV and 193.00 mV, with the exception of day 20. Water for human consumption is expected to have a pH in the range of 6.50–9.00 [85] and a redox potential between 100 and 300.00 mV [86]. Thus, both the pH and the redox potential of the water treated with our installation fell within the acceptable range for drinking water.

3.8. Effectiveness of the Adsorption Module of the Pilot-Scale Installation

Granulated bog iron ores are characterized by high arsenic adsorption capacity (up to 5.72 mg kg^{-1}, depending on the adsorbate concentration), short residence time (20 min) [56], they display high chemical stability, and they are resistant to bioweathering processes [87]. These properties allow this material to function as an effective adsorbent for removal of arsenics from contaminated water in both passive and active remediation systems, as demonstrated in our earlier work [56]. Here, we have coupled the use of granulated bog iron ores as an input to the adsorption module as well as the microbiological module described above, as a way to ensure the efficient conversion of As(III) to As(V) by *Ensifer* sp. M14, followed by the removal of As(V) by the bog iron ores. The pre-conversion of As(III) to As(V) is important as bog iron ores saturated with As(V) display higher chemical stability than bog iron ores saturated with As(III) [87].

Treatment of the arsenic contaminated water with the pilot-scale installation led to a dramatic decrease in arsenic concentrations, going from 2400 µg L^{-1} in the raw water to less than 10 µg L^{-1} (Figure 5). Analysis of the breakthrough curves for each of the adsorption columns indicated that the adsorbent in none of the columns reached equilibrium saturation during the 40-day experiment (Figure 5). Equilibrium saturation is herein defined as the maximum adsorption capacity (full saturation) of the adsorbent at a given concentration of the adsorbate; i.e., when the arsenic concentration in the input and output water is equal. Upon reaching equilibrium saturation, the adsorbent would be completely consumed and unable to further remove arsenic from the water, and it would therefore require regeneration or replacement. As the total arsenic concentration in water after each column was lower than the water entering the column, none of the columns reached equilibrium saturation. Thus, under the tested environmental conditions, the pilot-scale installation is expected to have been able to effectively continue the bioremediation process for much longer than the 40 days of the experiment (during which, 8 m^3 of water flowed through the system).

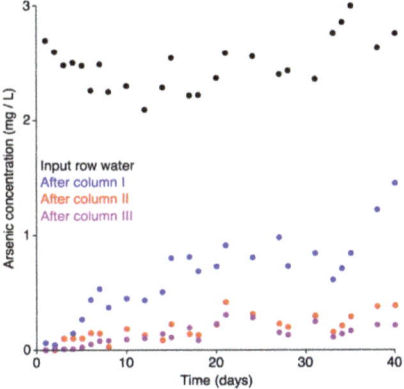

Figure 5. Arsenic adsorption breakthrough curves. The arsenic adsorption breakthrough curves for each column of the adsorption module are shown. Total arsenic concentrations in the raw water (black), and after column I (blue), column II (red), and column III (purple) are shown.

In Poland, the Regulation of the Polish Ministry of the Environment [88] currently sets the upper limit for arsenic contamination in water for use in technological purposes at 100 µg L^{-1}. In the experiment reported here, the arsenic concentration in the treated water remained below 100 µg L^{-1} for the first ten days of the experiment (Figure 5), and never exceeded 220 µg L^{-1} during the 40 day test. Thus, at least the first 2.0 m^3 of water treated by pilot-scale installation was below the Polish limit for use in technological purposes. However, if pooling the treated water (and thus averaging the arsenic concentration), it is likely that the cumulative concentration of arsenic in the 4.0 m^3 of water treated over the first 20 days remained below the limit.

The local adsorption capacity of the adsorbent varied between the columns and depended on the arsenic concentration of the inflowing water. The adsorbent from the first column was characterized by the highest adsorption capacity, which was 0.500 mg kg^{-1}. Adsorbent placed in the second and third columns had lower adsorption capacities of 0.031 and 0.021 mg kg^{-1}, respectively. Likely, these differences are due to the later columns adsorbing less arsenic and being farther from reaching equilibrium saturation. The adsorption capacities recorded in the current study were significantly lower than those described in our previous work, presumably due to the adsorbent not reaching equilibrium saturation [56].

4. Conclusions

Here, we reported the draft genome sequence of *Ensifer* sp. M14 in order to gain insights into the genomic adaptation of this organism to the stressful environment of the abandoned Zloty Stok gold mine from which it was isolated. In addition, we were interested in the genetic basis of the strains arsenic oxidation and resistance capabilities, resistance to arsenic and other heavy metals, and the biosafety of the strain for use in biotechnological applications. The results revealed hundreds of genes present in *Ensifer* sp. M14 that are not found in related species, and these genes are often colocalized in genomic islands. However, the majority of these genes encoded hypothetical proteins of unknown function. Based on the genome sequence, it appears that the majority of the genes have been acquired to deal with the hostile environment of the Zloty Stok gold mine, i.e., conferring resistance to heavy metals, and enabling arsenic oxidation, are located on the self-transmissible pSinA and pSinB megaplasmids. Additionally, analysis of the *Ensifer* sp. M14 genome suggested that this strain should be safe for use in biotechnology and bioremediation. However, it was noted that several putative antibiotic resistance genes are present in the genome, as is also true for the related strain *Ensifer adhaerens* OV14 that is used in biotechnological applications [89]. This property of *Ensifer* sp. M14 should be kept in mind during its application in order to limit the spread of antimicrobial resistance. The results of these genomic analyses provide hints into the genetic potential of *Ensifer* sp. M14. They will help focus future experimental research aimed at further characterizing the biology of this organism, and may contribute to the development of procedures for large-scale cultivation of this strain.

This study also reports the construction and validation of a pilot-scale installation designed for the remediation of arsenic contaminated waters. This novel installation couples a microbiological module, based on the arsenic oxidation abilities of *Ensifer* sp. M14, with an adsorption module, based on the use of granulated bog iron ores. The underlying principle is to use *Ensifer* sp. M14 to efficiently oxidize the As(III) ions to As(V), followed by the removal of the As(V) through adsorption by the bog iron ores. Characterization of the arsenic contaminated water following passage through the microbiological module generally revealed little to no As(III), consistent with the *Ensifer* sp. M14 generally ensuring effective conversion of As(III) to As(V). Additionally, a dramatic decrease (from 10-fold to greater than 250-fold) in the arsenic concentration of the water was observed following passage of this water through the adsorption module. These results therefore confirm the effectiveness of the tested installation for the remediation of arsenic contaminated waters, which pose risks to both the environmental and human health. Future work will be aimed at further developing and optimizing this system, which could involve, for example, the addition of beads to the reactor containing *Ensifer* sp. M14 biofilms.

Supplementary Materials: The following are available online at http://www.mdpi.com/2073-4425/9/8/379/s1. Table S1: Accession numbers for all genomes used in this work, Table S2: In silico test of the metabolic capacity of *Ensifer* sp. M14, Table S3: Orthologous groupings of six *Rhizobiaceae* strains, including three AOB and three strains that are not AOB, Table S4: COG category descriptions, Figure S1: Average nucleotide identity matrix. A matrix of the two-way ANI values for 46 *Sinorhizobium/Ensifer* strains is shown. Clustering was performed along both axes using hierarchical clustering with Pearson distance and average linkage, Figure S2: Average amino acid identity matrix. A matrix of the two-way AAI values for 46 *Sinorhizobium/Ensifer* strains is shown.

Clustering was performed along both axes using hierarchical clustering with Pearson distance and average linkage, Figure S3: Entner–Duodoroff pathway and the pentose phosphate pathway. A modified version of the KEGG pathway map ko00030 [90] displaying the Entner–Duodoroff pathway and the pentose phosphate pathway is shown. Reactions encoded by the *Ensifer* sp. M14 genome are colored green; those in white are missing. The figure was prepared using the KAAS webserver [91] using BLAST search with the bi-directional best hit assignment method, and with the default organism list for 'prokaryotes' plus *Sinorhizobium meliloti* Rm1021, Figure S4: Gluconeogenesis. A modified version of the KEGG pathway map ko00010 [90] displaying the pathway for gluconeogenesis is shown. Reactions encoded by the *Ensifer* sp. M14 genome are colored green; those in white are missing. The figure was prepared using the KAAS webserver [91] using BLAST search with the bi-directional best hit assignment method, and with the default organism list for 'prokaryotes' plus *Sinorhizobium meliloti* Rm1021, Figure S5: Tricarboxylic acid cycle. A modified version of the KEGG pathway map ko00020 [90] displaying the tricarboxylic acid (TCA) cycle is shown. Reactions encoded by the *Ensifer* sp. M14 genome are colored green; those in white are missing. The figure was prepared using the KAAS webserver [91] using BLAST search with the bi-directional best hit assignment method, and with the default organism list for 'prokaryotes' plus *Sinorhizobium meliloti* Rm1021, Data Set S1: *Sinorhizobium/Ensifer* gene presence and absence. This file contains the gene presence/absence output data from Roary for the pangenome analysis of 46 *Sinorhizobium/Ensifer* strains. Details on the information provided in the file is available at: https://sanger-pathogens.github.io/Roary/, Data Set S2: Metabolic reconstruction of *Ensifer* sp. M14. This archive contains the expanded, draft metabolic reconstruction of *Ensifer* sp. M14. The reconstruction is provided in COBRA format as a MATLAB file, as well as in a table within an Excel workbook. A readme file is included to explain the contents, Data Set S3: PhiSpy phage prediction. This file contains the PhiSpy phage prediction output for all *Ensifer* sp. M14 genes, as well as separate sheets for each of the putative prophage loci. Details on the information provided in the file is available at: https://github.com/linsalrob/PhiSpy, Data Set S4: Functional annotation of the *Ensifer* sp. M14 genome. This file contains the genome annotation and the eggNOG-mapper output for three sets of genes: (i) all genes in the *Ensifer* sp. M14 genome; (ii) all genes unique to the *Ensifer* sp. M14 genome; and (iii) all genes unique and common to the *Ensifer* sp. M14 and A49 genomes. Details on the eggNOG-mapper output provided in the file is available at: https://github.com/jhcepas/eggnog-mapper/wiki/Results-Interpretation.

Author Contributions: Conceptualization, G.C.d., K.D., A.M., L.D., and L.Dr.; Methodology, G.C.d., K.D., and J.K.; Software, G.C.d., A.M., C.F., and L.Dz.; Validation, G.C.d., K.D., A.M., L.Dz., and L.Dr.; Formal Analysis, G.C.d. and K.D.; Investigation, K.D., W.U., A.G., G.R., and T.B.; Resources, K.D., A.M., and L.Dr.; Data Curation, G.C.d., A.M., C.F., and L.Dz.; Writing-Original Draft Preparation, G.C.d., K.D., and L.Dr., Writing-Review & Editing, W.U., A.M., L.Dz., T.B., and G.R.; Visualization, G.C.d., K.D., and J.K.; Supervision, A.M. and L.Dr.; Project Administration, A.M. and L.Dr.; Funding Acquisition, K.D., A.M., and L.Dr.

Funding: This research was funded by National Centre for Research and Development (Poland) grant number LIDER/043/403/L-4/12/NCBR/2013, National Science Center (Poland) grant no. 2016/23/N/NZ9/01655 and intramural funding from the University of Florence, call "PROGETTI STRATEGICI DI ATENEO ANNO 2014".

Acknowledgments: G.C.d. was supported by a Post-Doctoral Fellowship from the Natural Sciences and Engineering Council of Canada. K.D. was supported by the European Molecular Biology Organization in the frame of the EMBO Short-Term Fellowship program [grant number 7376].

Conflicts of Interest: The authors declare no conflicts of interest.

References

1. Alisi, C.; Musella, R.; Tasso, F.; Ubaldi, C.; Manzo, S.; Cremisini, C.; Sprocati, A.R. Bioremediation of diesel oil in a co-contaminated soil by bioaugmentation with a microbial formula tailored with native strains selected for heavy metals resistance. *Sci. Total Environ.* **2009**, *407*, 3024–3032. [CrossRef] [PubMed]
2. Kuppusamy, S.; Thavamani, P.; Megharaj, M.; Lee, Y.B.; Naidu, R. Polyaromatic hydrocarbon (PAH) degradation potential of a new acid tolerant, diazotrophic P-solubilizing and heavy metal resistant bacterium *Cupriavidus* sp. MTS-7 isolated from long-term mixed contaminated soil. *Chemosphere* **2016**, *162*, 31–39. [CrossRef] [PubMed]
3. Yu, D.; Yang, J.; Teng, F.; Feng, L.; Fang, X.; Ren, H. Bioaugmentation treatment of mature landfill leachate by new isolated ammonia nitrogen and humic acid resistant microorganism. *J. Microbiol. Biotechnol.* **2014**, *24*, 987–997. [CrossRef] [PubMed]
4. Pepper, I.L.; Gentry, T.J.; Newby, D.T.; Roane, T.M.; Josephson, K.L. The role of cell bioaugmentation and gene bioaugmentation in the remediation of co-contaminated soils. *Environ. Health Perspect.* **2002**, *110* (Suppl. 6), 943–946. [CrossRef] [PubMed]
5. Feng, Z.; Li, X.; Lu, C.; Shen, Z.; Xu, F.; Chen, Y. Characterization of *Pseudomonas mendocina* LR capable of removing nitrogen from various nitrogen-contaminated water samples when cultivated with *Cyperus alternifolius* L. *J. Biosci. Bioeng.* **2012**, *114*, 182–187. [CrossRef] [PubMed]

6. Lee, P.K.H.; Warnecke, F.; Brodie, E.L.; Macbeth, T.W.; Conrad, M.E.; Andersen, G.L.; Alvarez-Cohen, L. Phylogenetic microarray analysis of a microbial community performing reductive dechlorination at a TCE-contaminated site. *Environ. Sci. Technol.* **2012**, *46*, 1044–1054. [CrossRef] [PubMed]
7. Drewniak, L.; Matlakowska, R.; Sklodowska, A. Arsenite and arsenate metabolism of *Sinorhizobium* sp. M14 living in the extreme environment of the Zloty Stok gold mine. *Geomicrobiol. J.* **2008**, *25*, 363–370. [CrossRef]
8. Drewniak, L.; Krawczyk, P.S.; Mielnicki, S.; Adamska, D.; Sobczak, A.; Lipinski, L.; Burec-Drewniak, W.; Sklodowska, A. Physiological and metagenomic analyses of microbial mats involved in self-purification of mine waters contaminated with heavy metals. *Front. Microbiol.* **2016**, *7*, 1252. [CrossRef] [PubMed]
9. Debiec, K.; Krzysztoforski, J.; Uhrynowski, W.; Sklodowska, A.; Drewniak, L. Kinetics of arsenite oxidation by *Sinorhizobium* sp. M14 under changing environmental conditions. *Int. Biodeterior. Biodegrad.* **2017**, *119*, 476–485. [CrossRef]
10. Drewniak, L.; Dziewit, L.; Ciezkowska, M.; Gawor, J.; Gromadka, R.; Sklodowska, A. Structural and functional genomics of plasmid pSinA of *Sinorhizobium* sp. M14 encoding genes for the arsenite oxidation and arsenic resistance. *J. Biotechnol.* **2013**, *164*, 479–488. [CrossRef] [PubMed]
11. Dziewit, L.; Bartosik, D. Comparative analyses of extrachromosomal bacterial replicons, ientification of chromids, and experimental evaluation of their indispensability. *Methods Mol. Biol.* **2015**, *1231*, 15–29. [CrossRef] [PubMed]
12. Romaniuk, K.; Dziewit, L.; Decewicz, P.; Mielnicki, S.; Radlinska, M.; Drewniak, L. Molecular characterization of the pSinB plasmid of the arsenite oxidizing, metallotolerant *Sinorhizobium* sp. M14—Insight into the heavy metal resistome of sinorhizobial extrachromosomal replicons. *FEMS Microbiol. Ecol.* **2017**, *93*, fiw215. [CrossRef] [PubMed]
13. Doyle, J.; Doyle, J. A rapid DNA isolation procedure for small quantities of fresh leaf tissue. *Phytochem. Bull.* **1987**, *19*, 11–15.
14. Joint Genome Institute JGI Bacterial DNA Isolation CTAB Protocol. Available online: https://jgi.doe.gov/user-program-info/pmo-overview/protocols-sample-preparation-information/jgi-bacterial-dna-isolation-ctab-protocol-2012/ (accessed on 20 July 2018).
15. Bankevich, A.; Nurk, S.; Antipov, D.; Gurevich, A.A.; Dvorkin, M.; Kulikov, A.S.; Lesin, V.M.; Nikolenko, S.I.; Pham, S.; Prjibelski, A.D.; et al. SPAdes: A new genome assembly algorithm and its applications to single-cell sequencing. *J. Comput. Biol.* **2012**, *19*, 455–477. [CrossRef] [PubMed]
16. Vasilinetc, I.; Prjibelski, A.D.; Gurevich, A.; Korobeynikov, A.; Pevzner, P.A. Assembling short reads from jumping libraries with large insert sizes. *Bioinformatics* **2015**, *31*, 3262–3268. [CrossRef] [PubMed]
17. Jain, C.; Rodriguez-R, L.M.; Phillippy, A.M.; Konstantinidis, K.T.; Aluru, S. High-throughput ANI analysis of 90K prokaryotic genomes reveals cear species boundaries. *bioRxiv* **2017**, 225342. [CrossRef]
18. Bosi, E.; Donati, B.; Galardini, M.; Brunetti, S.; Sagot, M.-F.; Lió, P.; Crescenzi, P.; Fani, R.; Fondi, M. MeDuSa: A multi-draft based scaffolder. *Bioinformatics* **2015**, *31*, 2443–2451. [CrossRef] [PubMed]
19. Seemann, T. Prokka: Rapid prokaryotic genome annotation. *Bioinformatics* **2014**, *30*, 2068–2069. [CrossRef] [PubMed]
20. Hyatt, D.; Chen, G.-L.; LoCascio, P.F.; Land, M.L.; Larimer, F.W.; Hauser, L.J. Prodigal: Prokaryotic gene recognition and translation initiation site identification. *BMC Bioinform.* **2010**, *11*, 119. [CrossRef] [PubMed]
21. Laslett, D.; Canback, B. ARAGORN, a program to detect tRNA genes and tmRNA genes in nucleotide sequences. *Nucleic Acids Res.* **2004**, *32*, 11–16. [CrossRef] [PubMed]
22. Kolbe, D.L.; Eddy, S.R. Fast filtering for RNA homology search. *Bioinformatics* **2011**, *27*, 3102–3109. [CrossRef] [PubMed]
23. Kalvari, I.; Argasinska, J.; Quinones-Olvera, N.; Nawrocki, E.P.; Rivas, E.; Eddy, S.R.; Bateman, A.; Finn, R.D.; Petrov, A.I. Rfam 13.0: Shifting to a genome-centric resource for non-coding RNA families. *Nucleic Acids Res.* **2018**, *46*, D335–D342. [CrossRef] [PubMed]
24. Huerta-Cepas, J.; Forslund, K.; Coelho, L.P.; Szklarczyk, D.; Jensen, L.J.; von Mering, C.; Bork, P. Fast genome-wide functional annotation through orthology assignment by eggNOG-mapper. *Mol. Biol. Evol.* **2017**, *34*, 2115–2122. [CrossRef] [PubMed]
25. Page, A.J.; Cummins, C.A.; Hunt, M.; Wong, V.K.; Reuter, S.; Holden, M.T.G.; Fookes, M.; Falush, D.; Keane, J.A.; Parkhill, J. Roary: Rapid large-scale prokaryote pan genome analysis. *Bioinformatics* **2015**, *31*, 3691–3693. [CrossRef] [PubMed]

26. Löytynoja, A. Phylogeny-aware alignment with PRANK. *Methods Mol. Biol.* **2014**, *1079*, 155–170. [CrossRef] [PubMed]
27. Stamatakis, A. RAxML version 8: A tool for phylogenetic analysis and post-analysis of large phylogenies. *Bioinformatics* **2014**, *30*, 1312–1313. [CrossRef] [PubMed]
28. Letunic, I.; Bork, P. Interactive tree of life (iTOL) v3: An online tool for the display and annotation of phylogenetic and other trees. *Nucleic Acids Res.* **2016**, *44*, W242–W245. [CrossRef] [PubMed]
29. Buchfink, B.; Xie, C.; Huson, D.H. Fast and sensitive protein alignment using DIAMOND. *Nat. Methods* **2015**, *12*, 59–60. [CrossRef] [PubMed]
30. Luo, C.; Rodriguez-R, L.M.; Konstantinidis, K.T. MyTaxa: An advanced taxonomic classifier for genomic and metagenomic sequences. *Nucleic Acids Res.* **2014**, *42*, e73. [CrossRef] [PubMed]
31. Hao, X.; Lin, Y.; Johnstone, L.; Liu, G.; Wang, G.; Wei, G.; McDermott, T.; Rensing, C. Genome sequence of the arsenite-oxidizing strain *Agrobacterium tumefaciens* 5A. *J. Bacteriol.* **2012**, *194*, 903. [CrossRef] [PubMed]
32. Henkel, C.V.; den Dulk-Ras, A.; Zhang, X.; Hooykaas, P.J.J. Genome sequence of the octopine-type *Agrobacterium tumefaciens* strain Ach5. *Genome Announc.* **2014**, *2*, e00225-14. [CrossRef] [PubMed]
33. Rudder, S.; Doohan, F.; Creevey, C.J.; Wendt, T.; Mullins, E. Genome sequence of *Ensifer adhaerens* OV14 provides insights into its ability as a novel vector for the genetic transformation of plant genomes. *BMC Genom.* **2014**, *15*, 268. [CrossRef] [PubMed]
34. Österman, J.; Marsh, J.; Laine, P.K.; Zeng, Z.; Alatalo, E.; Sullivan, J.T.; Young, J.P.W.; Thomas-Oates, J.; Paulin, L.; Lindström, K. Genome sequencing of two *Neorhizobium galegae* strains reveals a *noeT* gene responsible for the unusual acetylation of the nodulation factors. *BMC Genom.* **2014**, *15*, 500. [CrossRef] [PubMed]
35. Andres, J.; Arsène-Ploetze, F.; Barbe, V.; Brochier-Armanet, C.; Cleiss-Arnold, J.; Coppée, J.-Y.; Dillies, M.-A.; Geist, L.; Joublin, A.; Koechler, S.; et al. Life in an arsenic-containing gold mine: Genome and physiology of the autotrophic arsenite-oxidizing bacterium *Rhizobium* sp. NT-26. *Genome Biol. Evol.* **2013**, *5*, 934–953. [CrossRef] [PubMed]
36. Arkin, A.P.; Cottingham, R.W.; Henry, C.S.; Harris, N.L.; Stevens, R.L.; Maslov, S.; Dehal, P.; Ware, D.; Perez, F.; Canon, S.; et al. KBase: The United States Department of Energy Systems Biology Knowledgebase. *Nat. Biotechnol.* **2018**, *36*, 566–569. [CrossRef] [PubMed]
37. Li, L.; Stoeckert, C.J.; Roos, D.S. OrthoMCL: Identification of ortholog groups for eukaryotic genomes. *Genome Res.* **2003**, *13*, 2178–2189. [CrossRef] [PubMed]
38. Li, X.; Hu, Y.; Gong, J.; Lin, Y.; Johnstone, L.; Rensing, C.; Wang, G. Genome sequence of the highly efficient arsenite-oxidizing bacterium *Achromobacter arsenitoxydans* SY8. *J. Bacteriol.* **2012**, *194*, 1243–1244. [CrossRef] [PubMed]
39. Muller, D.; Médigue, C.; Koechler, S.; Barbe, V.; Barakat, M.; Talla, E.; Bonnefoy, V.; Krin, E.; Arsène-Ploetze, F.; Carapito, C.; et al. A tale of two oxidation states: Bacterial colonization of arsenic-rich environments. *PLoS Genet.* **2007**, *3*, e53. [CrossRef] [PubMed]
40. Li, X.; Gong, J.; Hu, Y.; Cai, L.; Johnstone, L.; Grass, G.; Rensing, C.; Wang, G. Genome sequence of the moderately halotolerant, arsenite-oxidizing bacterium *Pseudomonas stutzeri* TS44. *J. Bacteriol.* **2012**, *194*, 4473–4474. [CrossRef] [PubMed]
41. O'Leary, N.A.; Wright, M.W.; Brister, J.R.; Ciufo, S.; Haddad, D.; McVeigh, R.; Rajput, B.; Robbertse, B.; Smith-White, B.; Ako-Adjei, D.; et al. Reference sequence (RefSeq) database at NCBI: Current status, taxonomic expansion, and functional annotation. *Nucleic Acids Res.* **2016**, *44*, D733–D745. [CrossRef] [PubMed]
42. Akhter, S.; Aziz, R.K.; Edwards, R.A. PhiSpy: A novel algorithm for finding prophages in bacterial genomes that combines similarity- and composition-based strategies. *Nucleic Acids Res.* **2012**, *40*, e126. [CrossRef] [PubMed]
43. McArthur, A.G.; Waglechner, N.; Nizam, F.; Yan, A.; Azad, M.A.; Baylay, A.J.; Bhullar, K.; Canova, M.J.; De Pascale, G.; Ejim, L.; et al. The comprehensive antibiotic resistance database. *Antimicrob. Agents Chemother.* **2013**, *57*, 3348–3357. [CrossRef] [PubMed]
44. Kahlmeter, G.; Brown, D.F.J.; Goldstein, F.W.; MacGowan, A.P.; Mouton, J.W.; Odenholt, I.; Rodloff, A.; Soussy, C.-J.; Steinbakk, M.; Soriano, F.; et al. European Committee on Antimicrobial Susceptibility Testing (EUCAST) Technical notes on antimicrobial susceptibility testing. *Clin. Microbiol. Infect.* **2006**, *12*, 501–503. [CrossRef] [PubMed]

45. EUCAST European Committee on Antimicrobial Susceptibility Testing. Available online: http://www.eucast.org (accessed on 20 July 2018).
46. Eddy, S.R. A new generation of homology search tools based on probabilistic inference. *Genome Inform.* **2009**, *23*, 205–211. [CrossRef] [PubMed]
47. Finn, R.D.; Coggill, P.; Eberhardt, R.Y.; Eddy, S.R.; Mistry, J.; Mitchell, A.L.; Potter, S.C.; Punta, M.; Qureshi, M.; Sangrador-Vegas, A.; et al. The Pfam protein families database: Towards a more sustainable future. *Nucleic Acids Res.* **2016**, *44*, D279–D285. [CrossRef] [PubMed]
48. Haft, D.H.; Selengut, J.D.; Richter, R.A.; Harkins, D.; Basu, M.K.; Beck, E. TIGRFAMs and genome properties in 2013. *Nucleic Acids Res.* **2012**, *41*, D387–D395. [CrossRef] [PubMed]
49. Keating, S.M.; Bornstein, B.J.; Finney, A.; Hucka, M. SBMLToolbox: An SBML toolbox for MATLAB users. *Bioinformatics* **2006**, *22*, 1275–1277. [CrossRef] [PubMed]
50. Bornstein, B.J.; Keating, S.M.; Jouraku, A.; Hucka, M. LibSBML: An API library for SBML. *Bioinformatics* **2008**, *24*, 880–881. [CrossRef] [PubMed]
51. Schellenberger, J.; Que, R.; Fleming, R.M.T.; Thiele, I.; Orth, J.D.; Feist, A.M.; Zielinski, D.C.; Bordbar, A.; Lewis, N.E.; Rahmanian, S.; et al. Quantitative prediction of cellular metabolism with constraint-based models: The COBRA Toolbox v2.0. *Nat. Protoc.* **2011**, *6*, 1290–1307. [CrossRef] [PubMed]
52. DiCenzo, G.; Mengoni, A.; Fondi, M. Tn-Core: Context-specific reconstruction of core metabolic models using Tn-seq data. *bioRxiv* **2017**, 221325. [CrossRef]
53. DiCenzo, G.C.; Checcucci, A.; Bazzicalupo, M.; Mengoni, A.; Viti, C.; Dziewit, L.; Finan, T.M.; Galardini, M.; Fondi, M. Metabolic modelling reveals the specialization of secondary replicons for niche adaptation in *Sinorhizobium meliloti*. *Nat. Commun.* **2016**, *7*, 12219. [CrossRef] [PubMed]
54. DiCenzo, G.C.; Benedict, A.B.; Fondi, M.; Walker, G.C.; Finan, T.M.; Mengoni, A.; Griffitts, J.S. Robustness encoded across essential and accessory replicons of the ecologically versatile bacterium *Sinorhizobium meliloti*. *PLoS Genet.* **2018**, *14*, e1007357. [CrossRef] [PubMed]
55. Blin, K.; Wolf, T.; Chevrette, M.G.; Lu, X.; Schwalen, C.J.; Kautsar, S.A.; Suarez Duran, H.G.; de los Santos, E.L.C.; Kim, H.U.; Nave, M.; et al. antiSMASH 4.0—Improvements in chemistry prediction and gene cluster boundary identification. *Nucleic Acids Res.* **2017**, *45*, W36–W41. [CrossRef] [PubMed]
56. Debiec, K.; Rzepa, G.; Bajda, T.; Uhrynowski, W.; Sklodowska, A.; Krzysztoforski, J.; Drewniak, L. Granulated bog iron ores as sorbents in passive (bio)remediation systems for arsenic removal. *Front. Chem.* **2018**, *6*, 54. [CrossRef] [PubMed]
57. Tardy, V.; Casiot, C.; Fernandez-Rojo, L.; Resongles, E.; Desoeuvre, A.; Joulian, C.; Battaglia-Brunet, F.; Héry, M. Temperature and nutrients as drivers of microbially mediated arsenic oxidation and removal from acid mine drainage. *Appl. Microbiol. Biotechnol.* **2018**, *102*, 2413–2424. [CrossRef] [PubMed]
58. Williams, L.E.; Baltrus, D.A.; O'Donnell, S.D.; Skelly, T.J.; Martin, M.O. Complete genome sequence of the predatory bacterium *Ensifer adhaerens* Casida A. *Genome Announc.* **2017**, *5*, e01344-17. [CrossRef] [PubMed]
59. Grissa, I.; Vergnaud, G.; Pourcel, C. CRISPRFinder: A web tool to identify clustered regularly interspaced short palindromic repeats. *Nucleic Acids Res.* **2007**, *35*, W52–W57. [CrossRef] [PubMed]
60. Lycus, P.; Lovise Bøthun, K.; Bergaust, L.; Peele Shapleigh, J.; Reier Bakken, L.; Frostegård, Å. Phenotypic and genotypic richness of denitrifiers revealed by a novel isolation strategy. *ISME J.* **2017**, *11*, 2219–2232. [CrossRef] [PubMed]
61. Casida, L.E. *Ensifer adhaerens* gen. nov., sp. nov.: A bacterial predator of bacteria in soil. *Int. J. Syst. Bacteriol.* **1982**, *32*, 339–345. [CrossRef]
62. Krzywinski, M.; Schein, J.; Birol, I.; Connors, J.; Gascoyne, R.; Horsman, D.; Jones, S.J.; Marra, M.A. Circos: An information aesthetic for comparative genomics. *Genome Res.* **2009**, *19*, 1639–1645. [CrossRef] [PubMed]
63. Geddes, B.A.; Oresnik, I.J. Physiology, genetics, and biochemistry of carbon metabolism in the alphaproteobacterium *Sinorhizobium meliloti*. *Can. J. Microbiol.* **2014**, *60*, 491–507. [CrossRef] [PubMed]
64. Silver, S.; Phung, L.T. Genes and enzymes involved in bacterial oxidation and reduction of inorganic arsenic. *Appl. Environ. Microbiol.* **2005**, *71*, 599–608. [CrossRef] [PubMed]
65. Hughes, M.F. Arsenic toxicity and potential mechanisms of action. *Toxicol. Lett.* **2002**, *133*, 1–16. [CrossRef]
66. Voegele, R.T.; Bardin, S.; Finan, T.M. Characterization of the *Rhizobium (Sinorhizobium) meliloti* high- and low-affinity phosphate uptake systems. *J. Bacteriol.* **1997**, *179*, 7226–7232. [CrossRef] [PubMed]
67. Yuan, Z.-C.; Zaheer, R.; Finan, T.M. Regulation and properties of PstSCAB, a high-affinity, high-velocity phosphate transport system of *Sinorhizobium meliloti*. *J. Bacteriol.* **2006**, *188*, 1089–1102. [CrossRef] [PubMed]

68. Elias, M.; Wellner, A.; Goldin-Azulay, K.; Chabriere, E.; Vorholt, J.A.; Erb, T.J.; Tawfik, D.S. The molecular basis of phosphate discrimination in arsenate-rich environments. *Nature* **2012**, *491*, 134–137. [CrossRef] [PubMed]
69. Pickering, B.S.; Oresnik, I.J. Formate-dependent autotrophic growth in *Sinorhizobium meliloti*. *J. Bacteriol.* **2008**, *190*, 6409–6418. [CrossRef] [PubMed]
70. Fuchs, G. Alternative pathways of carbon dioxide fixation: Insights into the early evolution of life? *Annu. Rev. Microbiol.* **2011**, *65*, 631–658. [CrossRef] [PubMed]
71. Xiu, A.; Kong, Y.; Zhou, M.; Zhu, B.; Wang, S.; Zhang, J. The chemical and digestive properties of a soluble glucan from *Agrobacterium* sp. ZX09. *Carbohydr. Polym.* **2010**, *82*, 623–628. [CrossRef]
72. Padan, E.; Tzubery, T.; Herz, K.; Kozachkov, L.; Rimon, A.; Galili, L. NhaA of *Escherichia coli*, as a model of a pH-regulated Na+/H+antiporter. *Biochim. Biophys. Acta Bioenerg.* **2004**, *1658*, 2–13. [CrossRef] [PubMed]
73. Rathore, D.S.; Lopez-Vernaza, M.A.; Doohan, F.; Connell, D.O.; Lloyd, A.; Mullins, E. Profiling antibiotic resistance and electro-transformation potential of *Ensifer adhaerens* OV14; a non-*Agrobacterium* species capable of efficient rates of plant transformation. *FEMS Microbiol. Lett.* **2015**, *362*, fnv126. [CrossRef] [PubMed]
74. Nikaido, H. Multidrug resistance in bacteria. *Annu. Rev. Biochem.* **2009**, *78*, 119–146. [CrossRef] [PubMed]
75. Xiong, X.H.; Han, S.; Wang, J.H.; Jiang, Z.H.; Chen, W.; Jia, N.; Wei, H.L.; Cheng, H.; Yang, Y.X.; Zhu, B.; et al. Complete genome sequence of the bacterium *Ketogulonicigenium vulgare* Y25. *J. Bacteriol.* **2011**, *193*, 315–316. [CrossRef] [PubMed]
76. Martin, F.A.; Posadas, D.M.; Carrica, M.C.; Cravero, S.L.; O'Callaghan, D.; Zorreguieta, A. Interplay between two RND systems mediating antimicrobial resistance in *Brucella suis*. *J. Bacteriol.* **2009**, *191*, 2530–2540. [CrossRef] [PubMed]
77. Barakat, M.A. New trends in removing heavy metals from industrial wastewater. *Arab. J. Chem.* **2011**, *4*, 361–377. [CrossRef]
78. Lièvremont, D.; N'negue, M.A.; Behra, P.; Lett, M.C. Biological oxidation of arsenite: Batch reactor experiments in presence of kutnahorite and chabazite. *Chemosphere* **2003**, *51*, 419–428. [CrossRef]
79. Hong, J.; Silva, R.A.; Park, J.; Lee, E.; Park, J.; Kim, H. Adaptation of a mixed culture of acidophiles for a tank biooxidation of refractory gold concentrates containing a high concentration of arsenic. *J. Biosci. Bioeng.* **2016**, *121*, 536–542. [CrossRef] [PubMed]
80. Kamde, K.; Pandey, R.A.; Thul, S.T.; Dahake, R.; Shinde, V.M.; Bansiwal, A. Microbially assisted arsenic removal using *Acidothiobacillus ferrooxidans* mediated by iron oxidation. *Environ. Technol. Innov.* **2018**, *10*, 78–90. [CrossRef]
81. Wang, G.; Xie, S.; Liu, X.; Wu, Y.; Liu, Y.; Zeng, T. Bio-oxidation of a high-sulfur and high-arsenic refractory gold concentrate using a two-stage process. *Miner. Eng.* **2018**, *120*, 94–101. [CrossRef]
82. Wang, H.; Zhang, X.; Zhu, M.; Tan, W. Effects of dissolved oxygen and carbon dioxide under oxygen-rich conditions on the biooxidation process of refractory gold concentrate and the microbial community. *Miner. Eng.* **2015**, *80*, 37–44. [CrossRef]
83. Wang, S.; Zhao, X. On the potential of biological treatment for arsenic contaminated soils and groundwater. *J. Environ. Manag.* **2009**, *90*, 2367–2376. [CrossRef] [PubMed]
84. Katsoyiannis, I.A.; Zouboulis, A.I. Application of biological processes for the removal of arsenic from groundwaters. *Water Res.* **2004**, *38*, 17–26. [CrossRef] [PubMed]
85. Polish Minister of Health. *Regulation of the Polish Minister of Health on the Quality of Water Intended for Human Consumption*; No. 1989; Polish Minister of Health: Warsaw, Poland, 2015.
86. Goncharuk, V.V.; Bagrii, V.A.; Mel'nik, L.A.; Chebotareva, R.D.; Bashtan, S.Y. The use of redox potential in water treatment processes. *J. Water Chem. Technol.* **2010**, *32*, 1–9. [CrossRef]
87. Debiec, K.; Rzepa, G.; Bajda, T.; Zych, L.; Krzysztoforski, J.; Sklodowska, A.; Drewniak, L. The influence of thermal treatment on bioweathering and arsenic sorption capacity of a natural iron (oxyhydr)oxide-based adsorbent. *Chemosphere* **2017**, *188*, 99–109. [CrossRef] [PubMed]
88. Polish Ministry of the Environment. *Regulation on the Conditions to Be Met during Introducing Sewage into Waters or into the Ground, and in on Substances Particularly Harmful to the Aquatic Environment*; No. 06.137.984; Polish Ministry of the Environment: Warsaw, Poland, 2006.
89. Zuniga-Soto, E.; Mullins, E.; Dedicova, B. *Ensifer*-mediated transformation: An efficient non-*Agrobacterium* protocol for the genetic modification of rice. *Springerplus* **2015**, *4*, 600. [CrossRef] [PubMed]

90. Kanehisa, M.; Sato, Y.; Kawashima, M.; Furumichi, M.; Tanabe, M. KEGG as a reference resource for gene and protein annotation. *Nucleic Acids Res.* **2016**, *44*, D457–D462. [CrossRef] [PubMed]
91. Moriya, Y.; Itoh, M.; Okuda, S.; Yoshizawa, A.C.; Kanehisa, M. KAAS: An automatic genome annotation and pathway reconstruction server. *Nucleic Acids Res.* **2007**, *35*, W182–W185. [CrossRef] [PubMed]

© 2018 by the authors. Licensee MDPI, Basel, Switzerland. This article is an open access article distributed under the terms and conditions of the Creative Commons Attribution (CC BY) license (http://creativecommons.org/licenses/by/4.0/).

Article

Possible Role of Envelope Components in the Extreme Copper Resistance of the Biomining *Acidithiobacillus ferrooxidans*

Nia Oetiker [1], Rodrigo Norambuena [1], Cristóbal Martínez-Bussenius [1], Claudio A. Navarro [1], Fernando Amaya [2], Sergio A. Álvarez [2], Alberto Paradela [3] and Carlos A. Jerez [1,*]

1. Laboratory of Molecular Microbiology and Biotechnology, Department of Biology, Faculty of Sciences, University of Chile, Santiago 7800003, Chile; nia.oetiker.mancilla@gmail.com (N.O.); rodrigoanv93@gmail.com (R.N.); cm.bussenius@gmail.com (C.M.-B.); clnavarrol@gmail.com (C.A.N.)
2. Department of Biochemistry and Molecular Biology, Faculty of Chemical and Pharmaceutical Sciences, University of Chile, Santiago 7800003, Chile; fernando.amaya@ug.uchile.cl (F.A.); salvarez@uchile.com (S.A.Á.)
3. Proteomics Laboratory, National Biotechnology Center, CSIC, 28049 Madrid, Spain; alberto.paradela@cnb.csic.es
* Correspondence: cjerez@uchile.cl; Tel.: +56-2-29787376

Received: 30 May 2018; Accepted: 3 July 2018; Published: 10 July 2018

Abstract: *Acidithiobacillus ferrooxidans* resists extremely high concentrations of copper. Strain ATCC 53993 is much more resistant to the metal compared with strain ATCC 23270, possibly due to the presence of a genomic island in the former one. The global response of strain ATCC 53993 to copper was analyzed using iTRAQ (isobaric tag for relative and absolute quantitation) quantitative proteomics. Sixty-seven proteins changed their levels of synthesis in the presence of the metal. On addition of CusCBA efflux system proteins, increased levels of other envelope proteins, such as a putative periplasmic glucan biosynthesis protein (MdoG) involved in the osmoregulated synthesis of glucans and a putative antigen O polymerase (Wzy), were seen in the presence of copper. The expression of *A. ferrooxidans mdoG* or *wzy* genes in a copper sensitive *Escherichia coli* conferred it a higher metal resistance, suggesting the possible role of these components in copper resistance of *A. ferrooxidans*. Transcriptional levels of genes *wzy*, *rfaE* and *wzz* also increased in strain ATCC 23270 grown in the presence of copper, but not in strain ATCC 53993. Additionally, in the absence of this metal, lipopolysaccharide (LPS) amounts were 3-fold higher in *A. ferrooxidans* ATCC 53993 compared with strain 23270. Nevertheless, both strains grown in the presence of copper contained similar LPS quantities, suggesting that strain 23270 synthesizes higher amounts of LPS to resist the metal. On the other hand, several porins diminished their levels in the presence of copper. The data presented here point to an essential role for several envelope components in the extreme copper resistance by this industrially important acidophilic bacterium.

Keywords: *Acidithiobacillus ferrooxidans*; copper resistance; biomining; envelope components; proteomics; lipopolysaccharide

1. Introduction

Acidithiobacillus ferrooxidans is a gram-negative, acidophilic, chemolithoautotrophic bacterium able to use ferrous iron, reduced species of sulfur or metal sulfides as energy sources [1–5]. These bacteria are able to grow at high concentrations of several metals. This is an important property since they are used in biomining processes where copper concentrations are in the range of 15 to 100 mM [6–8]. Furthermore, these microorganisms can be used to exploit these natural resources sustainably [9].

Current knowledge indicates that *A. ferrooxidans* uses key elements involved in copper resistance in all bacteria [10–12], but in addition, it may have a broader repertoire of these known copper resistance determinants [13,14].

In the biomining environment, copper and other toxic metals are present in concentrations that are one to two orders of magnitude greater than those tolerated by neutrophils [6,15–18]. Most likely, the microorganisms forming part of the biomining consortium have developed additional strategies to resist the harsh conditions in which they live, and their study is therefore of great interest [19,20].

A. ferrooxidans ATCC 53993 is much more resistant to copper and other metals than *A. ferrooxidans* ATCC 23270. Both strains have the same copper resistance determinants but strain ATCC 53993 contains a genomic island (GI) having 160 extra genes, some of which code for additional copies of proteins involved in copper tolerance [13,14].

The response to copper of both *A. ferrooxidans* ATCC 53993 and ATCC 23270 was previously compared at 40 mM $CuSO_4$ [14,21,22]. These preliminary studies were done by using ICPL (isotope-coded protein labeling) quantitative proteomics [22] and showed that strain ATCC 23270 synthesized much more oxidative-stress-related proteins than strain 53993 in response to copper, clearly indicating that the former strain is much more sensitive to the metal [22]. A high overexpression of RND (Resistance-Nodulation-Division) efflux systems and copper periplasmic chaperones CusF were seen in both strains subjected to copper. However, in strain ATCC 53993 both of its additional genes present in its genomic island were also overexpressed. This behavior suggested a possible explanation for the much higher copper resistance of strain ATCC 53993. In addition, changes in the levels of the respiratory system copper-binding proteins AcoP, Rus and several other proteins with predicted functions suggested that numerous metabolic changes are involved in controlling the effects of the toxic metal in strain ATCC 53993 [22].

To understand in more detail the reason by which *A. ferrooxidans* ATCC 53993 stands higher copper concentrations compared with strain 23270, iTRAQ (isobaric tag for relative and absolute quantitation) proteomics, transcriptional expression of genes of interest and functional assays were used in the current report. Increased levels of novel possible copper resistance determinants present in the envelope of *A. ferrooxidans* ATCC 53993 such as outer membrane proteins, the periplasmic glucans synthesizing protein MdoG and proteins involved in lipopolysaccharide (LPS) synthesis, amongst others, were found in cells grown in the presence of copper. In addition, determination of the relative amounts of LPS present in the cells of each *A. ferrooxidans* strain also supports the idea that these polymers may also have an important role in copper resistance in these biomining bacteria.

2. Materials and Methods

2.1. Bacterial Strains and Growth Conditions

A. ferrooxidans strains ATCC 53993 and ATCC 23270 were grown at 30 °C in liquid 9 K medium containing ferrous sulfate (33.33 g/L) with an initial pH of 1.45 as previously described [23] and in absence or presence of $CuSO_4$. Copper concentrations between 40 and 300 mM were used depending on the experiment. In some experiments, concentrations of 100 or 200 mM were used for *A. ferooxidans* ATCC 53993 without prior adaptation since under these two conditions, similar cells numbers to control cells (in absence of copper) were obtained at their respective late exponential growth phases. At 300 mM copper strain ATCC 53993 required previous adaptation. On the other hand, to compare the effect of the metal at the same copper concentration in both strains, it was necessary to adapt strain ATCC 23270 to grow at 100 mM copper. For LPS determinations, strain ATCC 23270 was adapted to grow at 100 mM and strain ATCC 53993 to 200 mM copper. These adaptations were done starting with cells grown at 50 mM copper by increasing 5 mM copper in each successive culture until the desired concentrations were reached. After cells attained late exponential growth phases they were collected and triplicate separate cultures were employed for all experiments. Bacterial growth was determined by measuring the increase in cell numbers by using an Olympus BX50 optical microscope (Olympus, Tokyo, Japan) and a Petroff–Hausser counting chamber (Horsham, PA, USA).

2.2. Preparation of Total Protein Extracts for iTRAQ Analysis

A. ferrooxidans ATCC 53993 was grown with ferrous iron as oxidizable substrate until late exponential phase in absence or presence of 100 or 200 mM $CuSO_4$. Cells were harvested by centrifugation (4000× g for 15 min) and washed three times by centrifugation at 4 °C with dilute sulfuric acid (pH 1.5). This was followed by three washes with 50 mM sodium citrate, pH 7.0 by centrifugation at 4 °C to remove any minor ferrous iron remaining and at the same time, to neutralize the pH before cell rupture by sonic oscillation. Cells were then resuspended in sonication buffer (50 mM Tris-HCl pH 8.0, 1 mM ethylenediaminetetra-acetic acid (EDTA) containing phenylmethylsulfonyl fluoride (PMSF) as protease inhibitor (100 μg/mL) and were disrupted by sonic oscillation during 25 min on ice by using successive 5 s pulses and pauses. Finally, the lysate was centrifuged at 10,000× g for 10 min to remove unbroken cells and cell debris and the total protein amount in the cell-free extract was determined [21].

2.3. Protein Digestion and Tagging with iTRAQ-8-Plex® Reagent

Total protein concentration was determined using microBCA protein assay kit (Pierce, Appleton, WI, USA). For digestion, 50 μg of protein from each condition was precipitated by the methanol/chloroform method. Protein pellets were resuspended and denatured in 20 μL of 7 M urea, 2 M thiourea, 100 mM TEAB (triethylammonium bicarbonate), reduced with 2 μL of 50 mM Tris 2-carboxyethyl phosphine (TCEP) (AB SCIEX, Foster City, CA, USA), pH 8.0, at 37 °C for 60 min and followed by 2 μL of 200 mM cysteine-blocking reagent methyl methanethiosulfonate (MMTS) (Pierce) for 10 min at room temperature. Samples were diluted up to 120 μL with 50 mM TEAB to reduce the concentration of urea. Two μg of sequence grade-modified trypsin (Sigma-Aldrich, St. Louis, MO, USA) was added to each sample (ratio 1:25 enzyme:sample, which were then incubated at 37 °C overnight on a shaker. After digestion, samples were dried in a SpeedVac (Thermo Scientific, Waltham, MA, USA).

Each sample was reconstituted with 180 μL of 70% ethanol/50 mM TEAB, the different versions of the iTRAQ reagent 8-plex (AB SCIEX) were added in additional 20 μL and the mixture was incubated for 2 h at room temperature, according to the following labeling scheme: iTRAQ 113/117 reagent: control 1 and control 2 *A. ferrooxidans*; iTRAQ 115/119 reagent: *A. ferrooxidans* grown in 100 mM $CuSO_4$, 1 and grown in 100 mM $CuSO_4$, 2; iTRAQ 116/121 reagent: *A. ferrooxidans* grown in 200 mM $CuSO_4$, 1 and grown in 200 mM $CuSO_4$, 2. Two biological replicas were used in each case. After labeling, samples were combined and the reaction was stopped by evaporation in the SpeedVac.

2.4. Liquid Chromatography and Mass Spectrometry Analysis

A 2-μg aliquot of the combined sample was subjected to 2D-nano Liquid Chromatography-Electrospray Ionization Tandem Mass Spectrometry LC ESI-MSMS analysis using a nano liquid chromatography system nanoLC Ultra 1D plus, (Eksigent Technologies, AB SCIEX) coupled to a Quadrupole time of flight (QTOF) type, high speed Triple TOF 5600 mass spectrometer (AB SCIEX) equipped with a nanospray source. Injection volume was 5 μL and three independent technical replicas were analyzed. The analytical column used was a silica-based reversed phase Acquity UPLC Peptide BEH C18 column 75 μm × 15 cm, 1.7 μm particle size and 130 Å pore size (Waters, Dublin, Ireland). The trap column was a C18 Acclaim PepMap (Eksigent Technologies, AB SCIEX), 100 μm × 2 cm, 5 μm particle diameter, 100 Å pore size, switched on-line with the analytical column. The loading pump delivered a solution of 0.1% formic acid in water at 2 μL/min. The nano-pump provided a flow-rate of 300 nL/min and was operated under gradient elution conditions, using 0.1% formic acid in water as mobile phase A, and 0.1% formic acid in acetonitrile as mobile phase B. Gradient elution was performed according to the following scheme: Isocratic conditions of 96% A: 4% B for 5 min, a linear increase to 40% B in 205 min, then a linear increase to 90% B for 15 additional minutes, isocratic conditions of 90% B for 10 min and return to initial conditions in 2 min. Total gradient length was 250 min.

Data acquisition was performed with a TripleTOF 5600 System (AB SCIEX). Ionization occurred under the following conditions: Ionspray voltage floating (ISVF) 2800 V, curtain gas (CUR) 20, interface heater temperature (IHT) 150, ion source gas 1 (GS1) 20, declustering potential (DP) 85 V. All data was acquired using information-dependent acquisition (IDA) mode with Analyst TF 1.5 software (AB SCIEX). For IDA parameters, 0.25 s MS survey scan in the mass range of 350–1250 Da were followed by 25 MS/MS scans of 150 ms in the mass range of 100–1500 (total cycle time: 4 s). Switching criteria were set to ions greater than mass to charge ratio (m/z) 350 and smaller than m/z 1250 with charge state of 2–5 and an abundance threshold of more than 90 counts (cps). Former target ions were excluded for 20 s. IDA rolling collision energy (CE) parameters script was used for automatically controlling the CE.

2.5. Data Analysis and Statistics

MS/MS spectra were exported to Mascot generic format (mgf) using Peak View v1.2.0.3 and searched using OMSSA 2.1.9, X!TANDEM 2013.02.01.1, Myrimatch 2.2.140 and MS-GF+ (Beta v10072) [24] against a composite target/decoy database built from the 2748 *A. ferrooxidans* sequences at UniprotKB (June 2014). Search engines were configured to match potential peptide candidates with mass error tolerance of 25 ppm and fragment ion tolerance of 0.02 Da, allowing for up to two missed tryptic cleavage sites and a maximum isotope error (13C) of 1, considering fixed MMTS modification of cysteine and variable oxidation of methionine, pyroglutamic acid from glutamine or glutamic acid at the peptide N-terminus, and modification of lysine and peptide N-terminus with iTRAQ 8-plex reagents. Score distribution models were used to compute peptide-spectrum match p-values [24], and spectra recovered by a false discovery rate (FDR) $\leq$ 0.01 (peptide-level) filter were selected for quantitative analysis. Approximately 5% of the signals with the lowest quality were removed prior to further analysis. Differential regulation was measured using linear models [25], and statistical significance was measured using q-values (FDR). All analyses were conducted using software from Proteobotics (Madrid, Spain) [24].

2.6. Extraction of Total RNA from Acidithiobacillus ferrooxidans and Complementary DNA Synthesis

To determine the effect of copper on the expression of some genes of interest, *A. ferrooxidans* ATCC 23270 and ATCC 53993 cells were grown in absence or presence of $CuSO_4$ until cells reached late exponential phase of growth. At this time, total RNA was extracted from each culture condition by lysing the cells as previously reported [26], except that TRIzol (Invitrogen, Carlsbad, CA, USA) was used for the extraction [27,28]. Between three to five biological replicas were used for each experimental condition. Any remaining DNA was eliminated from RNA preparations by addition of 4 U of TURBO DNA-free DNase (Ambion, Thermo Scientific) following manufacturer's instructions. For complementary DNA (cDNA) synthesis, 0.8 µg of total RNA was reverse transcribed for 1 h at 42 °C using ImProm-II (Promega, Madison, WI, USA) reverse transcription system, 0.5 µg of random hexamers (Promega) and 3 mM $MgCl_2$ [28].

2.7. Primer Design, Real-Time PCR and Cloning of A. ferrooxidans Genes

Primers for quantitative real time PCR (qRT-PCR) were designed using the Primer3 software [29]. After separating PCR products by electrophoresis in a 1% agarose gel (0.5× Tris–acetate–EDTA pH 8.0 buffer), no cross-amplification or non-specific bands were detected. Copper-resistance related gene expression was analyzed by qRT-PCR with either the Corbett Rotor Gene 6000 system as described previously [21] or with the 96-well PikoReal Real-Time PCR System and Thermo Scientific PikoReal Software 2.2. Efficiency of each primer pair was calculated from the average slope of a linear regression curve, which resulted from qPCRs using a 10-fold dilution series (10 pg–10 ng) of *A. ferrooxidans* DNA as template. Efficiencies between 90 and 110% were used. Quantification cycle (Cq) values were automatically determined by Real-Time Rotor-gene 6000 PCR software (Corbett Life Sciences, Thermo Scientific/Qiagen, Hilden, Germany) or by Thermo Scientific PikoReal Software 2.2.

For transcriptional analysis of the different genes studied, a relative quantification method was used which is based in the ratio between the transcripts of a study sample (in presence of copper) versus a control sample (no copper) [30]. 16S rRNA$_{Af}$ was selected as a reference gene since its expression was found to be the most stable under our experimental conditions. To carry out the real-time PCR, 0.5 µL of 1:20 diluted cDNA or 0.5 µL of 1:200 diluted 16S rRNA$_{Af}$, 0.2 µL of each primer (10 µM) and 5.0 µL of master mix Rotor-Gene SYBR Green PCR (Qiagen) in a final volume of 10 µL, completed with RNA-free water were used. The program used was 10 min at 95 °C followed by 40 cycles of 5 s at 95 °C and 20 s at 60 °C.

2.8. Cloning A. ferrooxidans Genes in an Expression Vector

The functionality of different putative copper resistance genes from *A. ferrooxidans* was tested by using heterologous expression in *Escherichia coli*. A copper-sensitive *E. coli* K-12 (ΔcopA/ΔcusCFBA/ΔcueO) mutant was transformed with vector pTrc-His2A (Invitrogen) containing the genes of interest under the control of a promoter induced by IPTG, and minimal inhibitory concentration (MIC) values of these transformants were determined as described before [14,28].

2.9. Lipopolysaccharide Extraction

A. ferrooxidans cells grown in absence or presence of CuSO$_4$ were harvested by centrifugation (10,000× *g* for 5 min, at 4 °C). Cell pellets were washed twice with sulfuric acid solution (pH 1.5) and twice with 10 mM sodium citrate (pH 7) by resuspension followed by centrifugation (9200× *g* for 1 min). Cells were then resuspended in sulfuric acid solution. To normalize the number of cells, optical density of cell suspensions was measured at 600 nm, adjusting them to an optical density of 2 (OD$_{600nm}$ = 2) in 1 mL of sulfuric acid solution. Cell suspensions were then centrifuged at 10,000× *g* for 5 min. A partially modified Hitchcock & Brown method for LPS extraction was used [31]. The cell pellet was resuspended in 90 µL of lysis buffer solution (2% Sodium dodecyl sulfate (SDS); 4% 2-ME; 0.5 M Tris-HCl, pH 9.0). The suspension was heated for 30 min at 100 °C. Lysed cells were then digested with 100 µg/mL of DNase I (Ambion) for 90 min at 37 °C. Samples were thereafter treated with 1 mg/mL of Proteinase K (Sigma-Aldrich) for 90 min at 60 °C. Finally, samples were dialyzed for 30 min against nano-pure water using a nitrocellulose membrane, 0.025 µm pore size. Dialyzed samples were finally stored at 4 °C for further analysis.

2.10. Lipopolysaccharide Quantification

Extracted LPS was quantified by purpald assay [32]. Unsubstituted terminal vicinal glycol (UTVG) groups of the sugar residues such as Kdo and heptose in LPS can be subjected to periodate oxidation, yielding quantitative formaldehyde measurable by the purpald reagent. This assay provides the molarity of the UTVG present in LPS. LPS molarity can be found by dividing the molarity of the UTVG by the theoretical number of UTVG per LPS molecule. The numbers of UTVG present in LPS of *A. ferrooxidans* is unknown. Therefore, LPS concentration was expressed in relation to the molarity of UTVG present in each sample. The experimental procedure was carried out as previously described [33].

3. Results and Discussion

3.1. Proteomic Analysis of the Copper Response of A. ferrooxidans ATCC 53993

Proteins of cells grown in ferrous iron and in presence of 100 or 200 mM CuSO$_4$ were analyzed by quantitative iTRAQ proteomics. In cells subjected to 100 mM of copper 1656 proteins were identified, of which 28 changed their levels compared to control cells grown in absence of copper. Of these proteins, 11 had higher levels and 17 showed lower amounts (Table S1). On the other hand, in cells grown in 200 mM copper 1567 proteins were identified and 59 of these showed changes in their levels compared to the control. Seventeen showed higher levels than the control and 42 lower amounts (Table S1). This corresponds to about 2-fold more proteins changing at 200 mM than at 100 mM copper (Table S1). Most of the proteins changing at 100 mM copper were also seen to vary at 200 mM of the metal (Tables S2 and S3). The functional categories of all proteins changing in *A. ferrooxidans* 53993 are shown in Table S1 and Figure S1 and the data obtained is seen in Tables S2 and S3. Although at 200 mM copper ATCC 53993 cells grew reaching similar numbers to the control, they were apparently more affected than cells subjected to 100 mM copper since a greater number of proteins related to metabolism and protein biosynthesis decreased their levels whereas others related to energy production and copper resistance increased their amounts. Nevertheless, at 200 mM copper cells are still actively expressing the proteins related to the RND efflux systems (Table 1), as seen before at 40 mM copper sulfate [22]. In addition, an interesting group of proteins that also form part of the cell envelope changed their synthesis levels in presence of the metal (Table 1). Most of these proteins may be new possible copper resistance determinants present in *A. ferrooxidans* ATCC 53993.

Table 1. Levels of some selected known and new possible copper resistance determinants in *A. ferrooxidans* ATCC 53993 grown in the presence of 200 mM CuSO$_4$.

Function/Similarity	ORF	Name	q Value (FDR)	Coverage (%)	Peptide Number	Log$_2$ Fold Change (Cu 200/0 mM)
Outer membrane efflux protein	Lferr_1619	CusC1	0.001	45	9	1.258
Efflux transporter, RND family, MFP subunit	Lferr_1618	CusB1	0.001	63.3	15	0.859
Uncharacterized protein	Lferr_2057	CusF2	0.001	60	3	1.92
Uncharacterized protein	Lferr_0174	CusF3	0.001	60	3	1.63
Heavy metal efflux pump, CzcA family	Lferr_0172	CusA3	0.001	39.1	10	1.019
Outer membrane efflux protein	Lferr_2062	CusC2	0	45	12	1.084
Heavy metal efflux pump, CzcA	Lferr_1617	CusA1	0.002	35	9	0.855
Efflux transporter, RND family, MFP subunit	Lferr_2061	CusB2	0.003	68.3	6	0.968
Heavy metal efflux pump, CzcA family	Lferr_2060	CusA2	0.003	38.3	9	0.846
Periplasmic glucan biosynthesis protein MdoG	Lferr_1075	MdoG	0.009	48.1	15	0.415
Carbohydrate-selective porin OprB	Lferr_1898	OprB	0.005	36.46	10	−0.635
O-antigen polymerase	Lferr_0408	Wzy	0.026	3.19	1	1.795

3.2. Overexpression of Resistance-Nodulation-Division Efflux Transporters and Possible Generation of Excess Acidity

Transcriptional levels of some genes coding for possible components of the RND family of efflux transporters [10] are shown in Figure 1. These correspond to most of the genes coding for Cus system components in both *A. ferrooxidans* ATCC 53993 [20,22] and ATCC 23270 [28]. All these genes showed increased transcriptional levels in cells grown at the indicated copper concentrations. These Cus transporter systems are widely present in bacteria to remove copper from the cell [10]. *A. ferrooxidans* lives at an acid external pH (1–3) and its cytoplasmic pH is up to 5 units higher than external pH. This generates an elevated pH gradient across the cytoplasmic membrane that contributes to the proton motive force (PMF) comprising membrane potential ($\Delta\Psi$) and transmembrane pH difference

(ΔpH) [22,34]. RND type transporters are antiporters taking advantage of the proton gradient to efflux copper with the concomitant protons entrance to the cytoplasm. Due to its economy from the energetic point of view, these systems would be used preferentially by acidophilic microorganisms to remove intracellular copper.

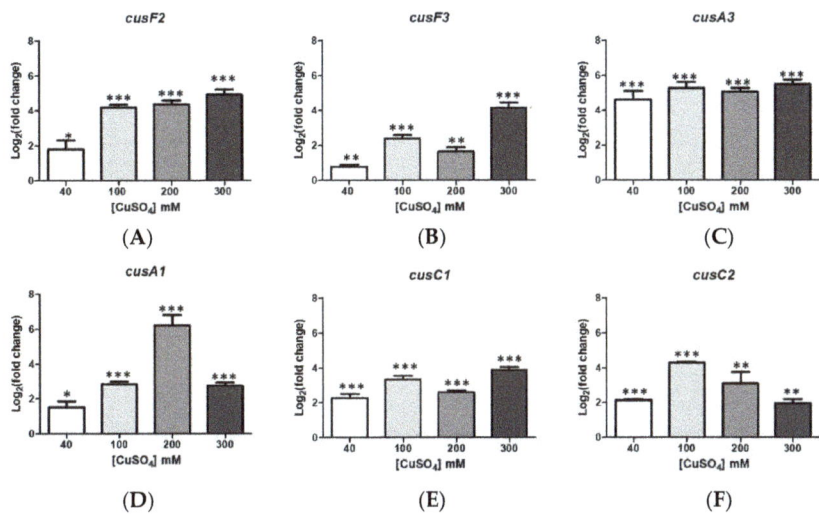

Figure 1. Transcriptional levels of several genes coding for Cus system components in *Acidithiobacillus ferrooxidans* ATCC 53993. The transcriptional levels of genes (**A**) *cusF2*; (**B**) *cusF3*; (**C**) *cusA3*; (**D**) *cusA1*; (**E**) *cusC1* and (**F**) *cusC2* were determined at the indicated copper concentrations as described in Material and Methods section. Error bars indicate the standard deviations based on three different experimental values. Application of *t*-Student test were: *** $p \leq 0.001$ ** $p \leq 0.01$ and * $p \leq 0.05$.

A possible cytoplasmic acidification would be expected if these efflux pumps were excessively used by *A. ferrooxidans* in presence of high metal concentrations. Conversely, as previously pointed out [22], this acidification could be diminished by the energetic metabolism of the bacterium, since oxidation of Fe(II) by molecular oxygen as the final electron acceptor consumes protons [35]. Still, RND systems may introduce an excess of protons from the acid culture medium to the cell during copper detoxification. This idea has not been demonstrated in *A. ferrooxidans*. However, a possible increase in the extracellular pH of the growth medium could be expected during growth in the presence of copper. Figure 2A shows growth curves of *A. ferrooxidans* ATCC 53993 and how cell growth was affected at the indicated copper concentrations. Initially, there was a strong partial inhibition of growth only at 300 mM copper.

Figure 2B clearly shows an increase in external pH of the growth medium (reaching around 0.6 pH units at 300 mM copper). On the contrary, pH values of control media containing different copper concentrations and no cells inoculated showed only very minor pH variations. Clearly, whether this interesting preliminary observation is due to intracellular acidification remains to be demonstrated.

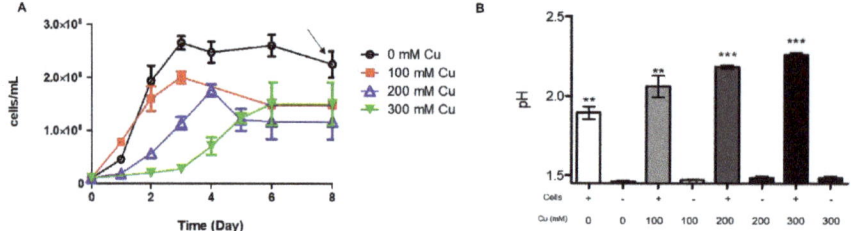

Figure 2. Growth medium pH changes of *A. ferrooxidans* ATCC 53993 grown in absence or presence of copper. (**A**) Cells were grown in ferrous iron medium at the indicated copper concentrations. Once cells reached stationary phase of growth (indicated by the arrow), aliquots of the cultures were taken and centrifuged to remove cells; (**B**) pH values of the media supernatants were determined and compared with pH changes of the medium containing the same copper concentrations but in absence of inoculated cells. Error bars indicate standard deviations based on three different experimental values. Application of *t*-Student test were: *** $p \leq 0.001$ ** $p \leq 0.01$ and * $p \leq 0.05$.

3.3. Changes of Several Additional Envelope Components Occur in Presence of High Copper Concentrations

Another protein with increased levels in cells grown in presence of copper was a putative periplasmic glucan biosynthesis protein MdoG coded by *Lferr_1075* (Table 1). This protein is involved in the synthesis of ramifications present in the osmoregulated periplasmic glucans (OPGs) in bacteria. These OPGs are present in all known proteobacteria and are formed by 5–24 D-glucose molecules bound by means of β-glycosidic bonds. The concentration of these glucans has been reported to change with variations in periplasmic osmolarity [36]. Due to their big size, OPGs are trapped in the periplasm, being unable to diffuse to the outside of cells. In *E. coli*, the carbon skeleton is synthesized by proteins coded by genes *opgG* (*mdog* orthologous) and *opgH*. OpgH is a glucosyl transferase that synthesizes the lineal skeleton of glucose units by means of β-1,2 bonds. In *E. coli*, MdoG is a 56 kDa periplasmic protein necessary for the polymerization of sugar molecules, although its function has not being completely established [36]. In *A. ferrooxidans*, putative MdoG (57.4 kDa) has been previously identified as a component of its periplasm [37]. In *E. coli* both *opgG* and *opgH* genes form part of the same operon. By analyzing the genomic context of *A. ferrooxidans* instead, it was found that *mdoG* and *mdoH* are separated by an open reading frame (ORF) coding for a protein of unknown function. On the other hand, protein MdoH did not change its levels in the results obtained here. The system for OPGs synthesis in *E. coli* involves four additional proteins (OpgD, OpgB, OpgC and OpgE) whose equivalent genes are absent in *A. ferrooxidans* genome. Nevertheless, only OpgG and OpgH are strictly necessary for the OPGs synthesis in *E. coli* [38]. OPGs biosynthesis starts with glucose transport to form glucose-6P, which is used to generate (uridine diphosphate glucose) UDP-glucose for production of OPGs via OpgH/OpgG [36]. It is known this molecule is formed by a glucose 1-phosphate uridil transferase that catalyzes the UTP and a proton addition to D-glucose-1-phosphate to generate UDP-D-glucose [39]. It can be suggested that generation of OPGs in *A. ferrooxidans* would involve also a higher UDP-glucose synthesis. Since this process consumes protons, it should alleviate excessive entrance of these cations when RND efflux pumps are heavily used to remove copper.

The CusA proton/Cu antiporter system is overexpressed in *A. ferrooxidans* subjected to copper as already seen in Figure 1, and under those conditions, a higher number of protons would be expected to enter the cytoplasm from the growth medium, as already suggested by the results shown in Figure 2. Thus, a higher synthesis of OPGs would also consume protons, in favor of keeping the normal cytoplasmic pH. Furthermore, it has been shown that an OPGs preparation acts as a blocker and a regulator of an OMPC-like porin channel selective of cations in *E. coli* [40]. On the other hand, cells unable to form OPGs showed an increased synthesis of OmpC [41]. It has also been documented that porins mediate copper entrance in *Mycobacterium tuberculosis* [42]. The existence of a relationship between both copper entrance and porins closing or decreasing their levels of synthesis is possible then, as seen here for OprB in *A. ferrooxidans* ATCC 53993 (Table 1). Examples of this behavior were previously reported for the major *A. ferrooxidans* porin Omp40 (Afe_2741) and OmpA (Afe_2685) in *A. ferrooxidans* ATCC 23270 [21].

To support proteomic results, transcriptional levels of genes coding for proteins MdoG and porins, were also determined in cells grown at different copper concentrations as shown in Figure 3. The results clearly indicate increasing levels of synthesis of mRNA coding for MdoG and decreasing levels of messenger RNAs for porin genes *omp40*, *oprB* and *ompA*, confirming the proteomic results already discussed. It is, therefore, possible that lower levels of porins, together with higher OPGs amounts, constitute a defense response to extreme copper conditions as seen here in *A. ferrooxidans* ATCC 53993, an idea that should be proven.

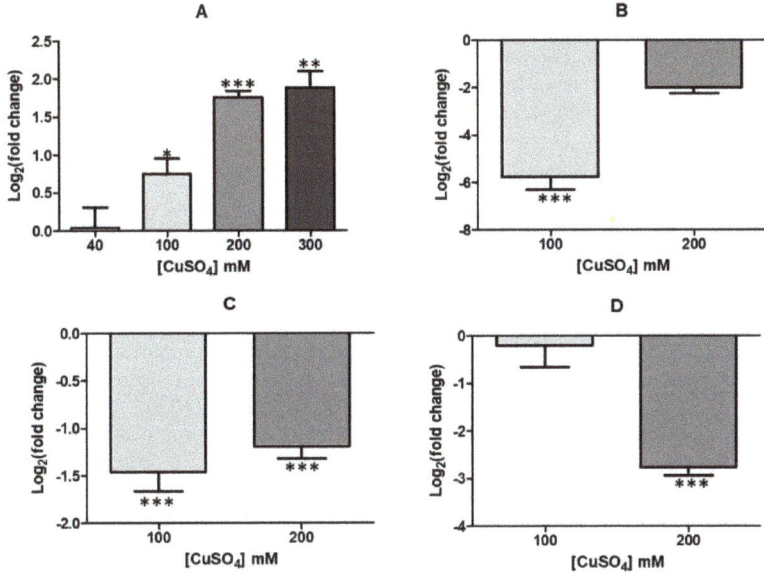

Figure 3. Transcriptional levels of selected envelope genes. (**A**) *mdoG*; (**B**) *oprB*, (**C**) *ompA* and (**D**) *omp40* in *A. ferrooxidans* ATCC 53993 grown in different copper concentrations. Error bars indicate standard deviations based on three different experimental values. Application of *t*-Student test were: *** $p \leq 0.001$ ** $p \leq 0.01$ and * $p \leq 0.05$.

Currently there are no efficient and easy to reproduce methods to generate knock-outs of genes in *A. ferrooxidans* [43]. Therefore, to ascertain whether the *mdoG* gene confers Cu-resistance to a heterologous host, it was expressed in *E. coli* as described in Materials and Methods. As seen in Figure 4, *A. ferrooxidans* putative *mdoG* gene conferred resistance to Cu when expressed in *E. coli* due to the increase of its MIC value from 1.0 to 3.0 mM copper. This result supports the possibility of MdoG being a copper resistance determinant in this acidophilic microorganism. In addition, the effect of overexpressing *mdoG* in *E. coli* was also tested in cells grown in the presence of Zn or Ni as shown in Figure 4.

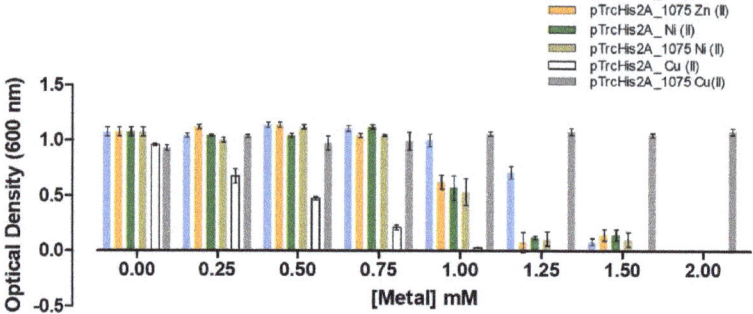

Figure 4. Heterologous functional analysis of the overexpression of *A. ferrooxidans mdoG* (*Lferr_1075*) gene in the Cu-sensitive *Escherichia coli* (K-12 ΔcopA/ΔcusCFBA/ΔcueO) grown in zinc (Zn), nickel (Ni) and copper (Cu). pTrcHis2A, empty vector; pTrcHis2A_1075 contains *mdoG* gene. Error bars indicate standard deviations based on three different experimental values.

Compared with copper, MdoG did not confer tolerance to Ni and Zn. Interestingly, a proteomic analysis of the response of *Rhodobacter sphaeroides* to high cobalt concentrations has been reported [44]. One of the changing proteins in presence of the metal was MdoG. It was previously suggested that cobalt would generate an alteration of permeability of the envelope, periplasm or cell wall as a possible resistance mechanism in this microorganism [44]. Whether the effect of copper is rather specific for MdoG from *A. ferrooxidans* remains to be elucidated. Another interesting protein found to be overexpressed in presence of copper was *Lferr_0408* (Wzy) (Table 1), a protein involved in O-antigen biosynthesis, the most external segment of LPS [45,46]. By expressing gene *wzy* from *A. ferrooxidans* in the Cu-sensitive *E. coli* (K-12 ΔcopA/ΔcusCFBA/ΔcueO) strain already used for *mdoG* gene, the results seen in Figure 5 were obtained. Once again, it is clear that expressing *wzy* gene in the heterologous host confers it a higher copper resistance.

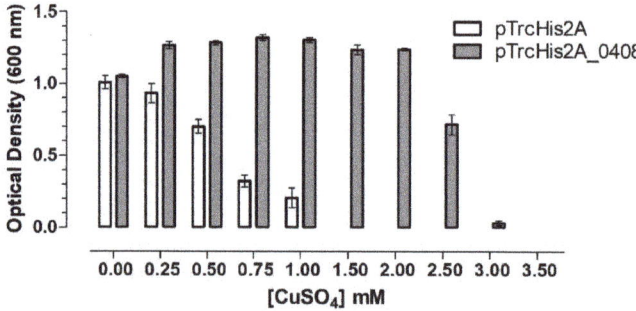

Figure 5. Heterologous functional analysis of overexpression of *A. ferrooxidans* antigen-O polymerase gene *wzy* (*Lferr_0408*) in Cu-sensitive *E. coli* (K-12 ΔcopA/ΔcusCFBA/ΔcueO) grown in copper. pTrcHis2A, empty vector; pTrcHis2A_0408 contains gene *wzy*. Error bars indicate standard deviations based on three different experimental values.

To support this result, the analysis of transcriptional expression of this and other genes involved in LPS generation was carried out. Figure 6 shows the levels of transcriptional expression of *wzy*, *wzz* and *rfaE* genes, all involved with LPS synthesis, in both *A. ferrooxidans* strains grown in the absence or presence of copper.

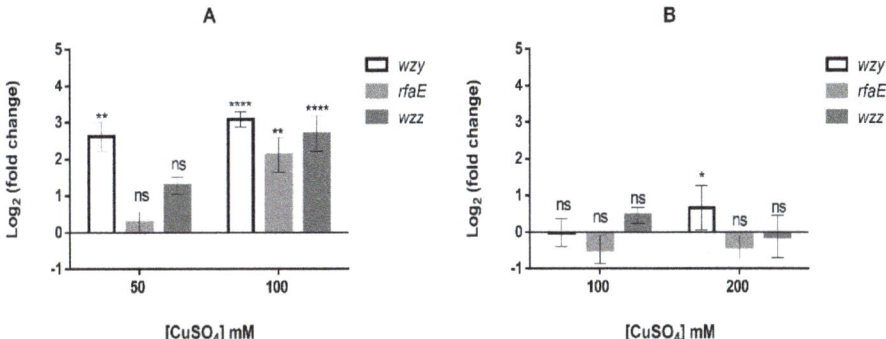

Figure 6. Transcriptional levels of genes *wzy*, *wzz* and *rfaE* related to lipopolysaccharides (LPS) synthesis in *A. ferrooxidans* exposed to copper. (**A**) Strain ATCC 23270; (**B**) Strain ATCC 53993. Values were obtained from three biological replicates. Error bars indicate standard deviations based on three different experimental values. Application of *t*-Student test were: **** $p \leq 0.0001$ ** $p \leq 0.01$ and * $p \leq 0.05$.

Strain ATCC 23270 clearly showed an increased level in expression of tested genes in presence of the metal (Figure 6A). On the contrary, the same genes did not show significant changes in their expression when strain ATCC 53993 was grown in presence of copper (Figure 6B). These results strongly suggest strain ATCC 23270 could synthesize higher amounts of LPS in presence of copper compared to ATCC 53993. However, when the amounts of LPS were determined in both strains in absence of the metal, ATCC 53993 showed about 3-fold higher amounts of LPS compared with ATCC 23270 (Figure 7A). This result indicates that normally, strain ATCC 53993 in addition of having extra copper resistance determinants in its genomic island, contains higher LPS levels compared with strain ATCC 23270. This could explain in part the higher copper tolerance of the former strain.

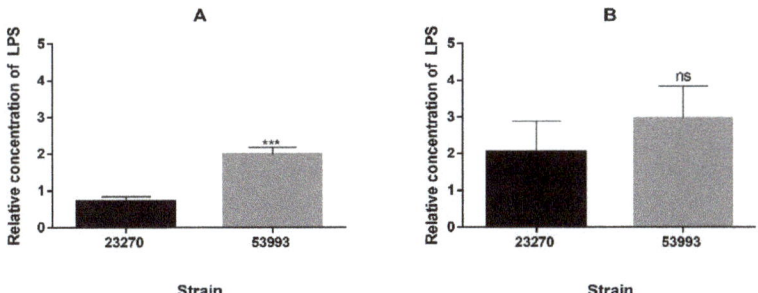

Figure 7. Relative LPS concentration of *A. ferrooxidans* ATCC 23270 and ATCC 53993 grown at different concentrations of CuSO$_4$. (**A**) Cells grown in absence of copper; (**B**) Cells grown in 100 mM CuSO$_4$. Values were obtained from three biological replicates. Error bars represent standard deviations for each condition. A *t*-Student statistic analysis was performed, where: *** indicates $p \leq 0.001$ and ns indicates $p > 0.05$.

Nonetheless, in presence of 100 mM copper both strains showed similar LPS levels (Figure 7B), suggesting that strain 23270 increases its LPS levels in presence of the metal, in agreement with results of the transcriptional expression of its genes in the presence of copper seen in Figure 6. Previously, *A. ferrooxidans* ATCC 53993 subjected to 40 mM Cu showed an increased level of protein RfaE possibly involved in LPS synthesis [22]. Apparently, LPS could bind metals in the cell surface depending on the composition of the polymers [47]. A summary of the main results obtained is shown in the working model of Figure 8.

Remarkably, it has been reported that *A. ferrooxidans* adapted to high copper and zinc ions concentrations showed changes in the surface chemical properties of this bacterium. Under these conditions, their surface negative charge was decreased due to changes in the structure of its surface layers [48].

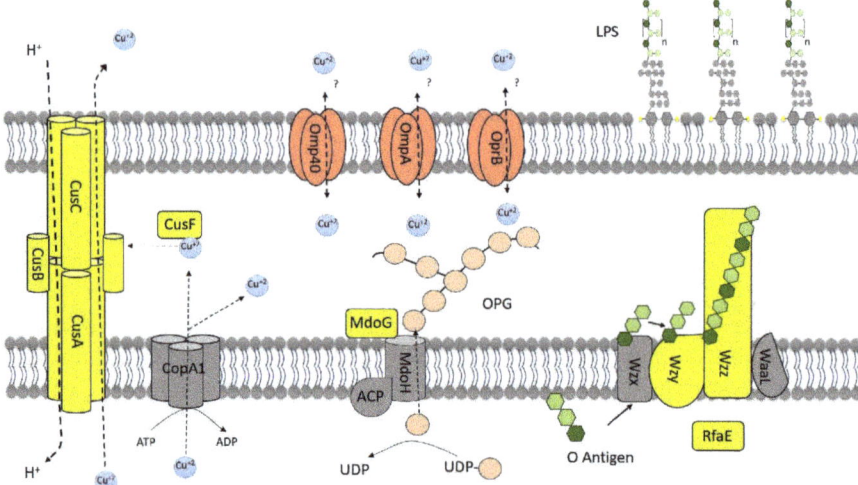

Figure 8. Summary working model of some proteins in *A. ferrooxidans* adapted to grow in presence of copper mentioned in this study. Yellow, proteins that increase their synthesis or transcript levels in presence of copper. Pink, proteins down-regulated in cells subjected to copper. OPG stands for osmoregulated periplasmic glucans. The locations and order in which proteins are illustrated are arbitrary.

4. Conclusions

The results presented here clearly indicate that several envelope components from *A. ferrooxidans* such as RND efflux pumps, LPS, porins, and periplasmic protein MdoG may be of great relevance for both, copper resistance and/or tolerance in their environment. Similar roles for these components in other members of the biomining bacterial consortia are also possible and their study may be of importance for industrial bioleaching operations.

Supplementary Materials: The following are available online at http://www.mdpi.com/2073-4425/9/7/347/s1. Figure S1: Functional categories of *A. ferrooxidans* proteins changing their synthesis levels in cells grown in presence of 100 and 200 mM $CuSO_4$. Table S1: Functional categories and numbers of *A. ferrooxidans* ATCC 53993 proteins changing their synthesis levels in cells grown in presence of 100 and 200 mM $CuSO_4$. Table S2: Proteins with increased levels in *A. ferrooxidans* ATCC 53993 grown in presence of 100 and 200 mM $CuSO_4$. Table S3: Proteins with lower levels of synthesis in *A. ferrooxidans* ATCC 53993 grown in presence of 100 and 200 mM $CuSO_4$.

Author Contributions: N.O., C.A.N., S.A.Á., A.P., and C.A.J. conceived and designed the experiments. N.O., R.N., F.A., and C.M.-B. performed the experiments. N.O., R.N., A.P. and C.A.J. wrote the paper. All authors revised the manuscript.

Funding: This research was funded by FONDECYT grant number 1150791 to Carlos A. Jerez.

Acknowledgments: This work was part of the Ph.D. thesis of Nia Oetiker supported by CONICYT-PCHA/Doctorado Nacional/2013- 21130264.

Conflicts of Interest: The authors declare no conflict of interest.

References

1. Vera, M.; Schippers, A.; Sand, W. Progress in bioleaching: Fundamentals and mechanisms of bacterial metal sulfide oxidation—Part A. *Appl. Microbiol. Biotechnol.* **2013**, *97*, 7529–7541. [CrossRef] [PubMed]
2. Rawlings, D.E. Characteristics and adaptability of iron- and sulfur-oxidizing microorganisms used for the recovery of metals from minerals and their concentrates. *Microb. Cell Fact.* **2005**, *4*, 1–15. [CrossRef] [PubMed]
3. Valenzuela, L.; Chi, A.; Beard, S.; Orell, A.; Guiliani, N.; Shabanowitz, J.; Hunt, D.F.; Jerez, C.A. Genomics, metagenomics and proteomics in biomining microorganisms. *Biotechnol. Adv.* **2006**, *24*, 197–211. [CrossRef] [PubMed]
4. Jerez, C.A. The use of genomics, proteomics and other OMICS technologies for the global understanding of biomining microorganisms. *Hydrometallurgy* **2008**, *94*, 162–169. [CrossRef]
5. Jerez, C.A. Bioleaching and biomining for the industrial recovery of metals. *Compr. Biotechnol.* **2011**, *3*, 717–729.
6. Navarro, C.A.; von Bernath, D.; Jerez, C.A. Heavy metal resistance strategies of acidophilic bacteria and their acquisition: Importance for biomining and bioremediation. *Biol. Res.* **2013**, *46*, 363–371. [CrossRef] [PubMed]
7. Watling, H.R. The bioleaching of sulphide minerals with emphasis on copper sulphides—A review. *Hydrometallurgy* **2006**, *84*, 81–108. [CrossRef]
8. Watkin, E.L.J.; Keeling, S.E.; Perrot, F.A.; Shiers, D.W.; Palmer, M.L.; Watling, H.R. Metals tolerance in moderately thermophilic isolates from a spent copper sulfide heap, closely related to *Acidithiobacillus caldus*, *Acidimicrobium ferrooxidans* and *Sulfobacillus thermosulfidooxidans*. *J. Ind. Microbiol. Biotechnol.* **2009**, *36*, 461–465. [CrossRef] [PubMed]
9. Jerez, C.A. Biomining of metals: How to access and exploit natural resource sustainably. *Microb. Biotechnol.* **2017**, *10*, 1191–1193. [CrossRef] [PubMed]
10. Franke, S.; Grass, G.; Rensing, C.; Nies, D.H. Molecular analysis of the copper-transporting efflux system CusCFBA of *Escherichia coli*. *J. Bacteriol.* **2003**, *185*, 3804–3812. [CrossRef] [PubMed]
11. Padilla-Benavides, T.; Thompson, A.M.G.; McEvoy, M.M.; Argüello, J.M. Mechanism of ATPase-mediated Cu^+ export and delivery to periplasmic chaperones: The interaction of *Escherichia coli* CopA and CusF. *J. Biol. Chem.* **2014**, *289*, 20492–20501. [CrossRef] [PubMed]
12. Argüello, J.M.; Raimunda, D.; Padilla-Benavides, T. Mechanisms of copper homeostasis in bacteria. *Front. Cell. Infect. Microbiol.* **2013**, *3*, 1–14. [CrossRef] [PubMed]
13. Cárdenas, J.P.; Quatrini, R.; Holmes, D.S. Genomic and metagenomic challenges and opportunities for bioleaching: A mini-review. *Res. Microbiol.* **2016**, *167*, 529–538. [CrossRef] [PubMed]
14. Orellana, L.H.; Jerez, C.A. A genomic island provides *Acidithiobacillus ferrooxidans* ATCC 53993 additional copper resistance: A possible competitive advantage. *Appl. Microbiol. Biotechnol.* **2011**, *92*, 761–767. [CrossRef] [PubMed]
15. Remonsellez, F.; Orell, A.; Jerez, C.A. Copper tolerance of the thermoacidophilic archaeon *Sulfolobus metallicus*: Possible role of polyphosphate metabolism. *Microbiology* **2006**, *152*, 59–66. [CrossRef] [PubMed]
16. Dopson, M.; Holmes, D.S. Metal resistance in acidophilic microorganisms and its significance for biotechnologies. *Appl. Microbiol. Biotechnol.* **2014**, *98*, 8133–8144. [CrossRef] [PubMed]
17. Orell, A.; Navarro, C.A.; Arancibia, R.; Mobarec, J.C.; Jerez, C.A. Life in blue: Copper resistance mechanisms of bacteria and Archaea used in industrial biomining of minerals. *Biotechnol. Adv.* **2010**, *28*, 839–848. [CrossRef] [PubMed]
18. Orell, A.; Navarro, C.A.; Rivero, M.; Aguilar, J.S.; Jerez, C.A. Inorganic polyphosphates in extremophiles and their possible functions. *Extremophiles* **2012**, *16*, 573–583. [CrossRef] [PubMed]
19. Martínez-Bussenius, C.; Navarro, C.A.; Jerez, C.A. Microbial copper resistance: Importance in biohydromrtallurgy. *Microb. Biotechnol.* **2017**, *10*, 279–295. [CrossRef] [PubMed]
20. Dopson, M.; Ossandon, F.J.; Lövgren, L.; Holmes, D. Metal resistance or tolerance? Acidophiles confront high metal loads via both abiotic and biotic mechanisms. *Front. Microbiol.* **2014**, *5*, 1–4. [CrossRef] [PubMed]
21. Almárcegui, R.J.; Navarro, C.A.; Paradela, A.; Albar, J.P.; von Bernath, D.; Jerez, C.A. New copper resistance determinants in the extremophile *Acidithiobacillus ferrooxidans*: A quantitative proteomic analysis. *J. Proteome Res.* **2014**, *7*, 946–960. [CrossRef] [PubMed]

22. Martínez-Bussenius, C.; Navarro, C.A.; Orellana, L.; Paradela, A.; Jerez, C.A. Global response of *Acidithiobacillus ferrooxidans* ATCC 53993 to high concentrations of copper: A quantitative proteomics approach. *J. Proteom.* **2016**, *145*, 37–45. [CrossRef] [PubMed]
23. Amaro, A.M.; Chamorro, D.; Seeger, M.; Arredondo, R. Effect of external pH perturbations on in vivo protein synthesis by the acidophilic bacterium *Thiobacillus ferrooxidans*. *J. Bacteriol.* **1991**, *173*, 910–915. [CrossRef] [PubMed]
24. Ramos-Fernández, A.; Paradela, A.; Navajas, R.; Albar, J.P. Generalized method for probability-based peptide and protein identification from tandem mass spectrometry data and sequencue database searching. *Mol. Cell. Proteom.* **2008**, *7*, 1748–1754. [CrossRef] [PubMed]
25. López-Serra, P.; Marcilla, M.; Villanueva, A.; Ramos-Fernández, A.; Palau, A.; Leal, L.; Wahi, J.E.; Setien-Baranda, F.; Szczesna, K.; Moutinho, C.; et al. A DERL3-associated defect in the degradation of SLC2A1 mediates the Warburg effect. *Nat. Commun.* **2014**, *5*, 3608. [CrossRef] [PubMed]
26. Vera, M.; Pagliai, F.; Guiliani, N.; Jerez, C.A. The chemolithoautotroph *Acidithiobacillus ferrooxidans* can survive under phosphate-limiting conditions by expressing a C-P lyase operon that allows it to grow on phosphonates. *Appl. Environ. Microbiol.* **2008**, *74*, 1829–1835. [CrossRef] [PubMed]
27. Alvarez, S.; Jerez, C.A. Copper ions stimulate polyphosphate degradation and phosphate efflux in *Acidithiobacillus ferrooxidans*. *Appl. Environ. Microbiol.* **2004**, *70*, 5177–5182. [CrossRef] [PubMed]
28. Navarro, C.A.; Orellana, L.H.; Mauriaca, C.; Jerez, C.A.; Navarro, C.A.; Orellana, L.H.; Mauriaca, C.; Jerez, C.A. Transcriptional and functional studies of *Acidithiobacillus ferrooxidans* genes related to survival in the presence of copper. *Appl. Environ. Microbiol.* **2009**, *75*, 6102–6109. [CrossRef] [PubMed]
29. Rozen, S.; Skaletsky, H. Primer3 on the WWW for general users and for biologist programmers. *Methods Mol. Biol.* **2000**, *132*, 365–386. [PubMed]
30. Pfaffl, M.W. A new mathematical model for relative quantification in real-time RT-PCR. *Nucleic Acids Res.* **2001**, *29*, e45. [CrossRef] [PubMed]
31. Hitchcock, P.J.; Brown, T.M. Morphological heterogeneity among *Salmonella* lipopolysaccharide chemotypes in silver-stained polyacrylamide gels. *J. Bacteriol.* **1983**, *154*, 269–277. [PubMed]
32. Quesenberry, M.S.; Lee, Y.C. A rapid formaldehyde assay using purpald reagent: Application under periodation conditions. *Anal. Biochem.* **1996**, *234*, 50–55. [CrossRef] [PubMed]
33. Lee, C.H.; Tsai, C.M. Quantification of bacterial lipopolysaccharides by the purpald assay: Measuring formaldehyde generated from 2-keto-3-deoxyoctonate and heptose at the inner core by periodate oxidation. *Anal. Biochem.* **1999**, *267*, 161–168. [CrossRef] [PubMed]
34. Baker-Austin, C.; Dopson, M. Life in acid: pH homeostasis in acidophiles. *Trends Microbiol.* **2007**, *15*, 165–171. [CrossRef] [PubMed]
35. Quatrini, R.; Appia-Ayme, C.; Denis, Y.; Jedlicki, E.; Holmes, D.S.; Bonnefoy, V. Extending the models for iron and sulfur oxidation in the extreme acidophile *Acidithiobacillus ferrooxidans*. *BMC Genom.* **2009**, *10*, 394. [CrossRef] [PubMed]
36. Bohin, J.P. Osmoregulated periplasmic glucans in Proteobacteria. *FEMS Microbiol. Lett.* **2000**, *186*, 11–19. [CrossRef] [PubMed]
37. Chi, A.; Valenzuela, L.; Beard, S.; Mackey, A.J.; Shabanowitz, J.; Hunt, D.F.; Jerez, C.A. Periplasmic proteins of the extremophile *Acidithiobacillus ferrooxidans*. *Mol. Cell. Proteom.* **2007**, *6*, 2239–2251. [CrossRef] [PubMed]
38. Bontemps-Gallo, S.; Bohin, J.P.; Lacroix, J.M. Osmoregulated periplasmic glucans. *EcoSal Plus* **2017**, *7*. [CrossRef] [PubMed]
39. Weissborn, A.C.; Liu, Q.; Rumley, M.K.; Kennedy, E.P. UTP-D-glucose-1-phosphate uridylyltransferase of *Escherichia coli*: Isolation and DNA sequence of the *galU* gene and purification of the enzyme. *J. Bacteriol.* **1994**, *176*, 2611–2618. [CrossRef] [PubMed]
40. Delcour, A.H.; Adler, J.; Kung, C.; Martinac, B. Membrane-derived oligosaccharides (MDO's) promote closing of an *E. coli* porin channel. *FEBS Lett.* **1992**, *304*, 216–220. [CrossRef]
41. Fiedler, W.; Rotering, H. Properties of *Escherichia coli* mutants lacking membrane-derived oligosaccharides. *J. Biol. Chem.* **1988**, *263*, 14684–14689. [PubMed]
42. Speer, A.; Rowland, J.L.; Haeili, M.; Niederweis, M. Porins increase copper susceptibility of *Mycobacterium tuberculosis*. *J. Bacteriol.* **2013**, *195*, 5133–5140. [CrossRef] [PubMed]

43. Gumulya, Y.; Boxall, N.J.; Khaleque, H.N.; Santala, V.; Carlson, R.P.; Kaksonen, A.H. In a quest for engineering acidophiles for biomining applications: Challenges and opportunities. *Genes* **2018**, *9*, 116. [CrossRef] [PubMed]
44. Pisani, F.; Italiano, F.; De Leo, F.; Gallerani, R.; Rinalducci, S.; Zolla, L.; Agostiano, A.; Ceci, L.R.; Trotta, M. Soluble proteome investigation of cobalt effect on the carotenoidless mutant of *Rhodobacter sphaeroides*. *J. Appl. Microbiol.* **2009**, *106*, 338–349. [CrossRef] [PubMed]
45. Lerouge, I.; Vanderleyden, J. O-antigen structural variation: Mechanisms and possible roles in animal/plant-microbe interactions. *FEMS Microbiol. Rev.* **2001**, *26*, 17–47. [CrossRef]
46. Snyder, D.S.; Brahamsha, B.; Azadi, P.; Palenik, B.; Acteriol, J.B. Structure of compositionally simple lipopolysaccharide from marine *Synechococcus*. *J. Bacteriol.* **2009**, *191*, 5499–5509. [CrossRef] [PubMed]
47. Langley, T.; Beveridge, T.J. Effect of O-side-chain-lipopolysaccharide chemistry on metal binding. *Appl. Environ. Microbiol.* **1999**, *65*, 489–498. [PubMed]
48. Vilinska, A.; Rao, K.H. Surface characterization of *Acidithiobacillus ferrooxidans* adapted to high copper and zinc ions concentration. *Geomicrobiol. J.* **2011**, *28*, 221–228. [CrossRef]

© 2018 by the authors. Licensee MDPI, Basel, Switzerland. This article is an open access article distributed under the terms and conditions of the Creative Commons Attribution (CC BY) license (http://creativecommons.org/licenses/by/4.0/).

MDPI
St. Alban-Anlage 66
4052 Basel
Switzerland
Tel. +41 61 683 77 34
Fax +41 61 302 89 18
www.mdpi.com

Genes Editorial Office
E-mail: genes@mdpi.com
www.mdpi.com/journal/genes

www.ingramcontent.com/pod-product-compliance
Lightning Source LLC
LaVergne TN
LVHW070427100526
838202LV00014B/1541